Graphene Materials: Surface Chemistry, Properties and Modification Techniques

Graphene Materials: Surface Chemistry, Properties and Modification Techniques

Editor: Chris Patterson

MURPHY & MOORE
www.murphy-moorepublishing.com

Published by Murphy & Moore Publishing
1 Rockefeller Plaza,
New York City, NY 10020, USA
www.murphy-moorepublishing.com

Graphene Materials: Surface Chemistry, Properties and Modification Techniques
Edited by Chris Patterson

International Standard Book Number: 978-1-63987-268-8 (Hardback)

Cataloging-in-Publication Data

Graphene materials : surface chemistry, properties and modification techniques / edited by Chris Patterson.
p. cm.
Includes bibliographical references and index.
ISBN 978-1-63987-268-8
1. Graphene. 2. Carbon. 3. Graphene--Properties. 4. Graphene--Technological innovations.
5. Nanostructured materials. 6. Materials science. I. Patterson, Chris.
TA455.G65 G733 2022
620.115--dc23

Contents

Preface

Graphene refers to a single layer of carbon atoms which are tightly bound in a hexagonal honeycomb structure. It is an allotropic form of carbon. Each carbon atom in graphene forms bond with three neighboring atoms. The three dimensional structure formed by stacking layers of graphene is called graphite. It is an excellent conductor of heat and electricity, and is capable of absorbing light of all visible wavelengths. Some common methods to produce graphene are chemical vapor deposition, graphite oxide reduction, nanotube slicing, electrochemical synthesis, etc. It finds extensive applications across various fields such as electronics, biomedical engineering and light processing. Future applications of graphene include manufacturing of supercapacitors. From theories to research to practical applications, case studies related to all contemporary topics of relevance to this field have been included in this book. Most of the topics introduced herein cover modification techniques and properties of graphene materials. As this field is emerging at a rapid pace, the contents of this book will help the readers understand the modern concepts and applications of the subject.

This book has been the outcome of endless efforts put in by authors and researchers on various issues and topics within the field. The book is a comprehensive collection of significant researches that are addressed in a variety of chapters. It will surely enhance the knowledge of the field among readers across the globe.

It gives us an immense pleasure to thank our researchers and authors for their efforts to submit their piece of writing before the deadlines. Finally in the end, I would like to thank my family and colleagues who have been a great source of inspiration and support.

Editor

1

Green Routes for Graphene Oxide Reduction and Self-Assembled Graphene Oxide Micro- and Nanostructures Production

Rebeca Ortega-Amaya, Yasuhiro Matsumoto ,
Esteban Díaz-Torres, Claudio Davet Gutierrez-Lazos,
Manuel Alejandro Pérez-Guzmán and
Mauricio Ortega-López

Abstract

Graphene-derived materials are currently studied because of their actual and projected applications. Among them, graphene oxide (GO) promises for outstanding applications as it can be prepared at large scale by simple, scalable, and low-cost techniques. The existent chemical methods based on the graphite exfoliation (phase solution and Hummers based) produce highly functionalized graphene, i.e., GO-like materials that converts into reduced GO (rGO) after a reduction treatment. The present work presents the current scenario on the GO green reduction methods, on the development of hierarchical carbon-based structures by the self-assembly of GO sheets at interfaces, and on rGO-based hybrid nanocomposites. It is worth noting that, to date, the production and application of graphene-related materials are the fastest-growing research areas.

Keywords: graphene oxide, reduced graphene oxide, green reduction, metal, composites, self-assemble

1. Introduction

Graphene, one-atom thick layer of densely packed carbon atoms into a honeycomb crystal lattice, is considered the key building block of graphite, carbon nanotubes, and fullerenes [1]. It is of current interest due to its remarkable physical and chemical properties, which makes it useful for theoretical studies for several technological applications. Current applications of graphene include flexible electronics, batteries, and so on [2]. Diverse methods have been proposed to produce high-quality single and few layer graphene films. Among them, graphite micromechanical cleavage, chemical vapor deposition, and graphitization of SiC have been the most utilized methods [3]. Although these methods produce high-quality graphene in a controlled way, they suffer from mass production scaling.

In the past years, graphene-derived materials, such as graphene oxide (GO), graphane (the hydrogenated version of graphene), graphene fluoride, and so on [4, 5] have been paid special interest because of their potential applications. Particularly, GO and its reduced version, reduced-graphene oxide (rGO), have emerged as a technologically important material by its their own right [6].

GO is mainly prepared through chemical methods and therefore achieves unique and useful physiochemical properties to prepare a variety of functional materials for a range of advanced applications, such as rGO self-assembled microstructures [7, 8] and, rGO-based composites with inorganic nanoparticles (metals, semiconductors, metal oxides). These GO-derived materials have successfully been tested in the technological areas of nanomedicine, electronics, environmental remediation, energy conversion, and others [7–9].

The chemical methods to prepare single-layer GO use graphite as the raw material, which is exfoliated either using strong oxidants in aqueous medium (based on Hummers' method) or using organic solvents (based on the solution-phase technique), among others [10]. During the graphite oxidation process, oxidative species intercalate into graphite galleries provoking the partial disruption of the graphene sp^2-hybridization and the covalent attachment of oxygen-rich species. This results on the weakening of the interlayer attractive force, so that single-layer GO sheets are easily obtained upon application of low power sonication in water [8].

From a structural point of view, GO is considered as a graphene sheet comprising in-plane undisturbed π-conjugated domains, and functionalized ones with covalently attached hydroxyl and epoxy groups, and additional carboxyl and carbonyl groups located at the sheet edge [11]. This chemical structure gives GO an amphiphilic character and then makes it dispersible in polar or nonpolar solvents [12]. This amphiphilic character preserves in rGO because it is obtained after the partial remotion of these functional groups by a reduction process.

Interestingly, rich oxygenated groups attached to the graphene structure makes GO and rGO highly hydrophilic and susceptible for further functionalization. Therefore, pristine or reduced GO can conveniently be functionalized to facilitate the interfacial interaction between GO and other materials including polymers, metal oxides, and inorganic nanoparticles to form GO-based composite materials, or to link the sheets together and then lead to macroscopic GO-based materials [13, 14].

Due to its multiple applications, GO is produced at an industrial level. Nowadays, worldwide research groups are looking for ways to find cost-effective and environment-friendly methods for graphene-derived materials' mass production. These include electrochemical, mechanical, and chemical exfoliation of graphite [15]. In general, these methods produce GO-like materials, i.e., functionalized graphene, and they may be further processed to produce rGO with multiple functionalities. To date, the phase solution graphite exfoliation-based methods have demonstrated their high versatility to fabricate bulk amounts of graphene-derived materials at relatively low cost [16].

There are diverse methods for GO reduction, such as thermal reduction, chemical reduction using toxic or green reductive reagents, and multistep reduction (either by combining chemical and thermal processes or by combining green and toxic reducers to get an effective reduction). Dangerous and toxic reagents such as hydrazine, oxalic acid, sodium hydrosulfite, and sodium borohydride were reported to reduce GO efficiently. On the other hand, GO environment-friendly reduction routes include flash photo reduction, hydrothermal dehydration, solvothermal reduction, catalytic reduction, and photocatalytic reduction. Furthermore, green reductants have also been essayed including vitamin C, alcohols, bovine serum albumin, ging-seng, bacteriorhodopsin, bacteria, and polyphenols (present in green tea and caffeic acid, among others) [17, 18].

This work presents an overview on the environmental-friendly methods to reduce GO and produce GO-based nanocomposites. A survey of their applications is also presented.

In addition, we present the mechanistic aspects on GO-based nanocomposites, as well as those associated in the formation of GO nano- and microstructures by self-assembly process.

2. Green methods to reduce GO

For applications where the exceptional electrical conductivity and transparency of graphene are demanded, GO can be subjected to an additional chemical treatment to detach the covalently oxygenated groups on graphene basal plane and restoring the sp^2-hybridization. As previously mentioned, the detected drawbacks for the chemical reductants, such as hydrazine, hydroquinone, and sodium borohydride [19], have fuelled the search for both environment-friendly methods and chemicals for GO reduction. The so-called "green technologies" satisfy both criteria, and the most reported green technologies may be classified, selected reducing agent, into four groups, as indicated in **Table 1**.

Green technologies	Reducer agent sources
Bioreduction	•Bacteria •Plants •Commercial biomolecules
Photoreduction	•Electromagnetic irradiation
Reduction by polymers	•Polyelectrolytes
Aided-metal reduction (or reduction by metals)	•Transition metals
Mechanochemical reduction	•Ball milling system
Electrochemical reduction	•Supporting electrolyte

Table 1. Representative green technologies for the GO reduction.

2.1. Green reducers and their effectiveness

2.1.1. Bacterial reducers

Bacteria are living beings capable of surviving under the most extreme conditions, i.e., in severe temperature and chemical composition. Bacteria have been found in the most warmed underwater pools, where tectonic plates emanate pernicious gasses and incandescent material or in lakes of extreme saline composition, surrounded by an environment that is highly concentrated in arsenic, such as those found in Mono Lake, California [20]. To survive, bacteria can take organic and inorganic molecules from the surrounding environment and transform them into the substance required to start the cellular process in which oxidation-reduction mechanism is employed to obtain an energy source [21, 22]. The overall redox process carried out by bacteria has been used in GO reduction by means of *Shewanella* [23], Bacillus subtilis [24], Extremophiles bacteria [25], *Escherichia coli* [26], and *Gluconacetobacter xylinus* [27].

Note that the involved reaction mechanisms depend on the bacteria cell structure, which determines the capacity for directly or indirectly hydrolyzing the acidic groups attached to the GO molecular structure, particularly, the groups that comprise oxygen atoms. Wang et al. [23] used *Shewanella* for reducing GO (**Figure 1**) through a mechanism that consists segregating the heme group proteins such as c-type cytochromes, through the membrane and these proteins act as electron intercessor [28].

Zhang et al. [24] reported that depending on the bacteria type, it is possible to select a process to efficiently reduce GO for specific applications of the final nanomaterial. It was also proposed that, in the bacteria-based reduction processes, the parallel action of different bacteria could increase the effectiveness of reduction process. Based on Zhang's observations, Raveendran et al. [25] achieved reducing GO using extremophiles bacteria, obtaining graphene with excellent conductive properties.

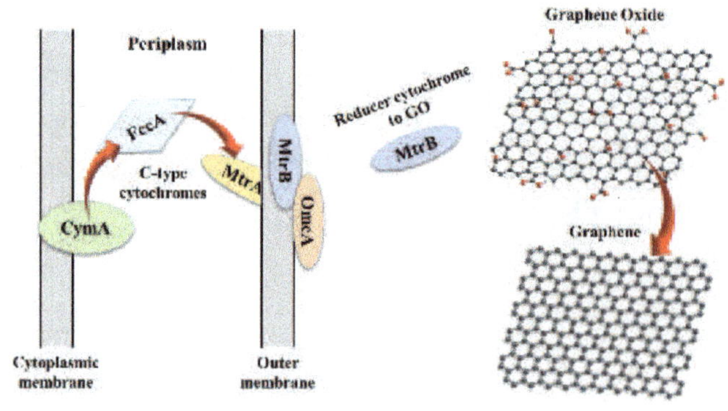

Figure 1. As bacteria, Shewanella oneidensis utilizes terminal electron acceptors during its respiratory metabolism. It transfers electrons from cell surface to any extracellular acceptor such as metal oxides or graphene oxide. It was proposed that GO reduction by Shewanella procceds via an electron exchange among MtrB, MtrC and OmcA cytochromes to finally transfer an electron to GO and then reduce it.

2.1.2. Biological reducers

The chemical compounds naturally existing in plants (phytochemicals) have been used for years as nutrients, drugs, etc. In the past few years, phytochemicals, such as vitamins, amino acids, saccharides, alkaloids, proteins, and enzymes [29, 30], have been studied as reductant precursors for metals and GO. The reported attempts for the GO reduction by using phytochemicals go into the employment of either laboratory-extracted (plant extracts) or commercial-purchased phytochemicals.

Herein, we present some relevant results emphasizing on the reductant chemical source.

2.1.2.1. Plant extract reducing

To date, the GO reduction by means of plant extract is intensively studied [31]. In this approach, the plant is chosen considering the antioxidant compound contents. For preparing the plant extract leaves, flowers, stems, and/or roots are refluxed in water, alcohol, or water-alcohol mixtures as solvents.

Green tea has proven to be an excellent source of antioxidant biomolecules. For example, it was successfully used for reduction of graphene oxide [32]. The reducing capacity of green tea is based on the antioxidant biomolecules extracted from emulsion, mainly polyphenols.

Extracts of chrysanthemum flower and lycium barbarum plants, used in the traditional Chinese medicine, were recently reported for GO reduction by Hou et al. [33, 34]. The extracts were obtained in aqueous media at boiling temperature and then filtered. Afterward, the extract was poured into the GO dispersion at the water boiling point for 24 h. The authors reported that the chemical composition of extracts, namely, chrysanthemum extract and flavonoids (diosmetin, luteolin, apigenin, and glucoside), were the predominant phytochemicals. Whereas the lycium barbarum extract comprised flavonoids, phenols, carotenoids, and polysaccharides as dominant phytochemicals.

The authors suggested that polyphenols present in chrysanthemum and lycium barbarum extracts transform to quinone releasing H^+ ions that interact with GO for reducing it. Importantly, chrysanthemum and lyceum barbarum plants hold promise to effectively reduce GO, because the C/O ratio values obtained by X-ray photoelectron spectroscopy (XPS) were 1.35, 4.96, and 6.5 for pristine GO, rGO-chrysanthemum, and rGO-lycium barbarum, respectively.

2.1.2.2. Commercial reductants

2.1.2.2.1. Vitamins

Vitamin C (L-ascorbic acid) has been widely used in GO reduction because of its reducing effectivity and is comparable to that of hydrazine, besides promoting highly stabilized dispersions of rGO sheets in water. It has been observed that oxidized L-ascorbic acid is unreactive and stable and does not provoke damage to living cells [31].

In some GO reduction reactions, L-tryptophan (an aromatic amino acid) has been considered as a stabilizing agent to prepare highly stable rGO aqueous dispersions [35]. It effectively prevents against agglomeration of the rGO sheets because it readily adsorbs on undisturbed π–π domains at the basal plane of the rGO chain, which minimizes the attractive π–π interaction. Furthermore, the remaining terminal carboxylate anion of L-tryptophan has provided an electrostatic repulsion between the individual graphene sheets.

The L-tryptophan-stabilized rGO dispersion prepared with vitamin C exhibited good electrical conductivity of 14.1 S/m (pristine GO: 5.72×10^{-10} S/m). The mechanistic aspects for the GO chemical reduction remains unknown, but a plausible reduction mechanism was proposed as comprising two-step S_N2 nucleophilic reactions. That is, L-ascorbic acid oxidizes into the dehydroascorbic anion releasing electrons and protons, which react with oxygenated groups on the GO sheet to reduce it.

The free-stabilizing agents including vitamin C-reduced rGO dispersions were also prepared by Zhang, who reported high stability for all the prepared samples. The electrical conductivity with a value of 800 S/m was obtained in the sample prepared for 48 h [36]. Fernandez-Merino et al. [37] reported that the reduction capability of vitamin C could be improved by increasing the alkali concentration into reducing solution; using this approach the reduction time was shortened to 15 min. Furthermore, rGO showed good dispersibility in polar organic solvents, with high C/O ratio (~12.5) as well as high electrical conductivity (7700 S/m). In addition, riboflavin (vitamin B2), phosphate salt of vitamin B2, and pyridoxine (vitamin B6) were used to reduce GO. These bioreductants have also been proven to successfully reduce GO [38].

2.1.2.2.2. Saccharides

Saccharides are nutrients that may be used as reducing agents; these are classified into four chemical groups: mono/di/oligo/polysaccharides. Monosaccharides, glucose, and fructose have demonstrated mild reductive ability and nontoxic property in GO reduction experiments. In general, their potential for reduction is closely related to the ease to form open-chained structures [31]. In the GO reduction, it was found that their oxidized products play an important role to stabilize rGO sheets in aqueous dispersions, i.e., they may act as capping agents. Both saccharides and their oxidized products are environmental friendly. Zhu et al. [39] used glucose, fructose, and sucrose in aqueous ammonia solution for the reduction of GO. They determined that the ammonia solution is useful for both completion and enhancement of the GO deoxygenation reaction rate. In addition, they found that the reduction capability of sucrose was weaker than that for the glucose and fructose, under similar reaction conditions. The resulting rGO powder was biocompatible and highly dispersible in water. Likewise, Akhavan et al. [40] found that glucose increases its power to reduce GO in the presence of an iron catalyst under neutral condition.

On the other hand, dextran (a polysaccharide) was tested as a GO reducer in aqueous ammonia [41]. However, the as-reduced rGO exhibited a rather low electrical conductivity (1.1 S/m) that can be notably improved (10,000 S/m) upon thermal annealing (500°C under Ar atmosphere).

2.1.2.2.3. Amino acids

L-Cysteine is a thiol-containing amino acid that is liable to oxidate to cystine. It inhibits oxidative properties because thiol groups can suffer redox reactions. Chen et al. [42] synthesized rGO using L-cysteine as reducing agent under mild conditions. They proposed the reduction pathway for GO by L-Cysteine might be like that observed in the GO reduction by vitamin C. That is, at first, the reactions comprehend nucleophilic attack by thiol groups, which develop upon proton releasing during the L-Cysteine oxidation process. Afterward, the released protons react with the oxygenated groups producing water and byproducts, inducing the GO reduction. The rGO suspension conductivity increases by about 10^6 times in comparison to that of pristine GO. Bose et al. [43] used other amino acid such as glycine for reducing GO. They found that glycine not only reduces the GO but also functionalized it, as a result amine group can covalently bound to a GO network. In other work, L-Lysine was successfully used for reduction of graphene oxide in the presence of carboxymethyl starch (CMS) as stabilizing agent. The rGO suspension exhibited good dispersion stability in water [44]. Furthermore, L-aspartic acid has been employed for synthesizing rGO, the product obtained by this process also presents uniform separation in water as well as good electrical conductivity of ~700 S/m [45]. Other studies have revealed that some amino acids such as tryptophan, arginine, and histidine reduce the GO and also augmented the consolidation of rGO–metal nanoparticles [46]

2.1.2.2.4. Gallic and citric acids

Gallic and citric acids are natural organic acids that have been tested as GO reductants. It was found that both acids could play the dual role as a reducing agent and a surfactant. Li et al. [47] found that the GO can be significantly reduced by gallic acid in aqueous ammonia, either at room temperature or under heating condition. Although, the reduction mechanism of GO by gallic acid has not been explored, it is expected that its three adjacent hydroxyl groups (pyrogallol moieties) interact with the GO in-plane oxygenated groups. The prepared rGO suspensions displayed excellent dispersibility in various solvents such as H_2O, N-Methyl-2-pyrrolidone (NMP), dimethylsulfoxide (DMSO), dimethylformamide (DMF), and methanol, probably due to adsorbed oxidized gallic acid.

On the other hand, citric acid has extensively been studied for the synthesis of silver and gold nanoparticles. Recently, Ortega-Amaya et al. [18] used the one pot approach to produce highly dispersible functionalized rGO by using citric acid. This process was made in aqueous medium at room temperature, under Ar atmosphere. To explain the dual role of citric acid as a reducer and a stabilizer, the authors assumed that protons released by the citric acid dissociation bind to epoxy or hydroxyl groups to form water molecules and an active carbocation at the GO network. Afterward, a di-ionized citrate $HCit^{2-}$ anion covalently binds to the carbocation to stabilize it. The whole effect was one of reduction by protons, and functionalization by HCi^{2-} anion. Last one being the predominant specie in the aqueous solution at pH 4.

2.1.3. Irradiation as reducer

UV, microwave, or ultrasound irradiation have been used for transforming colloidal GO to graphene with a similar quality as that produced by means of hydrazine. In acidic GO colloids, Lu et al. [48] obtained free contaminants rGO by microwave heating. First, an acidic GO colloid at pH 1, 3, or 5 was separately prepared by dropping a NaOH solution. Afterwards, each mixture was heated at 150°C under microwave irradiation, employing a power of 80 W, for 10 min. They monitored the GO reduction advance by visual observation of the color changes from brownish-yellow to black [49, 50].

A different method for the GO reduction based on electromagnetic irradiation was reported by Ding et al. [51]. The authors reported clean reduction of colloidal GO using the strong UV absorption property of water [52]. The UV radiation dissociates the water molecule into three radicals (hydrogen H_2, hydroperoxyl HO_2, and hydrated electrons e^-), each one retaining one of the earlier bounded electrons [53]. Then, hydrated electrons behave as a reducer to form rGO (**Figure 2**). Although the reduction process takes a long reaction time, it is possible to monitor the formation of rGO dispersions through UV-vis spectroscopy [54].

Another green processing by irradiation was published by Nyangiwe et al. [53], which is a very simple method and is described for the reduction of GO solution. By irradiating a GO sample dispersed in water with sunlight, the most oxygenic functional groups in GO were removed. The authors considered that photoreduction of GO by sunlight can be explained by a model proposed by Ji et al. [55], where the absorbed UV radiation in solvent excites the water molecule near its photoionization threshold (6.5 eV), generating solvated electrons, which will act like reducers. The complete process is described by the following equations [56]:

$$GO + hv \rightarrow GO(hole + e^{-}) \tag{2.1}$$

$$4holes + 2H_2O \rightarrow O_2 + 4H^{+} \tag{2.2}$$

$$4e^{-} + GO + 4H^{+} \rightarrow rGO + 2H_2O \tag{2.3}$$

2.1.4. Polymers as reducers

There are scarce reports on the GO reduction by polymers. Zang et al. [57] reported the GO reduction using poly(diallyldimethylammonium chloride) (PDDA) polyelectrolyte [57]. It has been reported that the addition of PDDA to a GO aqueous dispersion triggers a chemical reaction that promotes a color change in the GO dispersion, indicating that GO transforms to rGO. Although the mechanistic aspects of the GO reduction were not clearly explained, the PDDA-functionalized rGO exhibited an excellent dispersion in water. Therefore, polyelectrolyte might be used as a reducing agent as well as a stabilizer to prepare a colloidal suspension of graphene. This method is based on the Yang et al. report [56], where PDDA was adsorbed on the external surface of carbon nanotubes through π–π and electrostatic interactions [56, 58]. It was assumed that repulsive electrostatic interaction dominates to produce well dispersed PDDA-functionalized carbon nanotubes in water.

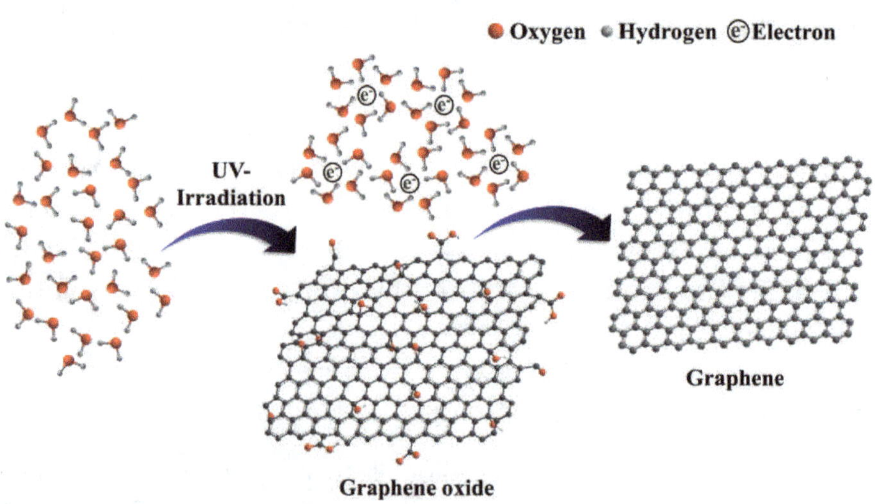

Figure 2. Schematic representation of the reduction of GO under UV irradiation.

2.1.5. *Transition metals as reducers*

An interestingly eco-friendly approach toward the GO reduction consists of using transition metals (e.g., Fe, Zn, Cu, and Co) as GO reducing agents. In this case, the reduction mechanism strongly depends on the experimental conditions (mainly pH and temperature) and it follows a frequently complex pathway. Some examples are described below.

GO was reduced by iron in aqueous medium by Fan et al. [59]. They studied the GO reduction by powdered iron (10 μm average size) in an acidic HCl-water mixture at room temperature. They proposed that H^+ interacts with the iron surface particle to bring forth the Fe/Fe^{2+} core/shell structure (iron particle with a thin sheet of charged Fe^{2+} ions). These positively charged Fe/Fe^{2+} species interacts with the functional groups on the GO sheets and after electron transport from Fe/Fe^{2+} to GO, the reduction of GO was achieved.

Experiments on GO reduction using Zn powder were essayed by Yang et al. [60]. To evaluate the Zn reduction capability and how it is affected by the solution pH and temperature, they prepared aqueous GO colloids with and without sodium hydroxide at room temperature and 100°C. They obtained lower reduction levels for all cases other than alkaline in 100°C conditions. The proposed reduction mechanism consists of an electron exchanging between Zn and GO to produce rGO and by products. The GO reduction using Zn powder were also carried out by Mei and Ouyang [61] and Liu et al. [62] under acidic conditions.

The formation of Cu_2O/rGO nanostructures during the GO reduction by Cu nanoparticles was described by Wu et al. [63]. They mixed polydisperse Cu nanoparticles with water-dispersed GO under neutral conditions. After sonication and heating at 95°C a composite comprising of rGO sheet decorated by Cu_2O nanoparticles was observed. Authors claimed that GO was reduced through a redox reaction between Cu an GO, in which Cu nanoparticles transformed to nanozised Cu_2O and GO was reduced. In addition, they reported that the GO reduction strongly depended on the Cu particle size, because experiments involving fine grained Cu powder was unable to effectively reduce GO.

GO reduction experiments using metal foils as a substrate were done by Cao et al. [64]. A number of metal foils (Cu, Ni, Co, Fe, and Zn) were separately immersed in a GO aqueous dispersion at pH = 6. After taking out and drying at ambient temperature, the metal foil was coated with a rGO film. It was assumed that the rGO film was developed by a self-assembly process of rGO nanosheets and that the GO was spontaneously reduced by direct transference of electrons from metal ions to GO. Some metal ions were found in the GO layers' galleries.

Hu et al. [65] used a GO dispersion at pH = 4 to immerse metallic foils (Cu, Fe, Zn, Co, and Al) and also a nonmetallic (carbon) film supported by Cu, and after (1–12 h) immersion, the metal foil was covered by self-assembled rGO multilayers and was dried at ambient conditions or freeze dried. They found that there had been electron transfer between the metal and GO propitiated for the acid condition (**Figure 3**). A significant difference is that they found metal oxide nanoparticles decorating the rGO; another important result was that the electron exchange to reduce GO had taken place even when a conductive layer (carbon or Au, Pt, Ag) covers the Cu substrate.

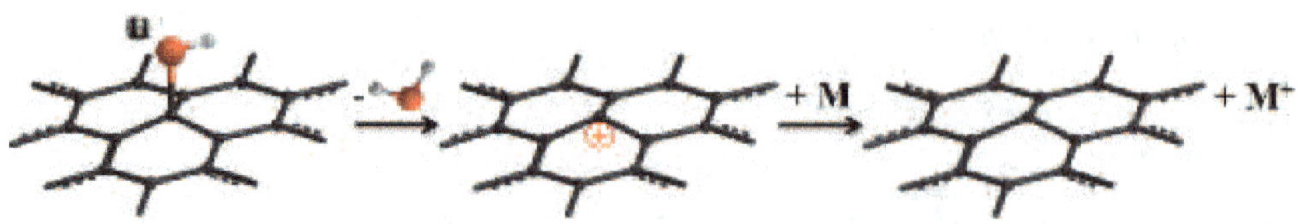

Figure 3. Scheme of GO reduction and metal oxidation. A hydroxyl group present at GO was protonated in acid conditions and then an electron transfer between metal GO took place, rGO was obtained, and oxidized metal.

2.1.6. Mechanochemical reducer: ball milling system

The mechanical reduction of GO into graphene was tested by a hydrogen-assisted ball-milling process [66]. The ball-milling process was carried out in a planetary micro ball-milling machine with a stainless steel chamber and stainless steel 5 mm diameter balls. First, 2.0 g of previously prepared GO powder by a modified Hummers' method [67] was loaded into the ball-milling chamber, and then filled with hydrogen gas. The chamber was rotated at 900 rpm for different times in the 30- to 240-min interval, to obtain a variety of ball-milled rGO samples. The GO reduction process with the milling time was visually verified by observing the GO color change from a brownish-yellow to black. The final powder, as analyzed by transmission electron microscopy, XPS, and infrared absorbance spectroscopy, consisted of well-exfoliated oxygen-free single-layer graphene [68, 69].

2.1.7. Electrochemical reduction

The electrochemical technique is widely used in thin film deposition on conductive substrates. The electrochemical reduction of GO develops either during the film deposition process or a preformed film as described in the comprehensive review [70]. The one-step and two-step approaches are usually employed to produce GO, and the reduction level can be controlled by varying the processing time, electrode material, on-off cycles, electrolyte type, and potential values. A variety of nontoxic electrolytes, such as NaCl, KCl, $NaHPO_4$, Na_2SO_4, KNO_3, and phosphate buffer solution (PSB), have been used. Furthermore, glass carbon, Au, Pt, Ag, and 3-aminopropyltriethoxysilane (APTES) have been tested as electrode materials [70].

In the one-step approach, GO sheets are dispersed into a mixture of electrolyte and buffer solution, and the power source is turned on and the GO thin film deposition and reduction occur simultaneously at the cathode surface material.

In the two-step approach, a thin film is deposited by some technique (drop-casting, spray pyrolysis, layer-by-layer, etc.) on an electrode of a three-electrode system (reference, working, and auxiliary electrodes), and then immersed into the electrolyte solution. Under controlled conditions of electrolyte temperature and composition, as well as electrodes potential, a rGO thin film is obtained. Recently, Fang et al. [71] used the two-step system to produce large area rGO and rGO/silk fibroin composites. They used a reference electrode of Ag/AgCl, auxiliary electrodes of ITO, Ag wire, and titanium and were tested individually and the working electrode GO materials, electrolytes of NH_4Cl, KCl, or [dmin][BF_4] were used. After the reduction process, the large area GO were tested for electrical properties, having 28,200 S/m conductivity after the reduction.

2.2. Summary of green reducing methods

A summary of the above-mentioned green reducing methods and its reduction rate is presented in **Table 2**.

Reducing agent of GO	Reducing grade reported, based on XPS measurements	References
Shewanella	C–C bonds increased from 28% in GO to 90–95% in rGO	[23]
Gluconacetobacter xylinus	C/O ratio increased from 1.8 in GO to 3.1 in rGO	[27]
Camellia sinensis (green tea)	C–C bonds increased from 58.9% in GO to 74.0% in rGO	[32]
Vitamin C	C/O ratio increased from 2.3 for GO to 12.5 for rGO. Samples reduced with (hydrazine was 5.5)	[37]
Microwave irradiation	C/O ratio is ~9.12, closed to value for rGO obtained by conventional hydrazine reduction ~10	[48]
UV irradiation	Functional groups that contain oxygen in the GO nanosheets were successfully removed. However, complete elimination of oxygen groups must be due to longer UV irradiation	[51]
Poly (diallyldimethylammonium chloride) (PDDA)	C–C bonds increased from 24.5% in GO to 52.7% in rGO	[57]
Metallic zinc	C/O atomic ratio increased from 1.19 in GO to 7.19 in rGO	[60]
Ball milling	In the presence of H_2, a dramatic decrease in the oxygen-bonded carbon components is observed	[66]

Table 2. Summary of reductive green methods.

3. Self-assembly of micro- and nanostructures

Due to their amphiphilic character GO sheets are valuable building blocks for preparing a variety of carbon-based nano- and microsized nanostructures by a self-assembly process. Because GO sheets are few nanometers in size, their self-assembly hierarchy proceeds to develop 1D, 2D, and3D nano- and microsized materials. It thus enables to use templates for directing the self-assembly of GO sheets into complex structures with the specific shape and morphology for a given application.

It is noteworthy that the GO sheets' self-assembly was observed to occur at interfaces such as solid-liquid and air-liquid [12], and hence some hybrid metal- or metal oxide-GO nanocomposites comprise GO-coated inorganic nanoparticles.

Regarding GO self-assembled micro- and nanostructures, the presence of functional groups on the GO sheet surface promotes the assembly of nanoscale GO sheets into macroscopic 2D structures (films or fibers) and 3D bulk graphene by GO solution filtration or hydrothermal treatment. The forces that drive the self-assembly process are quite like those forces that participate in the self-assembly process of colloidal nanocrystals. Potential applications for these emergent structures are flexible fiber-type actuators, robots, motors, photovoltaic cells, and supercapacitors.

3.1. Metal-rGO hybrid nanocomposites

3.1.1. Broad classification of nanocomposites

Graphene and derivative materials are widely used to develop novel nanocomposites when combined with polymers and/or nanoparticles (semiconductors, metals, or metal oxides) [72]. These materials display superior physicochemical properties than those of their individual components and are currently essayed for water remediation, sensing, catalysis, photovoltaic films, materials reinforcement, and biomedical applications.

At present, a number of nanocomposites have been prepared by diverse methods and with specific physicochemical properties for biomedical [73], energy conversion, environmental and electrochemical storage [12], and miscellaneous [9] applications, as reported in recent review articles.

Among the variety of chemical and physical synthesis methods reported in the literature, we just include representative examples of the four classes of nanocomposites as described below.

According to their final morphology, rGO hybrid nanocomposites are broadly classified as supported, encapsulated, incorporated, and multilayered composites [7, 13]. Schematic representation of nanocomposites is presented in **Figure 4**.

3.1.2. Processing methods

The primary nanostructure GO sheets and nanoparticles (metal, metal oxide, or semiconductor) to develop nanocomposite materials are mainly processed by chemical methods.

The supported-rGO nanoparticles can be prepared by either direct synthesis of inorganic nanoparticle in the rGO dispersion or by mixing of previously prepared rGO and nanoparticles colloids. In the first approach, precursors are first dissolved in a convenient solvent, and then poured into the rGO dispersion. For preparing rGO-metal or rGO-metal oxide nanocomposites, the preferred method consists in adding the metal precursors (chlorides, nitrates, etc.) and a reducing agent (vitamin C, citric acid, L-lauric acid, etc.) into a previously pristine GO colloid. The whole processing can be performed using a range of synthesis systems such as microwave oven, hydrothermal, electrodeposition, sonication, and so on.

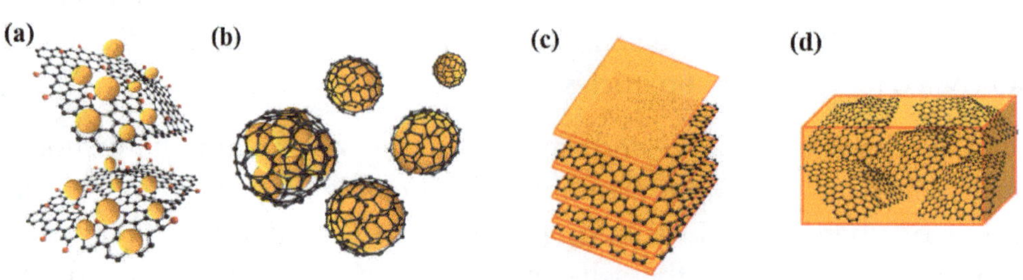

Figure 4. Kinds of metal-rGO nanocomposites. (a) Supported rGO surface is decorated by metal nanoparticles. (b) Encapsulated nanocomposites, few or multilayers of rGO are wrapping individual or clusters of MNP. (c) Incorporated layers of rGO are intercalated by metal layers. (d) rGO sheets are present in a metal matrix.

For instance, Kim et al. [74] used ascorbic acid to simultaneously reduce GO, Pd, Pt, Au, and Ag. An aqueous GO dispersion was kept at 100°C, then metal precursor and ascorbic acid solution were sequentially added. The final product consisted of rGO-supported nanosized noble metals. They used the rGO/Pd nanocomposite as catalytic material for Suzuki coupling reaction and observed that the nanocomposite catalytic activity was almost fully restored after five cycles.

An interesting version of supported-rGO nanoparticles approach is one in which rGO wraps the nanoparticle. In some cases, rGO sheets conformally enwrap the nanoparticle developing a hermetically sealed multilayer coating. The resultant composite comprises GO sheets decorated with a number of nanoparticle/rGO structures. These core/shell nanostructures are of current interest, since they are protected against oxidation or degradation. Chemical methods in solution and chemical vapor deposition have successfully been used to prepare such nanocomposites. They are used as lithium storage electrodes, high performance anodes, and biomedical applications.

Other interesting approach to obtain coated-rGO nanoparticles is the aerosol encapsulation technique reported by Chen et al. [75], to coat citric acid-stabilized Ag nanoparticles. These workers used an ultrasonic system to generate an aerosol composed of GO and Ag nanostructures, which was transported into a furnace at 600°C by using N_2 as the carrier gas. After draying, the individual drops transformed into a sample composed of Ag/rGO microstructures. This composite could be of interest for applications in tissue engineering, magnetic resonance imaging, X-ray computed tomography, and bioimaging contrast agent.

The layer-by-layer method is a thin film deposition technique in which alternating layers are successively deposited, and a film with a multilayered structure is obtained. Techniques such as immersion, spray coating, spin coating, and electrochemical are suitable ones to deposit multilayered nanocomposite films with large interfacial area. These kinds of nanocomposites are ideal for energy storage and generation.

Jang et al. [76] reported rGO/maghemite multilayered nanocomposite preparation. The GO was exfoliated by thermal expansion in vacuum at 200°C and then heated up to 300°C for 5 h. The powder of exfoliated GO, iron acetylacetonate, and oleic acid were mixed together and mechanically ground with a pestle and mortar. After heating at 600°C for 3 h, they obtained intercalated rGO-maghemite nanocomposite, and studied its performance as an anode material of Li ion batteries. In contrast with individually tested rGO and iron oxide samples, the nanocomposite displayed enhanced cycling stability and rate performance.

Incorporated nanocomposites have a low GO content (less than 1% vol) and are usually prepared by the ball-milling technique and a postsintering process. In this case, the properties to be exploited are mechanical and electrical properties, since they are applied in structural elements and implants.

Recently, the preparation of rGO incorporated in Al, Ni, Mg, and Cu matrices was reported [77]. It is expected that rGO sheets replace carbon nanotubes as a reinforcement material because rGO can be produced at large scale and at a lower cost. Zhang and Zhan [78] reported rGO-reinforced copper by ball milling and spark plasma sintering. They found that the presence of 0.1–1% vol rGO greatly enhances its mechanical properties (yield and tensile strengths) compared to those of pure Cu.

3.2. GO thin films and membrane

Currently, the GO self-assembled micro- and nanostructures are being essayed as semiconductor in thin film transistors, transparent electrode of solar cells, active material for chemical sensing, etc. These applications require paper-like and thin film of self-assembled GO nanosheets.

There are plenty of reports on the GO self-assembly into 2D and micro- and nanostructures [79, 80]. Shao et al. [14] did an exhaustive description of the mechanistic aspects of the GO self-assembly at diverse interfaces. Herein, we present the more recent findings on 2D microstructures, including thin films.

Until now several self-assembly mechanisms to form GO thin films have been proposed. The GO thin films formation frequently occurs at liquid-air-type interfaces by evaporation and Langmuir-Blodgett assembly [81, 82]. The techniques employed are simple in almost all cases and also let the assembly on the suspension surface, for instance, dip-coating, drop-casting, spin-coating, and spray-coating for mentioning some of them.

Langmuir-Blodgett assembly leads to the formation of GO very thin films and GO single layer, so when GO is obtained, it can be dispersed in a highly volatile organic solvent in the presence of little amounts of water, and then as the solvent evaporates, the GO begins to aggregate at the interface of water-air, forming a GO monolayer. The assembled material at the interfaces can be collected later by dipping a substrate [83]. The main advantage lies in the collected GO very thin films on the substrate, and they are a source to reduce the GO into graphene sheets by either chemical or thermal treatment [19, 84].

The resulting GO films have high transmittance, high surface chemical activity, and low sheet resistance. The morphology and dispersion degree can be modified by the pH of the synthesis solution; so, the pH modulates the amphiphilic nature of GO layers as evidenced by Cote et al. [85]. The self-assembled GO films are dominated by attractive forces as van der Waals forces and π–π interaction which lead to stacking of single GO layer.

In the evaporation case, the solvent is heated to accelerate its evaporation and the agglomeration of GO sheets in the interface of water-air, any solvent with relatively low evaporation point can be used to disperse and afterward evaporate it to promote the self-assembly on the solvent surface. In this way, it is possible to obtain both thin films and membranes. Similarly, the Langmuir-Blodgett assembly and evaporation induced lead to the formation of GO membranes by staking layer-by-layer (**Figure 5**). Cote and Shao [85, 86] provide examples of self-assembly using the Langmuir-Blodgett and evaporation-induced mechanisms.

As started below, the GO membrane can be considered as 2D microstructures with a morphology that depends either on the attractive or repulsive interaction among individual GO layers.

3.3. 1D and 2D microstructures (membranes and fibers) and 3D bulk structures

The GO sheet self-assembly can produce thin films and 2D membranes as aforementioned; however, this would occur at a liquid-air interface. If the interface is now liquid-solid type, a variety of microstructures of 2D and 3D can be obtained by self-assembly of single GO layers in the presence of a solid. The GO sheets' interactions with the solid surface involve π–π interaction, hydrogen bonding, electrostatic forces, and surface tension.

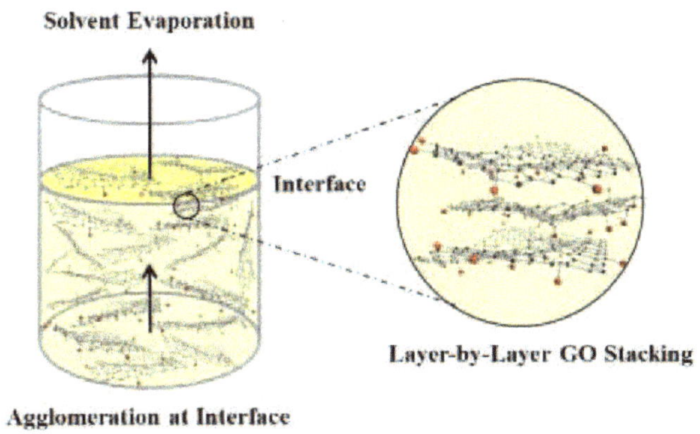

Figure 5. Self-assembly process into GO thin layers at the interface of liquid-air. The assembly is assisted by the solvent evaporation.

GO self-assembly driven by electrostatic forces produce several morphologies of 2D and 3D such as thin films, membranes, and capsules [87]. In this case, the ionization of COOH groups provides a negative charge distribution at the GO sheet edge. This charge distribution can be controlled by the pH of GO dispersion. An electric field applied in the GO solution is able to drift the negatively charged GO layers toward the positive electrode (solid element) dipped in the dispersion. GO sheets are agglomerated on the electrode surface and forced to assemble by staking. An interesting effect is presented during the GO sheets drift, since the GO layers that assemble are electrochemically reduced. So the electric field can remove the oxygen-based functional groups [88] and promote the assembly only by π–π interaction.

The presence of particles or nanoparticles in a GO dispersion can also drive the self-assembly, in this case foreign particles act as agglomeration centers, which destabilize the GO dispersion. This can be considered as a GO colloid in a disperse state [83, 89]. Therefore, attractive forces originated from particles overcome the electrostatic repulsion (colloidal stabilization) and lead to agglomeration and assembly.

Three-dimensional GO structures with polyhedral-like morphology were reported in Ref. [90]. In this case, the self-assembly process was observed at microscale. GO is synthesized by Hummers' method and the material is patterned on a silicon wafer. The patterned 2D GO membranes become the building blocks. A metallic frame is deposited around the GO membranes; it drives the folding of each membrane by surface tension forces. Considering the membrane size, the attractive forces as van der Waals are not manifested at this scale; therefore, surface tension forces are dominant here. These 3D cubes can be used as microcontainer of liquids and gases. Different assembly stages are shown in **Figure 6**, initially any material can be patterned as isolated blocks, **Figure 6(a)**, and then the metallic frame is deposited by a photolithographic process, **Figure 6(b)**. The metallic frame is constituted by two different metals, which linked the assembly blocks and allow the self-folding [91].

1D GO microfibers have been obtained by self-assembly of single GO layers; however, this is an example of microstructures' self-assembly at a liquid-air-type interface. As reported by Tian et al. [92], which these 1D fibers are formed by two forces combination, π–π interaction and van der Waals attractive forces are gradually manifested in the GO dispersion. These forces drive the single GO layers agglomeration toward the GO dispersion surface and staking layer

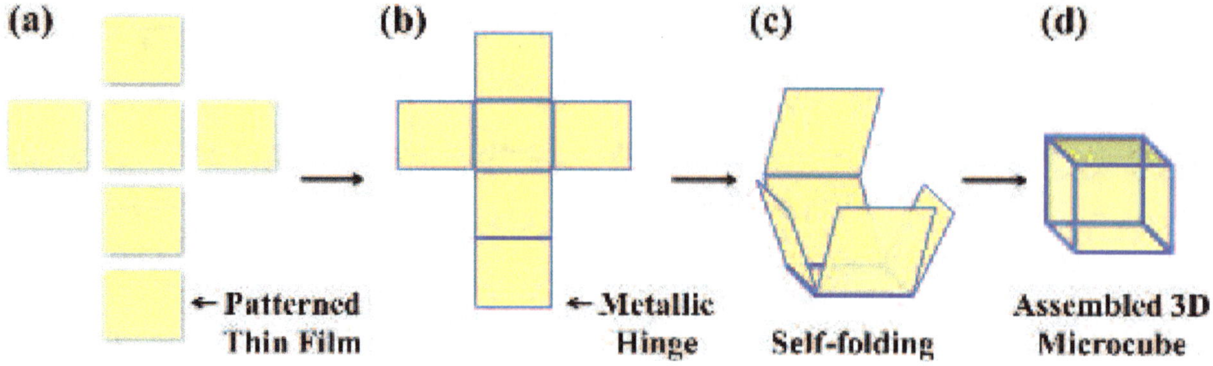

Figure 6. Self-assembly process of 3D microcubes, schematizing different stages by self-folding.

by layer. Then, progressive accumulation of GO sheets produces these 1D GO microfibers. The fibers are annealed later to obtain rGO. The authors in this case do not provide further information about the rolling up process that give place to the 1D fibers.

4. Summary

Synthesis, reduction, and advanced application of graphene oxide (GO) are fast growing research areas because there exist a great variety of preparation techniques for mass production, the chemical-based ones being the most promising. For its chemical richness, chemically obtained GO is an extraordinary product in various aspects. First, it can be obtained by means of scalable, simple, and low-cost techniques, which is important for gram- or kilogram-scale applications (e.g., rGO-metal-based composites for the lithium battery anode, rGO-based foams, water cleaning, etc.). Second, it has demonstrated to be an excellent precursor material for developing advanced materials, such as graphene, graphane when treated under hydrogen atmosphere, and Teflon-like materials when fluorinated.

This chapter presents an overview on the GO reduction by green methods, on the production methods of carbon-based structures by GO sheets self-assembly, and on preparation methods of GO-based metal nanocomposites.

The so-called green methods for GO reduction demand that both, starting chemicals and byproducts, are safe to handle and environmentally friendly. Technologies such as bioreduction, photoreduction, reduction by polymers, reduction by metals, mechanochemical reduction, and electrochemical reduction fulfill both criteria.

On the other hand, the amphiphilic character of GO sheets make them valuable as building blocks for preparing a variety of carbon-based structures produced by their self-assembly, as well as hybrid nanocomposites when combined with metal semiconductor nanoparticles. The self-assembled carbon structures and hybrid nanocomposites are currently essayed for water remediation, sensing, catalysis, photovoltaic films, materials reinforcement, and biomedical applications.

Author details

Rebeca Ortega-Amaya[1*], Yasuhiro Matsumoto[1,2], Esteban Díaz-Torres[2], Claudio Davet Gutierrez-Lazos[3], Manuel Alejandro Pérez-Guzmán[2] and Mauricio Ortega-López[1,2]

*Address all correspondence to: ortegaa@cinvestav.mx

1 SEES, Electrical Engineering Department, Center for Research and Advanced Studies of the National Polytechnic Institute, Mexico City, Mexico

2 Nanoscience and Nanotechnology Program, Center for Research and Advanced Studies of the National Polytechnic Institute, Mexico City, Mexico

3 Center for Research in Physical Mathematical Sciences, Faculty of Mathematical Physical Sciences, The Autonomous University of Nuevo León, San Nicolás de los Garza, Nuevo León, Mexico

References

[1] Novoselov K S, Geim A K, Morozov S V, Jiang D, Zhang Y, Dubonos S V, Grigorieva I V and Firsov A A. 2004. Electric field effect in atomically thin carbon films. Science 306; 666–9.

[2] Taiichi O, Tetsuya S, Akira S, Maki S, Eiichi S, Maxim R and Victor R. 2011. Graphene. CRC Press: Boca Raton, Florida. pp. 85–116.

[3] Huang X, Qi X, Boey F and Zhang H. 2012. Graphene-based composites. Chem. Soc. Rev. 41; 666–86.

[4] Geim A K and Grigorieva I V. 2013. Van der Waals heterostructures. Nature 499; 419–25.

[5] Pumera M and Wong C H A. 2013. Graphane and hydrogenated graphene. Chem. Soc. Rev. 42; 5987–95.

[6] Huh S H. 2011 Physics and Applications of Graphene Experiments, ed S. Mikhailov. InTech: Rijeka, Croatia.

[7] Bai H, Li C and Shi G. 2011. Functional composite materials based on chemically converted graphene. Adv. Mater. 23; 1089–115.

[8] Compton O C and Nguyen S T. 2010. Graphene oxide, highly reduced graphene oxide, and graphene: versatile building blocks for carbon-based materials. Small 6; 711–23.

[9] Khan M, Tahir M N, Adil S F, Khan H U, Siddiqui M R H, Al-warthan A A and Tremel W. 2015. Graphene based metal and metal oxide nanocomposites: synthesis, properties and their applications. J. Mater. Chem. A 3; 18753–808.

[10] Cai M, Thorpe D, Adamson D H and Schniepp H C. 2012. Methods of graphite exfoliation. J. Mater. Chem. 22; 24992–5002.

[11] Dreyer D R, Park S, Bielawski C W and Ruoff R S. 2010. The chemistry of graphene oxide. Chem. Soc. Rev. 39; 228–40.

[12] Li F, Jiang X, Zhao J and Zhang S. 2015. Graphene oxide: a promising nanomaterial for energy and environmental applications. Nano Energy 16; 488–515.

[13] Bai S and Shen X. 2012. Graphene-inorganic nanocomposites. RSC Adv. 2; 64–98.

[14] Shao J-J, Lv W and Yang Q-H. 2014. Self-assembly of graphene oxide at interfaces. Adv. Mater. 26; 5586–612.

[15] Santanu D and Wonbong C. 2011. Graphene. CRC Press: Boca Raton, Florida. pp. 27–64.

[16] Aunkor M T H, Mahbubul I M, Saidur R and Metselaar H S C. 2016. The green reduction of graphene oxide. RSC Adv. 6; 27807–28.

[17] Gurunathan S, Han J W, Eppakayala V, Dayem A A, Kwon D-N and Kim J-H. 2013. Biocompatibility effects of biologically synthesized graphene in primary mouse embryonic fibroblast cells. Nanoscale Res. Lett. 8; 393.

[18] Ortega-Amaya R, Matsumoto Y, Flores-Conde A, Pérez-Guzmán M A and Ortega-López M. 2016. In situ formation of rGO quantum dots during GO reduction via interaction with citric acid in aqueous medium. Mater. Res. Express 3; 105601.

[19] Pei S and Cheng H-M. 2012. The reduction of graphene oxide. Carbon 50; 3210–28.

[20] Wolfe-Simon F, Blum J S, Kulp T R, Gordon GW, Hoeft S E,Pett-Ridge J, Stolz J F, Webb S M, Weber P K, Davies P C W, Anbar A D and Oremland R S. 2011. A bacterium that can grow by using arsenic instead of phosphorus. Science 332; 1163–6.

[21] Agharkar M, Kochrekar S, Hidouri S and Azeez M A. 2014. Trends in green reduction of graphene oxides, issues and challenges: a review. Mater. Res. Bull. 59; 323–8.

[22] Salas E C, Sun Z, Lüttge A and Tour J M. 2010. Reduction of graphene oxide via bacterial respiration. ACS Nano; 4 4852–6.

[23] Wang G, Qian F, Saltikov C W, Jiao Y and Li Y. 2011. Microbial reduction of graphene oxide by *Shewanella*. Nano Res. 4; 563–70.

[24] Zhang H, Yu X, Guo D, Qu B, Zhang M, Li Q and Wang T. 2013. Synthesis of bacteria promoted reduced graphene oxide-nickel sulfide networks for advanced supercapacitors. ACS Appl. Mater. Interfaces 5; 7335–40.

[25] Raveendran S, Chauhan N, Nakajima Y, Toshiaki H, Kurosu S, Tanizawa Y, Tero R, Yoshida Y, Hanajiri T, Maekawa T, Ajayan P M, Sandhu A and Kumar D S. 2013. Ecofriendly route for the synthesis of highly conductive graphene using extremophiles for green electronics and bioscience Part. Part. Syst. Charact. 30; 573–8.

[26] Akhavan O and Ghaderi E. 2012. *Escherichia coli* bacteria reduce graphene oxide to bactericidal graphene in a self-limiting manner. Carbon 50; 1853–60.

[27] Nandgaonkar A G, Wang Q, Fu K, Krause W E, Wei Q, Gorga R and Lucia L A. 2014. A one-pot biosynthesis of reduced graphene oxide (RGO)/bacterial cellulose (BC) nanocomposites. Green Chem. 16; 3195–201.

[28] Schütz B, Seidel J, Sturm G, Einsle O and Gescher J. 2011. Investigation of the electron transport chain to and the catalytic activity of the diheme cytochrome c peroxidase CcpA of *Shewanella oneidensis*. Appl. Environ. Microbiol. 77; 6172–80.

[29] Iravani S. 2011. Green synthesis of metal nanoparticles using plants. Green Chem. 13; 2638–50.

[30] Thakur S and Karak N. 2012. Green reduction of graphene oxide by aqueous phytoextracts. Carbon 50; 5331–9.

[31] Thakur S and Karak N. 2015. Alternative methods and nature-based reagents for the reduction of graphene oxide: a review. Carbon 94; 224–42.

[32] Wang Y, Shi Z and Yin J. 2011. Facile synthesis of soluble graphene via a green reduction of graphene oxide in tea solution and its biocomposites. ACS Appl. Mater. Interfaces 3; 1127–33.

[33] Hou D, Liu Q, Cheng H, Li K, Wang D and Zhang H. 2016. Chrysanthemum extract assisted green reduction of graphene oxide. Mater. Chem. Phys. 183; 76–82.

[34] Hou D, Liu Q, Cheng H, Zhang H and Wang S. 2017. Green reduction of graphene oxide via *Lycium barbarum* extract. J. Solid State Chem. 246; 351–6.

[35] Gao J, Liu F, Liu Y, Ma N, Wang Z and Zhang X. 2010. Environment-friendly method to produce graphene that employs vitamin C and amino acid. Chem. Mater. 22; 2213–8.

[36] Zhang J, Yang H, Shen G, Cheng P, Zhang J and Guo S. 2010. Reduction of graphene oxide via L-ascorbic acid. Chem. Commun. 46; 1112–4.

[37] Fernández-Merino M J, Guardia L, Paredes J I, Villar-Rodil S, Solís-Fernández P, Martínez-Alonso A and Tascón J M D. 2010. Vitamin C is an ideal substitute for hydrazine in the reduction of graphene oxide suspensions. J. Phys. Chem. C. 114; 6426–32.

[38] Fernández-Merino MJ, Villar-Rodil S, Paredes J I, Solís-Fernández P, Guardia L, García R, Martínez-Alonso A and Tascón J M D. 2013. Identifying efficient natural bioreductants for the preparation of graphene and graphene-metal nanoparticle hybrids with enhanced catalytic activity from graphite oxide. Carbon 63; 30–44.

[39] Zhu C, Guo S, Fang Y and Dong S. 2010. Reducing sugar: new functional molecules for the green synthesis of graphene nanosheets. ACS Nano 4; 2429–37.

[40] Akhavan O, Ghaderi E, Aghayee S, Fereydooni Y and Talebi A. 2012. The use of a glucose-reduced graphene oxide suspension for photothermal cancer therapy. J. Mater. Chem. 221; 3773–81.

[41] Kim Y-K, Kim M-H and Min D-H. 2011. Biocompatible reduced graphene oxide prepared by using dextran as a multifunctional reducing agent. Chem. Commun. 47; 3195–7.

[42] Chen D, Li L, Guo L. 2011. An environment-friendly preparation of reduced graphene oxide nanosheets via amino acid. Nanotechnology 22; 325601.

[43] Bose S, Kuila T, Mishra A K, Kim N H and Lee J H. 2012. Dual role of glycine as a chemical functionalizer and a reducing agent in the preparation of graphene: an environmentally friendly method. J. Mater. Chem. 22; 9696–703.

[44] Ma J, Wang X, Liu Y, Wu T, Liu Y, Guo Y, Li R, Sun X, Wu F, Li C and Gao J. 2013. Reduction of graphene oxide with l-lysine to prepare reduced graphene oxide stabilized with polysaccharide polyelectrolyte. J. Mater. Chem. A 1; 2192–201.

[45] Tran D N H, Kabiri S and Losic D. 2014. A green approach for the reduction of graphene oxide nanosheets using non-aromatic amino acids. Carbon 76; 193–202.

[46] Xu C, Wang X and Zhu J. 2008. Graphene metal particle nanocomposites. J. Phys. Chem. C 112; 19841–5.

[47] Li J, Xiao G, Chen C, Li R and Yan D. 2013. Superior dispersions of reduced graphene oxide synthesized by using gallic acid as a reductant and stabilizer. J. Mater. Chem. A 1; 1481–7.

[48] Lu T, Pan L, Nie C, Zhao Z and Sun Z. 2011. A green and fast way for reduction of graphene oxide in acidic aqueous solution via microwave assistance. Physica Status Solidi(a) 208; 2325–7.

[49] Tien H N, Luan V H, Hoa L T, Lee T K, Kong B-S, Chung J S, Kim E J and Hur S H. 2013. The rapid and enhanced reduction of graphene oxide by microwave assisted acid catalyzed reaction. J. Nanosci. Nanotechnol. 13; 7104–7.

[50] Voiry D, Yang J, Kupferberg J, Fullon R, Lee C, Jeong H Y, Shin H S and Chhowalla M. 2016. High-quality graphene via microwave reduction of solution-exfoliated graphene oxide. Science 353; 1413–6.

[51] Ding Y H, Zhang P, Zhuo Q, Ren H M, Yang Z M and Jiang Y. 2011. A green approach to the synthesis of reduced graphene oxide nanosheets under UV irradiation. Nanotechnology 22; 215601.

[52] Quickenden T I and Irvin J A. 1980. The ultraviolet absorption spectrum of liquid water. J. Chem. Phys. 72; 4416–28

[53] Nyangiwe N N K M, Thema F T, Nukwa K, Kotsendi L and Mazza M. 2015. Free-green synthesis and dynamics of reduced graphene sheets via sun light irradiation. Graphene 4; 7

[54] Li D, Muller M B, Gilje S, Kaner R B and Wallace G G. 2008. Processable aqueous dispersions of graphene nanosheets. Nat Nano 3; 101–5

[55] Ji T, Hua Y, Sun M and Ma N. 2013. The mechanism of the reaction of graphite oxide to reduced graphene oxide under ultraviolet irradiation. Carbon 54; 412–8.

[56] Yang D-Q, Rochette J-F and Sacher E. 2005. Spectroscopic evidence for π-π interaction between poly(diallyl dimethylammonium) chloride and multiwalled carbon nanotubes. J. Phys. Chem. B 109; 4481–4.

[57] Zhang S, Shao Y, Liao H, Engelhard M H, Yin G and Lin Y. 2011. Polyelectrolyte-induced reduction of exfoliated graphite oxide: a facile route to synthesis of soluble graphene nanosheets. ACS Nano 5; 1785–91.

[58] Zhang S, Shao Y, Yin G and Lin Y. 2011. Self-assembly of Pt nanoparticles on highly graphitized carbon nanotubes as an excellent oxygen-reduction catalyst. Appl. Catal. B Environ. 102; 372–7.

[59] Fan Z-J, Kai W, Yan J, Wei T, Zhi L-J, Feng J, Ren Y-m, Song L-P and Wei F. 2011. Facile synthesis of graphene nanosheets via Fe reduction of exfoliated graphite oxide. ACS Nano 5; 191–8.

[60] Yang S, Yue W, Huang D, Chen C, Lin H and Yang X. 2012. A facile green strategy for rapid reduction of graphene oxide by metallic zinc. RSC Adv. 2; 8827–32.

[61] Mei X and Ouyang J. 2011. Ultrasonication-assisted ultrafast reduction of graphene oxide by zinc powder at room temperature. Carbon 49; 5389–97.

[62] Liu P, Huang Y and Wang L. 2013. A facile synthesis of reduced graphene oxide with Zn powder under acidic condition. Mater. Lett. 91; 125–8.

[63] Wu T, Gao J, Xu X, Wang W, Gao C and Qiu H. 2013. A new rapid chemical route to prepare reduced graphene oxide using copper metal nanoparticles. Nanotechnology 24; 215604.

[64] Cao X, Qi D, Yin S, Bu J, Li F, Goh C F, Zhang S and Chen X. 2013. Ambient fabrication of large-area graphene films via a synchronous reduction and assembly strategy. Adv. Mater. 25; 2957–62.

[65] Hu C, Zhai X, Liu L, Zhao Y, Jiang L and Qu L. 2013. Spontaneous reduction and assembly of graphene oxide into three-dimensional graphene network on arbitrary conductive substrates. Sci. Rep. 3; 2065.

[66] Chang D W, Choi H-J, Jeon I-Y, Seo J-M, Dai L and Baek J-B. 2014. Solvent-free mechanochemical reduction of graphene oxide. Carbon 77; 501–7.

[67] Kumar N A, Choi H-J, Shin Y R, Chang DW, Dai L and Baek J-B. 2012. Polyaniline-grafted reduced graphene oxide for efficient electrochemical supercapacitors. ACS Nano 6; 1715–23.

[68] Jeon I-Y, Choi H-J, Jung S-M, Seo J-M, Kim M-J, Dai L and Baek J-B. 2013. Large-scale production of edge-selectively functionalized graphene nanoplatelets via ball milling and their use as metal-free electrocatalysts for oxygen reduction reaction. J. Am. Chem. Soc. 135; 1386–93.

[69] Jeon I-Y, Shin Y-R, Sohn G-J, Choi H-J, Bae S-Y, Mahmood J, Jung S-M, Seo J-M, Kim M-J, Wook Chang D, Dai L and Baek J-B. 2012. Edge-carboxylated graphene nanosheets via ball milling. Proc. Natl. Acad. Sci. 109; 5588–93.

[70] Toh S Y, Loh K S, Kamarudin S K and Daud W R W. 2014. Graphene production via electrochemical reduction of graphene oxide: synthesis and characterization. Chem. Eng. J. 251; 422–34.

[71] Feng X, Chen W and Yan L. 2016. Electrochemical reduction of bulk graphene oxide materials. RSC Adv. 6; 80106–13.

[72] Singh V, Joung D, Zhai L, Das S, Khondaker S I and Seal S. 2011. Graphene based materials: past, present and future. Prog. Mater. Sci. 56; 1178–271.

[73] Yin P T, Shah S, Chhowalla M and Lee K-B. 2015. Design, synthesis, and characterization of graphene–nanoparticle hybrid materials for bioapplications. Chem. Rev. 115; 2483–531.

[74] Kim S-H, Jeong G H, Choi D, Yoon S, Jeon H B, Lee S-M and Kim S-W. 2013. Synthesis of noble metal/graphene nanocomposites without surfactants by one-step reduction of metal salt and graphene oxide. J. Colloid Interface Sci. 389; 85–90.

[75] Chen Y, Guo F, Jachak A, Kim S-P, Datta D, Liu J, KulaotsI, Vaslet C, Jang H D, Huang J, Kane A, Shenoy V B and Hurt R H. 2012. Aerosol synthesis of cargo-filled graphene nanosacks. Nano Lett. 12; 1996–2002.

[76] Jang B, Chae O B, Park S-K, Ha J, Oh S M, Na H B and Piao Y. 2013. Solventless synthesis of an iron-oxide/graphene nanocomposite and its application as an anode in high-rate Li-ion batteries. J. Mater. Chem. A 1; 15442–6.

[77] Bartolucci S F, Paras J, Rafiee M A, Rafiee J, Lee S, Kapoor D and Koratkar N. 2011. Graphene–aluminum nanocomposites. Mater. Sci. Eng. A 528; 7933–7.

[78] Zhang D and Zhan Z. 2016. Strengthening effect of graphene derivatives in copper matrix composites. J. Alloys Compd. 654; 226–33.

[79] Jin Y, Xue Q, Zhu L, Li X, Pan X, Zhang J, Xing W, Wu T and Liu Z. 2016. Self-assembly of hydrofluorinated janus graphene monolayer: a versatile route for designing novel janus nanoscrolls. Sci. Rep. 6; 26914.

[80] Sudeep P M, Narayanan T N, Ganesan A, Shaijumon MM, Yang H, Ozden S, Patra P K, Pasquali M, Vajtai R, Ganguli S, Roy A K, Anantharaman M R and Ajayan P M. 2013. Covalently interconnected three-dimensional graphene oxide solids. ACS Nano. 7; 7034–40.

[81] Zheng Q, Ip W H, Lin X, Yousefi N, Yeung K K, Li Z and Kim J-K. 2011. Transparent conductive films consisting of ultralarge graphene sheets produced by Langmuir–Blodgett assembly. ACS Nano 5; 6039–51.

[82] Cote L J, Kim F and Huang J. 2009. Langmuir Blodgett assembly of graphite oxide single layers. J. Am. Chem. Soc. 131; 1043–9.

[83] Kim J, Cote L J, Kim F, Yuan W, Shull K R and Huang J. 2010. Graphene oxide sheets at interfaces. J. Am. Chem. Soc. 132; 8180–6.

[84] Gao X, Jang J and Nagase S. 2010. Hydrazine and thermal reduction of graphene oxide: reaction mechanisms, product structures, and reaction design. J. Phys. Chem. C 114; 832–42.

[85] Cote L J, Kim J, Zhang Z, Sun C and Huang J. 2010. Tunable assembly of graphene oxide surfactant sheets: wrinkles, overlaps and impacts on thin film properties. Soft Matter 6; 6096–101.

[86] Shao J-J, Lv W, Guo Q, Zhang C, Xu Q, Yang Q-H and Kang F. 2012. Hybridization of graphene oxide and carbon nanotubes at the liquid/air interface. Chem. Commun. 48; 3706–8.

[87] Zou J and Kim F. 2012. Self-assembly of two-dimensional nanosheets induced by interfacial polyionic complexation. ACS Nano 6; 10606–13.

[88] An S J, Zhu Y, Lee S H, Stoller M D, Emilsson T, Park S, Velamakanni A, An J and Ruoff R S. 2010. Thin film fabrication and simultaneous anodic reduction of deposited graphene oxide platelets by electrophoretic deposition. J. Phys. Chem. Lett. 1; 1259–63.

[89] Leite E R and Ribeiro C. 2012. Crystallization and Growth of Colloidal Nanocrystals. Springer: New York, New York.

[90] Joung D, Gu T and Cho J H. 2016. Tunable optical transparency in self-assembled three-dimensional polyhedral graphene oxide. ACS Nano 10; 9586-94.

[91] Leong T G, Zaragshar A M and Gracias D H. 2010. Three-dimensional fabrication at small size scales. Small 6; 792-806.

[92] Tian Z, Xu C, Li J, Zhu G, Shi Z and Lin Y. 2013. Self-assembled free-standing graphene oxide fibers. ACS Appl. Mater. Interfaces 5; 1489–93.

2

Graphene Derivatives: Controlled Properties, Nanocomposites, and Energy Harvesting Applications

Ulises Antonio Méndez Romero,
Miguel Ángel Velasco Soto, Liliana Licea Jiménez,
Jaime Álvarez Quintana and
Sergio Alfonso Pérez García

Abstract

Graphene is a ground-breaking two-dimensional (2D) material that possesses outstanding electrical, optical, thermal, and mechanical properties and that promises a new generation of devices. Despite all these, some applications require graphene-based materials with different characteristics, such as good solubility in organic solvents and a specific band gap to be dispersible in polymer nanocomposite matrix and applied as active layer, electron transport layer (ETL) or hole transport layer (HTL) in organic photovoltaics. Chemically modified graphene derivatives are studied, searching for better dispersions and even more properties for different applications. Most of the attention has been drawn to dispersions of graphene oxides or highly reduced graphene oxides. Therefore, this allows an opportunity to study the characteristics of materials with intermediate oxidation degrees and its applications.

Keywords: graphene oxide, reduced graphene oxide, band gap modification, OPVs, reduced graphene oxide functionalized

1. Introduction

Graphene, a unique two-dimensional carbon material with exceptional properties as high electrical and thermal conductivity, zero band gap, high optical transmittance and superior mechanical performance than steel [1], is often regarded as the material of the twenty-first century. A great focus is being added to graphene in all scientific fields, such as physics, chemistry, and even medicine. Nanocomposites, dispersions, and films that include graphene in a polymer matrix are the trending topic of recent research studies. With graphene science on the growth, more investigation lines are getting opened, such as graphene synthesis and modification.

Certain applications require some characteristics that even graphene, with all its properties, cannot provide. This is the main reason for the continuous search for chemical modifications to graphene: graphene derivatives, which have an important role to play in recent research. It is a well-known fact that graphene oxide (GO) is one of the most studied derivatives. GO is graphene with carbon-oxygen bonds and has functionalities such as hydroxyl, carbonyl, and carboxyl groups (**Figure 1**). Therefore, GO is a highly hydrophilic molecule and, more important for electronics and energy-harvesting applications, this process provides a band gap (E_g). For example, the energy difference between the highest occupied molecular orbital (HOMO) and the lowest unoccupied molecular orbital (LUMO) in the resulting complete oxidized material is around 3 eV [2], which is also directly related to the carbon-oxygen ratio [3]. Oxidation reactions include strong acid attacks with H_2SO_4, HNO_3, H_2O_2, $KMnO_4$, or a mixture of them, being one of the most used by the Hummers method [4].

Most of the methods for quantification of carbon-oxygen functional groups include Boehm's titration [6], which uses different basic solutions to react in sequence with carboxyl, carbonyl, and hydroxyl moieties. Another quantification method is X-ray photoelectron spectroscopy (XPS), which uses the area result from fitting the signals corresponding to the binding energy of the sample for different carbon peaks. XPS can be used also to quantify sp^2 and sp^3 domains.

One of the most interesting applications for this type of materials is organic photovoltaics (OPVs). In this chapter, various approaches are explored to develop materials with a possible application in OPVs, based on modified GO into the active layer. It is well known that GO has a band gap of 3.1 eV, which can be further modulated by chemical reduction reactions [2]. The

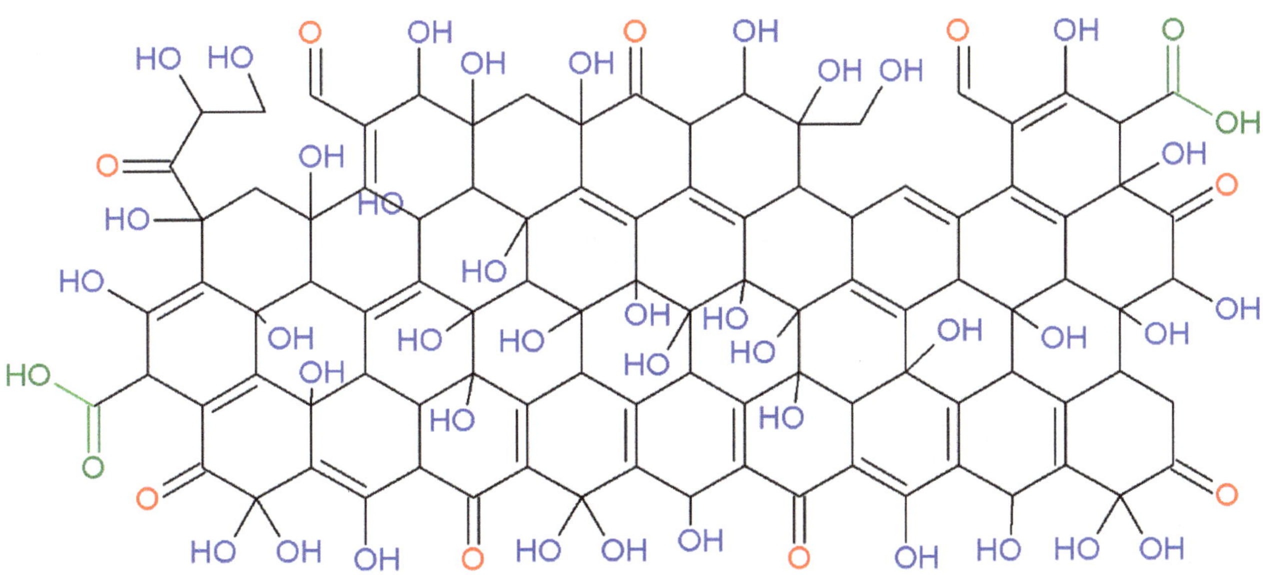

Figure 1. Graphene oxide structure in Lerf-klinowsky Model and different C-O functionalities. Hydroxyl in blue, carbonyl in red, and carboxyl in green [5].

reduction process can be carried out by innocuous agents, such as fructose, glucose, ascorbic acid, or ascorbic acid-6-palmitate, obtaining band gaps around 1.1 eV.

In recent years, alternatives have been explored with organic compounds (i.e., OPVs, also known as excitonic cells [7]), which have a great number of advantages compared with traditional silicon photovoltaics. Some of the main advantages are a lower weight, flexibility, simple mass production, and low cost. On the other hand, the OPVs have the limitation of low power conversion efficiency (PCE), compared to silicon based solar cells, with a maximum of 11% [8, 9], offering a big opportunity area for improvement. The PCE is the ratio between the maximum power in watts produced by the cell ($P_{\max}$), divided by the power in watts of the incident light (ϕ_e) cf. Eq. (1).

$$PCE = \frac{P_{\max}}{\phi_e} \tag{1}$$

The most efficient OPVs are based on the concept of bulk heterojunction (BHJ) [10]. In this type of solar cell, presented in **Figure 2**, where electron donor material (usually polymer) is mixed with an electron-accepting material (derived from fullerenes in most cases) in an organic solvent (1,2-dichlorobenzene, oDCB, for example) and subsequently deposited by spin coating on an indium-tin oxide (ITO) layer previously deposited on a glass or PET substrate. During evaporation of the solvent and subsequent treatments, an interconnected bicontinuous microphase is generated, which creates a large interfacial area between the donor material and the acceptor material (i.e., the BHJ), also called photoactive layer or simply active layer. The next layer deposited is lithium fluoride (LiF), which prevents the transportation of the holes, followed by a metal electrode (aluminum for example). These two layers are deposited by sputtering, inside a glove box in atmosphere of N_2 [11].

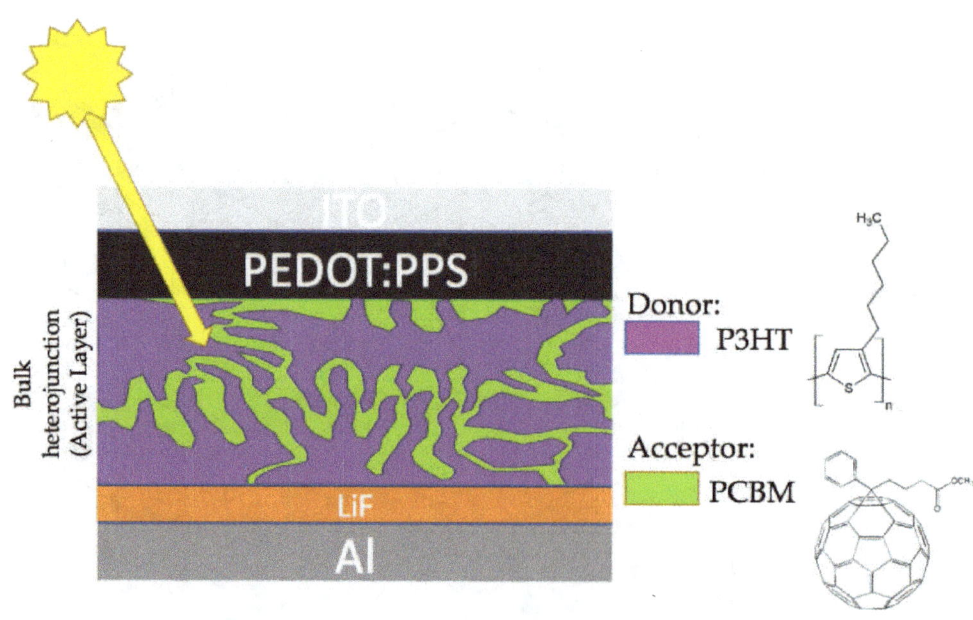

Figure 2. Typical diagram of an OPV-BHJ [12].

The heterojunction is important because of the large donor-acceptor interface that is created since it is precisely there, where the separation of charges takes place.

Another parameter used to describe OPVs is the external quantum efficiency (EQE), which is a parameter that indicates the ratio of the photogenerated electrons related to the incident photons. This depends on five factors: absorption of the light (η_A), diffusion of exciton (η_{diff}), dissociation of exciton (η_{diss}), transport of charges through films (η_{tr}), and charge collection on electrodes (η_{cc}) [13]. The equation of quantum efficiency is given below:

$$EQE = \eta_A \eta_{\mathrm{diff}} \eta_{\mathrm{diss}} \eta_{tr} \eta_{cc} \tag{2}$$

Finally, the fill factor (FF) is another term that characterizes the device and defines the efficiency of a cell and is given in Eq. (3).

$$FF = \frac{J_{sc} V_{oc}}{\phi_e} = \frac{J_m V_m}{J_{sc} V_{oc}} \tag{3}$$

Where J_{sc} is the short circuit current density, V_{oc} is the open circuit voltage, and J_m and V_m are the maximum current and voltage, respectively. These values are obtained by an electrical characterization of the device, using typical curves of current against voltage, through two measurements, one in darkness and the other under illumination.

The common feature of the materials used in the active layer of the OPVs is that they are small molecules or polymers with conjugated bonds (i.e., carbon-carbon bonds in alternating sp^2 hybridization), allowing the continuous transportation of electrons along the molecular structure. In addition, these materials must have adequate values of HOMO, LUMO, and E_g. Some examples of these materials are poly(3-hexylthiophene) (P3HT), [6,6]-phenyl C61 butyric acid methyl ester (PCBM), or [6,6]-phenyl C71 butyric acid methyl ester ($PC_{71}BM$).

A further aspect to consider is the solubility of graphene oxide in organic solvents, normally used for manufacturing OPVs such as 1,2-dichlorobenzene, chloroform, or toluene.

2. Graphene oxide

Graphene oxide (GO) is a molecule of graphene with oxygen functional groups (carbonyl, hydroxyl, epoxy, and carboxyl), synthesized from graphite, either by the method of Brodie, Staudenmaier, or Hummers [4, 14, 15] (structure as shown in **Figure 1**). GO has an E_g related to the stoichiometric ratio between carbon and oxygen, resulting in a zero band gap (0% oxygen) around 3 eV for graphene oxide with a proportion of 50% oxygen, according to the computational calculations reported elsewhere [3]. **Figure 3** shows the chemical reaction in Hummers method.

It is important to mention that **Figure 3** corresponds to the synthesis of graphite oxide, which is a bright yellow dispersion that needs a multiple step washing with DI H_2O in centrifuge and

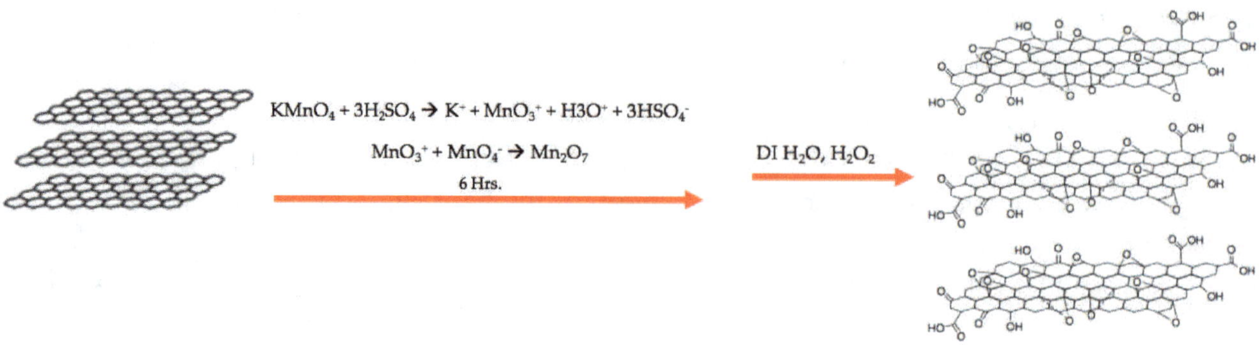

Figure 3. Chemical oxidation in Hummers' method [4].

exfoliation in ultrasonic bath in order to obtain the graphene oxide. The latter is a brown-reddish dispersion with a characteristic UV-vis spectrum, with $\pi \rightarrow \pi^*$ and $n \rightarrow \pi^*$ transitions for C—C and C=O bonding, respectively, shown in **Figure 4** [2].

A more detailed analysis of the GO is performed with XPS for C1s, presented in **Figure 5**, in order to determine the contribution for C=C bonding and C=O, which is 40 and 60%, respectively, for GO after 6 hours of oxidation in the presented Hummers reaction. This technique allows to specify the C=C and C—C, which is 21.3% and 18.73%. Also for the C=O=C and C=OH is 52.03%, C—O is 3.54%, and for COOH 4.4% [2].

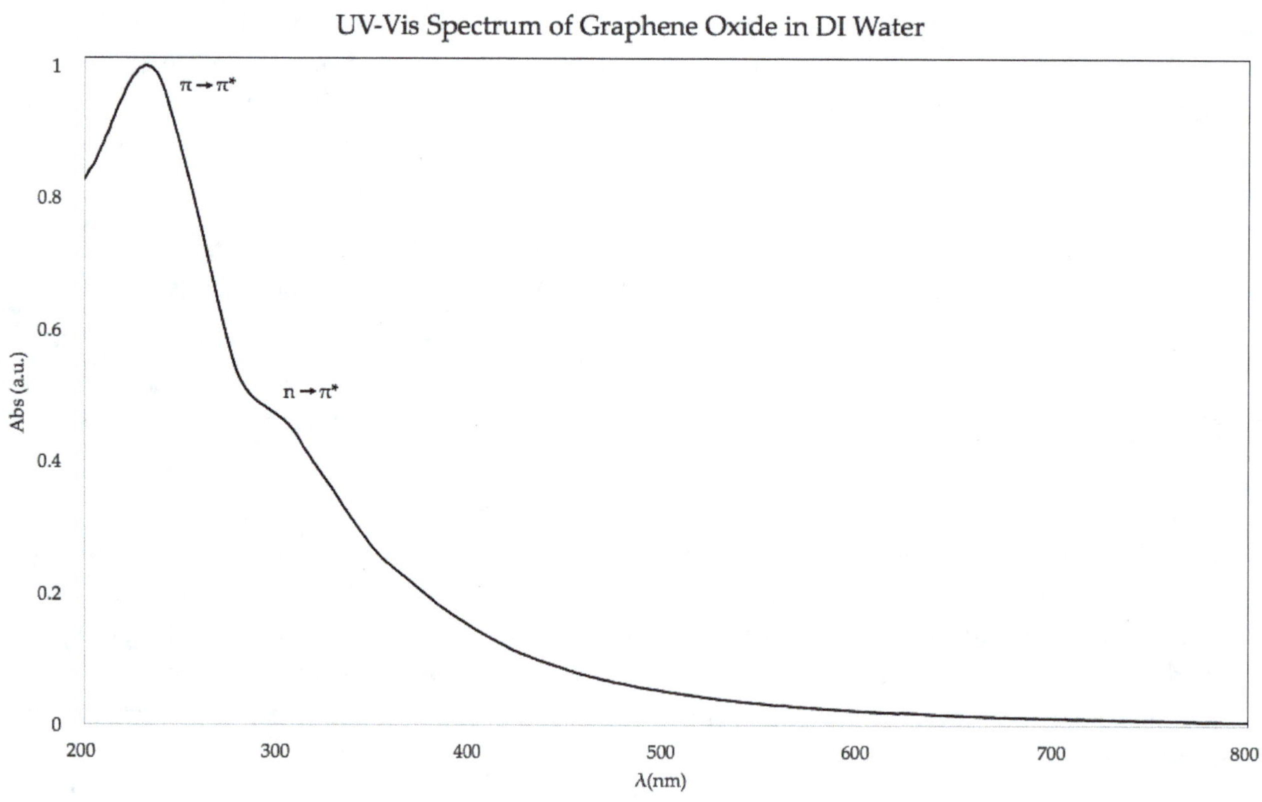

Figure 4. UV-vis spectrum of graphene oxide dispersion in DI H_2O [12].

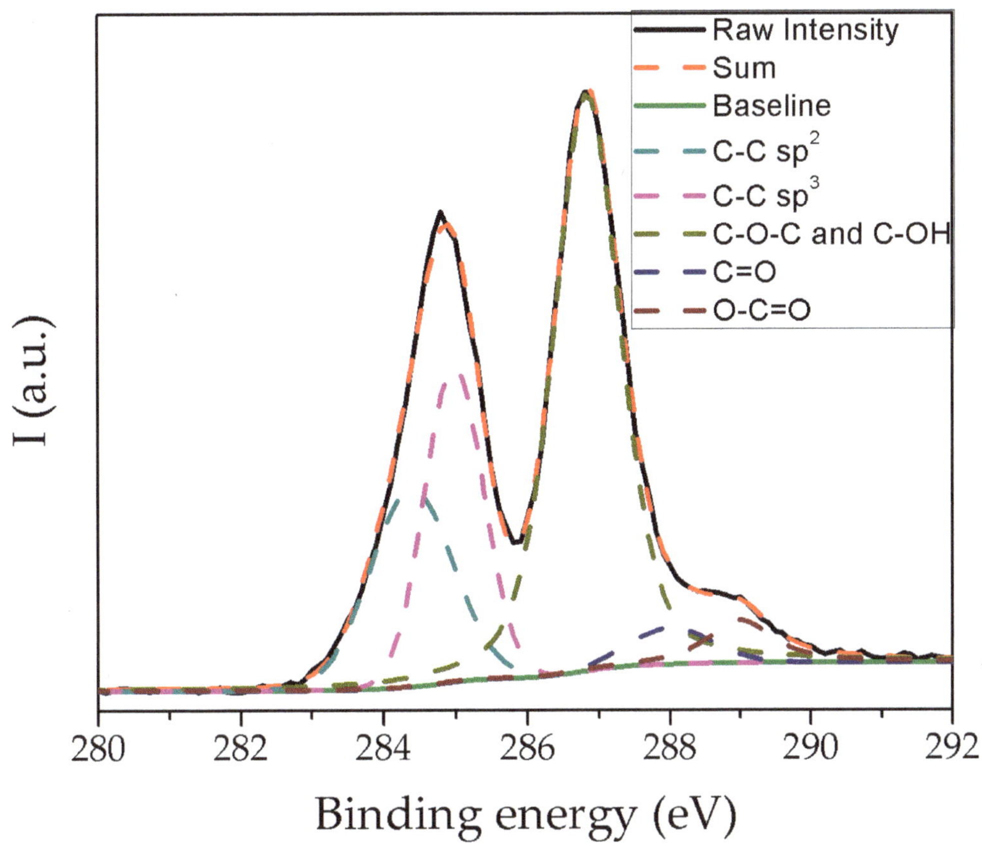

Figure 5. XPS for C1s [2].

3. Reduced graphene oxide

To apply these graphene-based materials, additional processes are required to obtain a material that meets the required characteristics. Graphite, as previously discussed in this chapter, must be oxidized (GO) in order to obtain a band gap. However, this molecule is extremely hydrophilic and therefore insoluble in organic solvents commonly used in OPVs, such as oDCB, toluene, and chloroform.

Nevertheless, the resulting E_g is too high—as we mentioned before—but because its relation with the percentage of oxygen, a common strategy used to modify that value consists in the chemical reduction (i.e., removing the oxygen atoms), obtaining values of 2.7–0.1 eV, when hydrazine is used. However, due its toxicity, it is also preferable to use harmless reducing agents such as fructose, glucose, and ascorbic acid in basic medium, obtaining values of 2.7–1.1 eV [2, 16]. The XPS analysis, plays a key role for the oxygen quantification, is shown in **Figure 6**.

To determine the real contribution of the mentioned reducing agents, an experiment with only NH_4OH was carried out, and an increase of up to 35% in sp^2 domains was observed. However, the contribution of carbonyl and carboxyl had a low modification, as seen in **Table 1**, demonstrating that NH_4OH alone is not able to decrease the most oxidized states of carbon. A

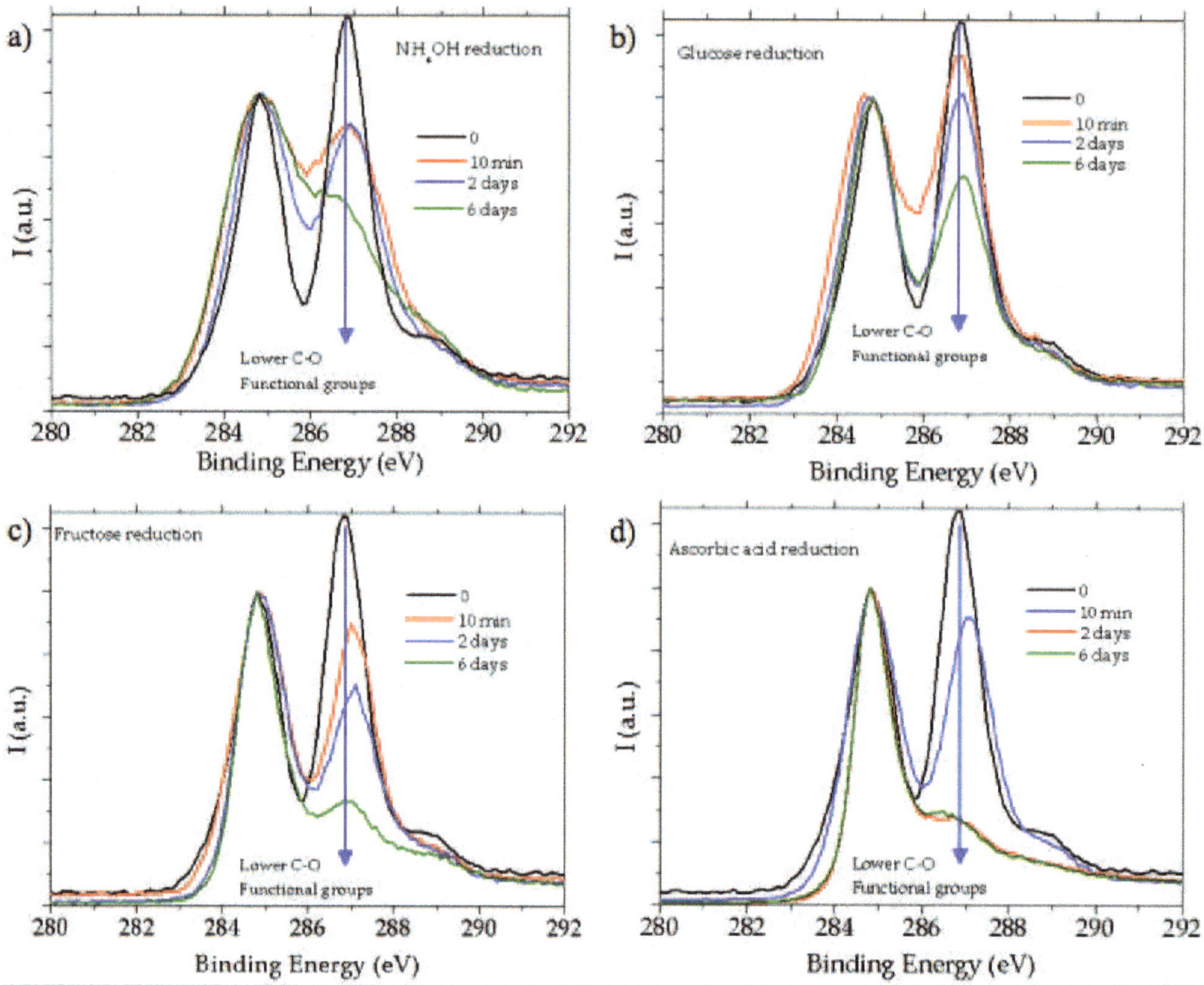

Figure 6. XPS C1s spectra for reduction with: a) NH4OH; b) glucose at pH 10; c) fructose at pH 10; and d) ascorbic acid at pH 10 after 10 min, 2 days, and 6 days [2].

comparable behavior is detected for the addition of glucose, where a similar result on the oxygen signal decrease can be observed (cf. **Figure 6b**).

Fructose reveals enhanced recovery of sp^2 contribution to 41.3% after 6 days. Among all reagents investigated, ascorbic acid provides the best recovery capability of sp^2 domains (~53% after 6 days) and elimination capability of epoxide and hydroxyl groups. A summary is presented in **Table 1** [2].

Time	C—C (sp^2)	C—C (sp^3)	C—O—C and C—OH	C=O	COOH
C1s contributions of GO after 6 hours of oxidation (%)					
0 min	18.73	21.30	52.03	3.54	4.40
Contributions after NH_4OH reduction (%)					
10 min	30.66	30.07	28.56	6.95	3.77
2 days	34.88	21.85	32.21	5.80	5.26
6 days	35.30	27.50	22.04	7.47	7.69

Time	C—C (sp^2)	C—C (sp^3)	C—O—C and C—OH	C=O	COOH
C1s contributions after glucose reduction (%)					
10 min	31.79	25.40	34.25	5.67	2.89
2 days	30.64	19.14	39.54	5.92	4.76
6 days	37.06	20.71	30.52	5.92	5.80
C1s contributions after fructose reduction (%)					
10 min	32.67	21.57	36.56	5.13	4.07
2 days	34.37	23.69	33.51	2.19	6.25
6 days	41.31	19.99	20.71	1.62	16.38
C1s contributions after ascorbic acid reduction (%)					
10 min	30.38	21.02	35.02	7.59	5.99
2 days	51.03	17.86	9.85	6.46	14.81
6 days	53.03	16.11	12.14	7.03	11.69

Reprinted from Ref. [2].

Table 1. Area percentage of functional groups after reduction.

Graphene oxides resulting from chemical reductions (rGO) continue to exhibit low dispersibility in organic solvents, so it is necessary to functionalize them with long chain hydrocarbon molecules.

4. Surface modification

The development of novel methods for surface modification of graphene is a challenge in the science and technology of nanocomposite materials. The functionalization of graphene derivatives should provide good stability and is necessary to improve their dispersion in any polymeric matrix. Therefore, many physical and chemical properties in nanocomposites obtained could be improved. The chapter aims to present the recent research advances in new methods of solution processable graphene derivatives, chemical modification, properties, and their incorporation in nanocomposites materials for applications in energy harvesting.

Graphene oxide has dispersion difficulties in nonpolar solvents due to its hydrophilic nature. A work [17] concerning the test of organic solvents for dispersion is the base for this analysis. The work suggests that the best solvents for this endeavor are those with high electrical dipole moment, including THF, ethylene glycol, and DMF. However, there are some exceptions for the dispersion and exfoliation of GO synthesized by Hummers method. One of these deviations from the rule is DMSO; in this solvent, dispersion was bad. The work is supported by photographs where sediments were visible on the bottom of the glass, making evident its bad dispersion.

In an unpublished work from our group, graphene oxide solubility with chlorinated solvents was studied, looking for the incorporation of graphene oxide to polymer solar cells to determine the solubility of different GOs with variable degrees of oxidation. **Figure 6** shows the reduction of GO with reducing agents in basic media. Carbon-oxygen ratios calculated from XPS survey are shown in **Figure 7**.

The solubility tests are shown in **Figure 8**. The maximum solubility achieved with this reducing method was less than 2 mg/mL. The trend shows that a minimum 3.5 carbon/oxygen ratio is required to disperse partially reduced graphene oxide.

For practical applications, a solubility of 10 mg/mL of electron acceptor is required; therefore, to increase solubility, surface modifications are required.

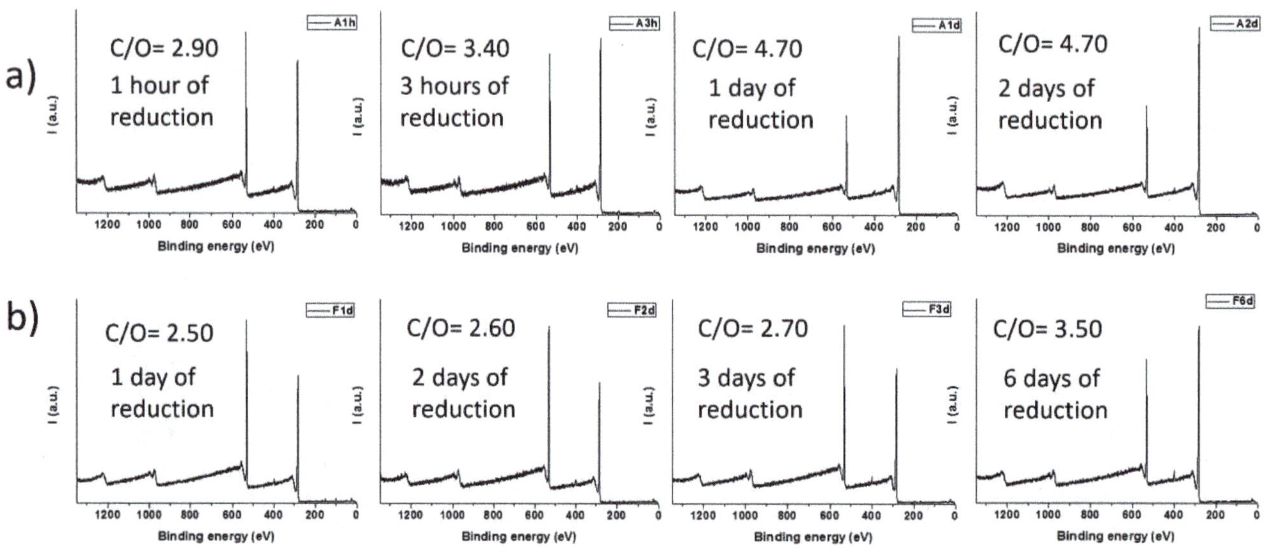

Figure 7. XPS survey of (a) GO with variable degrees of oxidation reduced with fructose; (b) reduced with ascorbic.

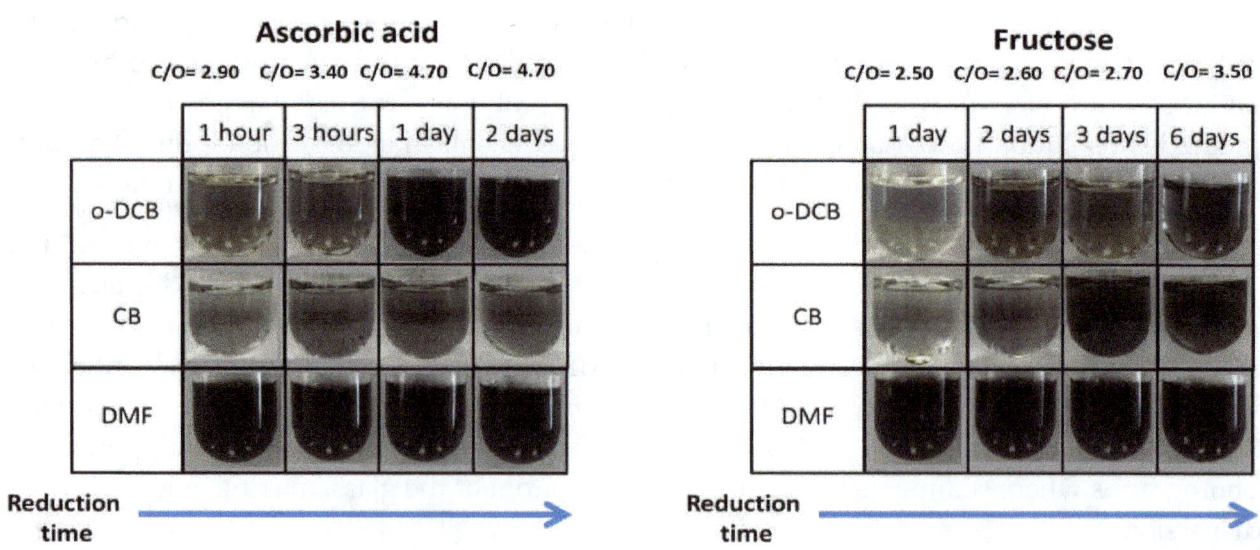

Figure 8. Dispersions of reduced GO with ascorbic acid and fructose in o-dichlorobenzene (o-DCB), chlorobenzene (CB) and dimethylformamide (DMF). For o-DCB 3.5 C/O ratio is required to have a good dispersion.

Graphene oxide is a family of decorated graphene structure with oxygen moieties that can be further functionalized with different polymer chains. Most reported functionalizations exploit the carboxylic endings for anchoring long alkyl chains through esterification or amidation.

Noncovalent functionalization can work via electrostatic interactions or π-π interactions, between the functionalizing agent and the graphene. This kind of functionalization is preferred

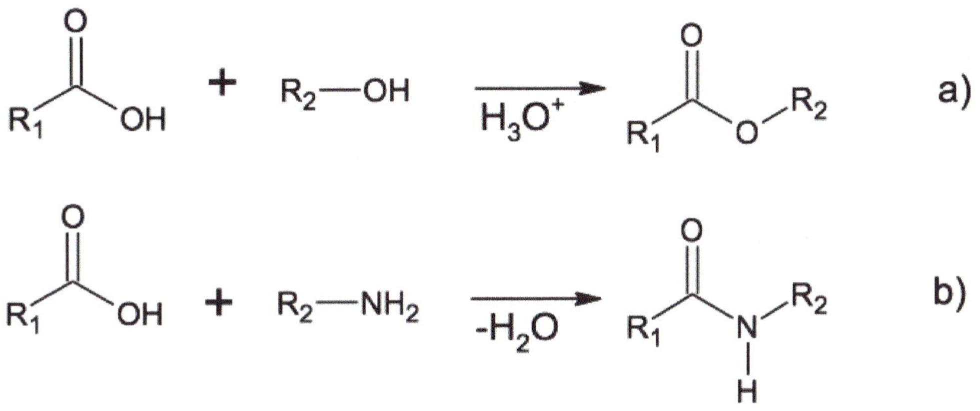

Figure 9. (a) Esterification reaction and (b) amidation reaction [5, 20].

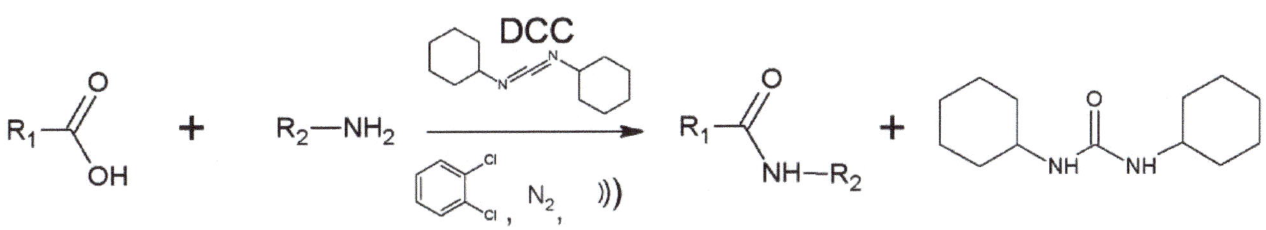

Figure 10. Amidation of a carboxylic acid using DCC in N_2 atmosphere [12].

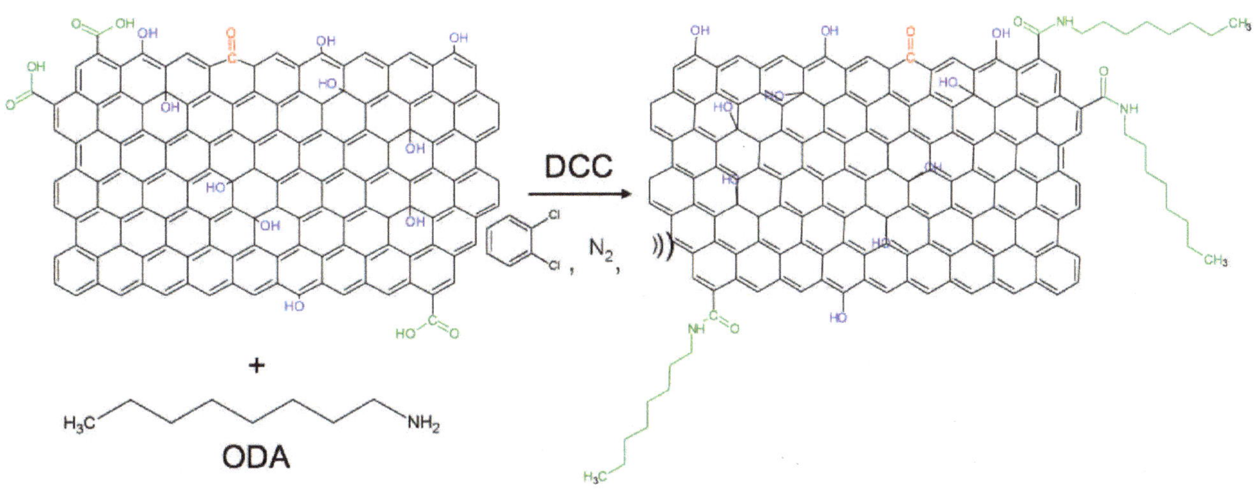

Figure 11. Surface modification of reduced graphene oxide with octadecylamine [12].

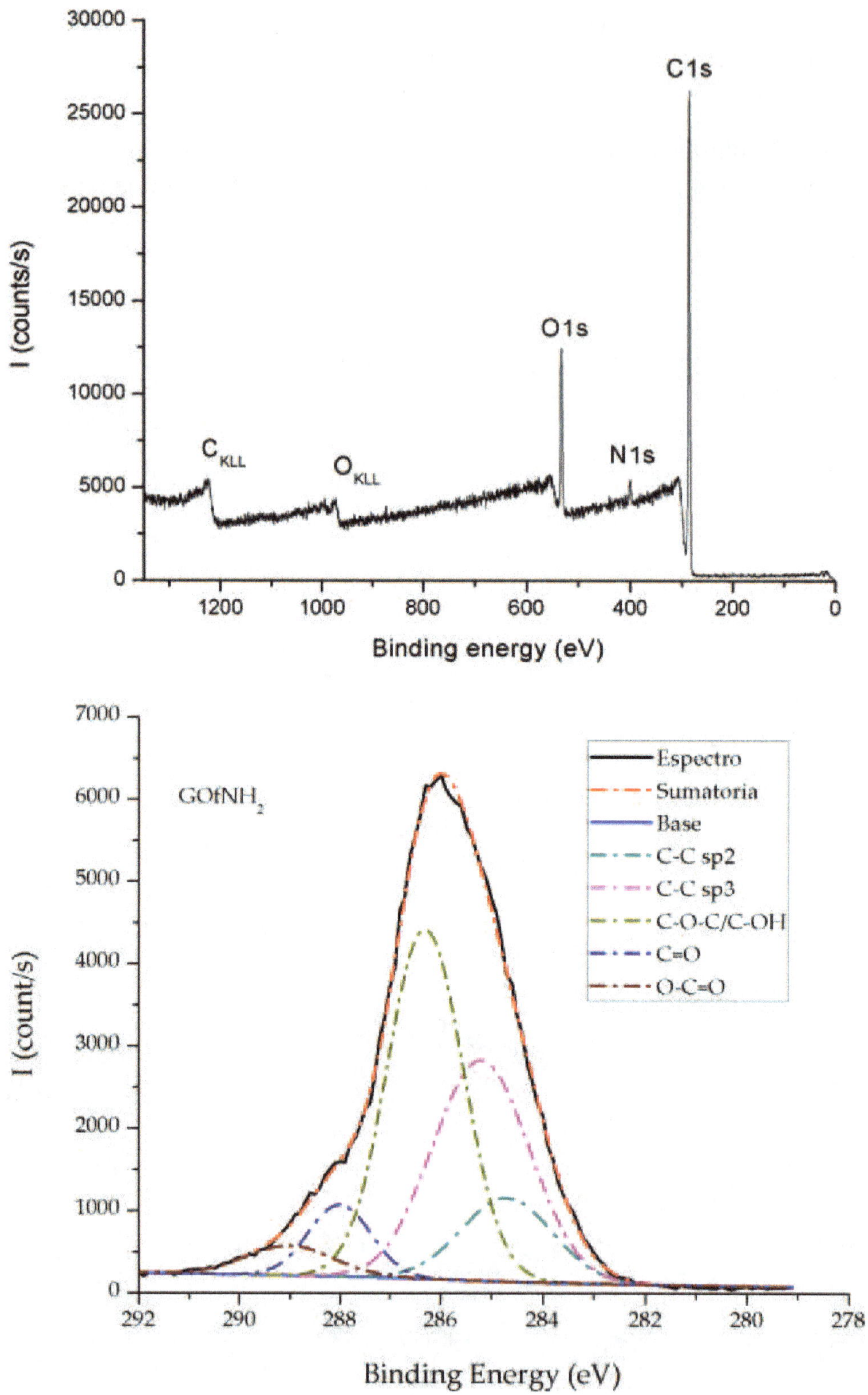

Figure 12. XPS survey and High resolution C1s of reduced graphene oxide functionalized with ODA [12].

for applications in which optical or electronic properties are critical. A disadvantage is the difficulty of removing the dispersant agent from the graphene surface due to high stability between them [18, 19].

Covalent functionalization is based on direct chemical reactions over functional groups of graphene oxide. The first step of the reaction will be referred as later proposed classification of graphene, because its similarities on reactions between carbon nanotubes (CNTs). These reactions are referred as first-generation (1G) modifications. They have the characteristic of introducing a functional group by covalently bonding to a carbon atom; frequently these reactions include strong oxidations.

Another kind of modification is the consecutive reaction of the already introduced 1G functional groups. Typically the reactions include esterifications and amidations, which appear in **Figure 9**. This is useful, not only for dispersing nanoparticles in solvent media but also for conferring affinity for polymer matrices because interactions between dispersant agent and the matrix occur. These reactions are referred as second-generation (2G) modifications.

However, the above representation is just a simplified way, as unusual it might seem. In order to carry out a successful amidation, a catalyst is often used, as dicyclohexylcarbodiimide, as shown in **Figure 10**.

With this procedure, it is possible to use the carboxylic acids at the edge of the reduced graphene oxide to make a surface modification not only to improve the solubility, but also to prevent the restacking of the layers by the steric hindrance using a long chain alkylamine as octadecylamine (ODA), given the molecular structure shown in **Figure 11**.

This modification can be observed in the XPS analysis for the peak attributed to N1s is shown in **Figure 12**, which was not present before the functionalization.

5. Band gap modification

As previously mentioned, the band gap of GO is related to the carbon-oxygen ratio, which implies that removing the number of oxygen present in the molecule leads to a band gap decrease. Moreover, gradual restoration of graphitic structure (sp^2) with time offers the possibility of tuning the band gap in a controlled way. The XPS is an amazing tool to determine the content of oxygen on graphene materials. However, the easiest way to follow the changes in band gap through chemical reduction reactions is from UV-vis spectrums and then analyzed by Tauc plot. The information obtained by the spectrum is the absorbance of light at every wave length (A_λ), which is related to the intensity of the incident light (I_0) in a logarithmic way, as shown in Eq. (4).

$$A_\lambda = -\mathrm{Log}_{10}\frac{I}{I_0} \tag{4}$$

It is a well-known fact that transmittance is the ratio between the intensity of the incident light (I_0) and transmitted (I), as described in Eq. (5). This is also related to the Lambert Beer's law and the absorption coefficient ($\boldsymbol{\alpha}$) and z is the optical path of the cell, usually equal of 1 cm.

$$T = \frac{I}{I_0} = e^{-\alpha z} \tag{5}$$

Which allows to relate the absorbance to transmittance as follow in Eq. (6) and finally, relate $\boldsymbol{\alpha}$ to T in Eq. (7).

$$T = 10^{-A_\lambda} \tag{6}$$

$$\alpha = -LnT \tag{7}$$

Finally, it is just needed to change the wavelength for energy (E) in eV, with Eq. (8).

$$E = \frac{hc}{\lambda} \tag{8}$$

For example, from **Figure 4**, using these simple equations, it is possible to make the Tauc plot for GO as shown in **Figure 13**, where it is very easy to determine the band gap value.

The reduced graphene oxide with ascorbic acid, for 1 hour, 3 hours, 1 day, and 2 days at room temperature, [2] shown in **Figure 14**, presents the reduction in C-O contribution at 286.6 eV for C1s in XPS and the Tauc plot for each sample, making clear the band gap modification, in accordance to the oxygen decrease.

This applies for the other samples, such as the reduced graphene oxide with fructose, presented in **Figure 15**, with the same behavior: the reduction in the oxygen content is directly related to the band gap modification.

As expected, the surface modification shown in **Figure 11** also gives a band gap modification by the incorporation of different atoms to carbon in accordance to some computational results [21]. The value of this band gap modification for functionalization was obtained experimentally from the Tauc plot and is presented in **Figure 16**.

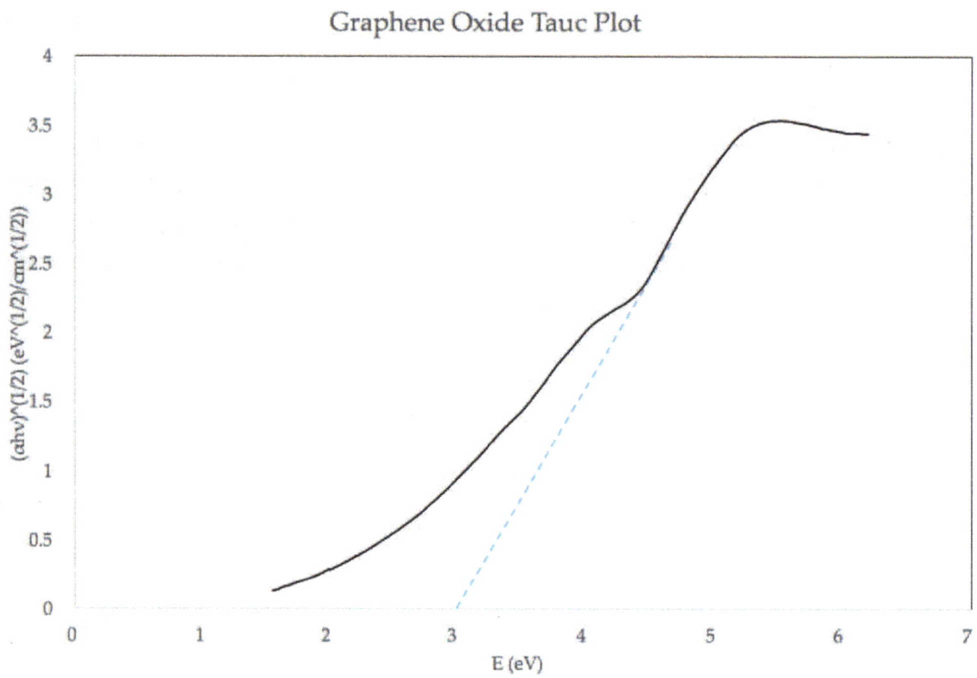

Figure 13. Tauc Plot for graphene oxide after 6 hours of oxidation [12].

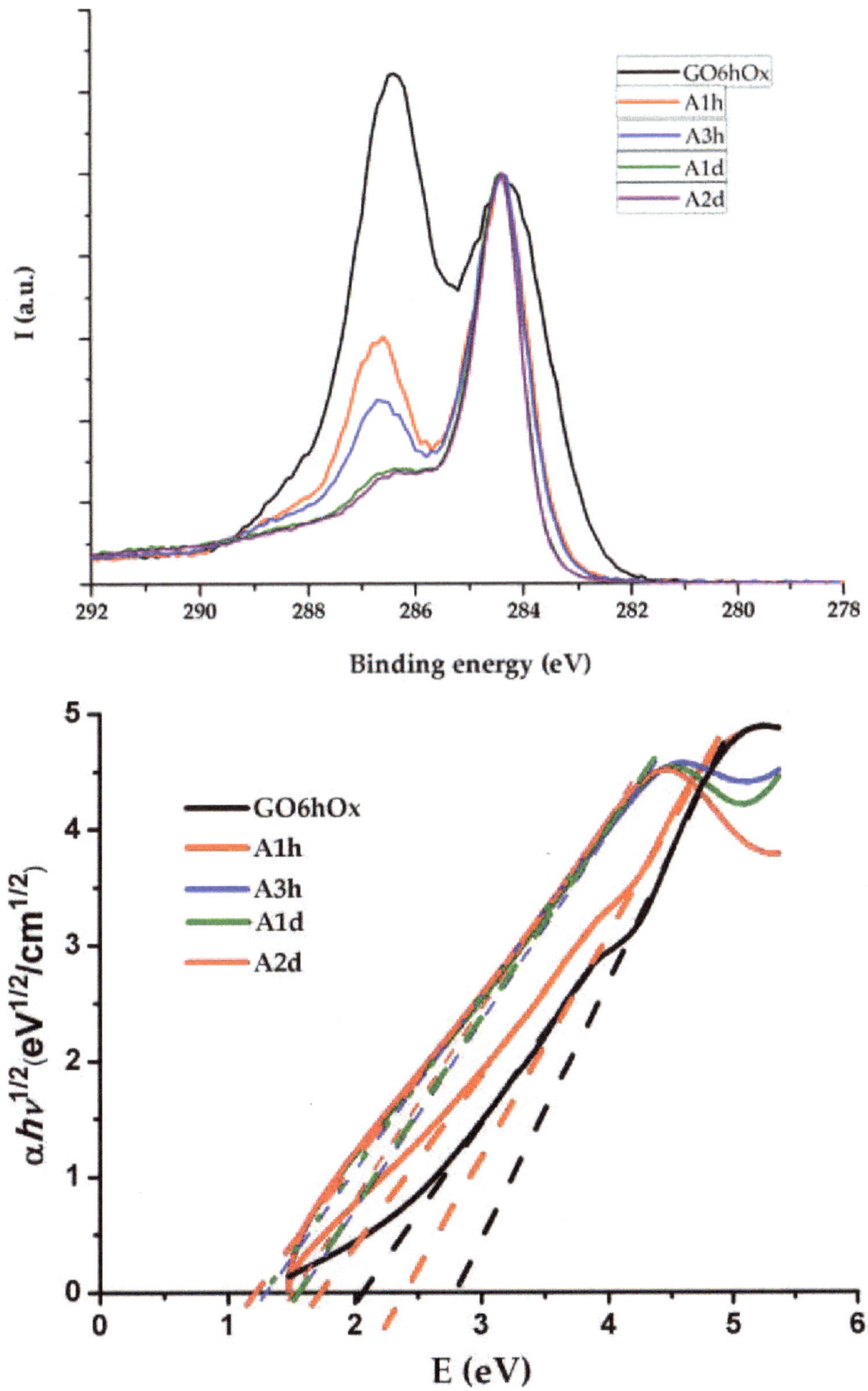

Figure 14. XPS C1s and Tauc plots for reduced graphene oxide with ascorbic acid. [2]

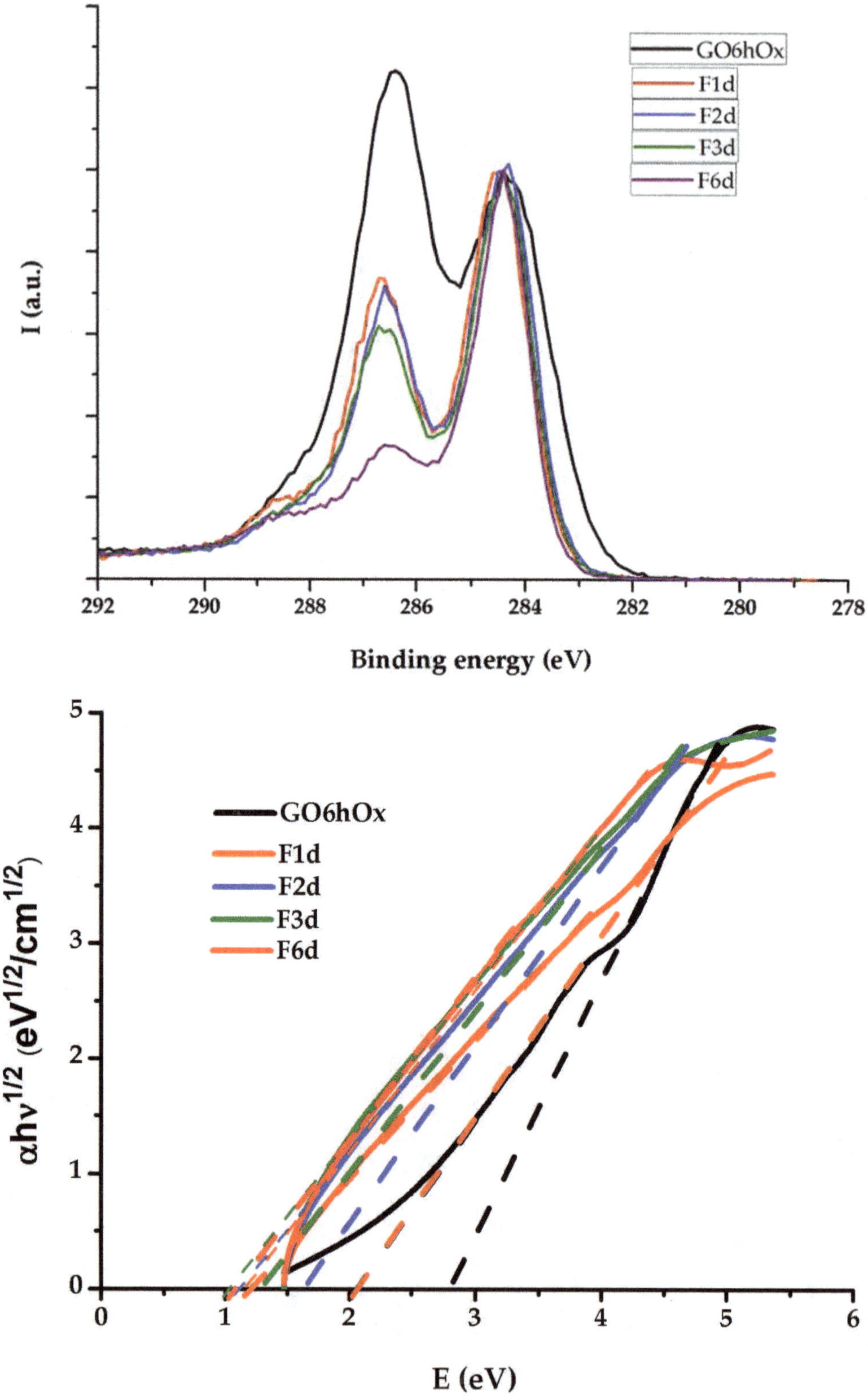

Figure 15. XPS C1s and Tauc plots for reduced graphene oxide with fructose. [2]

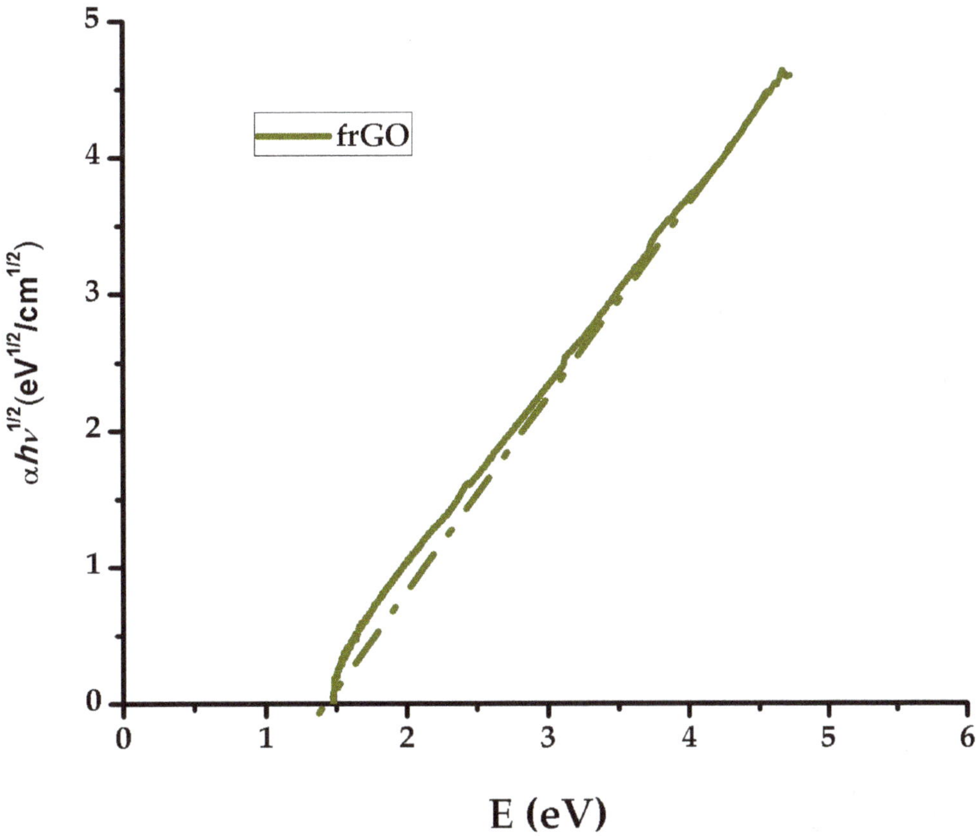

Figure 16. Band gap modification by surface modification of reduced graphene oxide [12].

6. Applications for energy harvesting devices

Energy harvesting is the process of capturing minute amounts of energy from one or more of these naturally occurring energy sources, accumulating them and storing them for further use. Graphene has the potential to address several scientific challenges, ranging from the need for more efficient alternative energy technologies. The unique properties of graphene previously mentioned along the present chapter shows the characteristics for energy-harvesting application, especially for its use in solar cells.

Since graphene monolayers were analyzed and obtained by micromechanical exfoliation in 2004 [22], graphene, as well as graphene oxide and reduced graphene oxide, has been added to the list of materials with potential application in OPVs in their different components, either in substitution of a previously used material or forming synergies that lead to an improvement in the devices. This is because of the three types of materials that have very different band gaps, from zero to 3 eV; HOMO values from −6.6 to −4.8 eV; LUMO values of −3.6 to −4 eV for GO and rGO; and Fermi level of −4.5 eV for graphene [23].

Based on the previously described characteristics of the graphene, it is expected that the first applications designed for it are in [1] substitution of the indium-tin oxide (ITO) electrodes, since this material is fragile, expensive, and with a transmittance of 90% [1]. It has been found that by replacing ITO with graphene, the efficiency of the device increases from 3.86 to 3.98% [24].

The main application of GO in OPVs, consists in the incorporation of this as a hole transport layer (HTL) between the active layer and the anode, replacing the PEDOT:PSS, achieving increases in the PCE of the order of 10% in a device with active layer of P3HT: PCBM [25, 26].

Moreover, it has been reported that incorporation of reduced and functionalized graphene oxide has successfully been carried out as an electron transport layer (ETL), between the active layer and the cathode, improving PCE by 15% for a device with a typical active layer of P3HT: PCBM [18, 27].

As expected, graphene and graphene oxide-based materials have also been used in the active layer of the OPVs, combined with P3HT: PCBM, obtaining a maximum PCE of 1.046% [28] or even without any polymer, only using PCBM/rGO/SWCNTs as active layer, obtaining a PCE of 1.3%, but with a significant improvement in thermal stability and resistance to aging by atmosphere with oxygen [29].

In addition, it has been reported that the incorporation of metallic nanoparticles in the organic photovoltaic devices leads to a significant increase in their efficiency. According to [30], the incorporation of reduced graphene oxide decorated with gold nanoparticles (rGO-AuNPs) instead of the typical PEDOT: PSS hollow conveyor layer, generates an increase of more than 25% in the efficiency of the device, this because the introduction of localized surface plasmon resonance (LSPR) generates a considerable increase in photocurrent. Following this trend, the incorporation of rGO-AgNPs into an active P3HT: PCBM blend layer, leads to a 24.7% increase in device efficiency, due to an increase in the photocurrent by the LSPR [31].

It has even been observed an increase of 116% in OPVs devices containing GO-AgNPs-PEDTO:PSS compared to those containing only GO-PEDOT:PSS as active layer [32].

7. Conclusion

The vast properties of graphene open endless possibilities for very different purposes. However, some applications require characteristics achieved through graphene derivatives such as graphene oxide, reduced graphene oxide, and reduced graphene oxide functionalized. Tailoring of the properties and surface chemistry of graphene lays on the control of its performance in a polymer matrix and a device.

Therefore, the capability of controlling these properties is of great importance. Moreover, a functionalization that makes reduced graphene oxide soluble in organic solvents must be successfully achieved, due to the varieties of applications and specific needs of the different energy harvesting devices' demands.

The chemical reduction of graphene oxide has been carried out by innocuous agents such as ascorbic acid, fructose, and glucose, controlling the band gap value of the resulting materials.

A process to decrease the band gap, by means, reducing the oxygen content in graphene oxide, i.e. rGO, was established. GO was synthesized through a modified Hummers method, and it has been found that NH_4OH, glucose, fructose, and ascorbic acid at pH 10 can be used to modify the band gap. The XPS C1s and UV-vis spectra indicate the gradual elimination of oxygen groups and restitution of graphitic structure, ensuing to a decreased value of band gap.

Ascorbic acid at pH 10 is more effective and fast for sp^2 restitution, resulting in an optical band gap of 1.55 eV. The chemical reduction by fructose and glucose at pH 10 is slower which could be preferred in order to have a more precise control of band gap, but the total sp^2 restoration is lower than that achieved by the ascorbic acid, nevertheless, it has the lowest band gap 1.15 eV with fructose at pH 10.

For application in organic solar cells, low band gaps are important because the PCE depends strongly on the effective generation, diffusion, and dissociation of the exciton. Because most of the solar energy is in the visible spectrum, a band gap in these values means more absorption of light and hence more exciton generation. Another aspect to consider is that restitution of sp^2 domains entails a better exciton diffusion, because of the high carrier mobility. Finally, considering the band gap as the energy difference between the HOMO and LUMO, a low value is advantageous because the difference of the LUMOs, both in the acceptor and donor molecules or polymers, must be adequate for the exciton bonding dissociation and therefore increase the PCE.

Acknowledgements

The authors are beholden for the technical support and facilities at CIMAV Monterrey, as well as to the Mexican National Research Council CONACyT for the scholarship of the students involved in this work. Lilia Magdalena Bautista Carrillo and Luis Gerardo Silva Vidaurri for UV–Vis spectra and XPS spectra, respectively, are also acknowledged.

Author details

Ulises Antonio Méndez Romero[1], Miguel Ángel Velasco Soto[1], Liliana Licea Jiménez[1,2], Jaime Álvarez Quintana[1,2] and Sergio Alfonso Pérez García[1,2]*

*Address all correspondence to: alfonso.perez@cimav.edu.mx

1 Center for Advanced Materials Research S.C. (CIMAV), Unidad Monterrey, Apodaca, NL, Mexico

2 GENES—Group of Embedded Nanomaterials for Energy Scavenging, Apodaca, NL, Mexico

References

[1] Garg, R., et al., Deposition methods of graphene as electrode material for organic solar cells. Advanced Energy Materials, 2016. 1601393. doi:10.1002/aenm.201601393

[2] Velasco-Soto, M.A., et al., Selective band gap manipulation of graphene oxide by its reduction with mild reagents. Carbon, 2015. **93**: pp. 967–973.

[3] Huang, H., et al., Oxygen density dependent band gap of reduced graphene oxide. Journal of Applied Physics, 2012. **111**(5): p. 054317.

[4] Hummers, W.S. and R.E. Offeman, Preparation of graphitic oxide. Journal of the American Chemical Society, 1958. **80**(6): pp. 1339–1339.

[5] Velasco-Soto, M.A., et al., Chapter 8 - Carbon Polymer Nanocomposites A2 - Domínguez, M. Sánchez, in Nanocolloids, C.R. Abreu, Editor. 2016, Elsevier: Amsterdam. pp. 265–297.

[6] Dreyer, D.R., et al., The chemistry of graphene oxide. Chemical Society Reviews, 2010. **39** (1): pp. 228–240.

[7] Thompson, B.C. and J.M.J. Fréchet, Polymer–fullerene composite solar cells. Angewandte Chemie International Edition, 2008. **47**(1): pp. 58–77.

[8] Green, M.A., et al., Solar cell efficiency tables (version 47). Progress in Photovoltaics: Research and Applications, 2016. **24**(1): pp. 3–11.

[9] Xue, J., et al., A hybrid planar–mixed molecular heterojunction photovoltaic cell. Advanced Materials, 2005. **17**(1): pp. 66–71.

[10] Yu, G., et al., Polymer photovoltaic cells: enhanced efficiencies via a network of internal donor-acceptor heterojunctions. Science, 1995. **270**(5243): pp. 1789–1791.

[11] Krebs, F.C., Polymer photovoltaics: a practical approach. Vol. 175. 2008: SPIE Press Bellingham, WA.

[12] Velasco-Soto, M.A., Desarrollo de Celdas Solares Poliméricas Usando Derivados de Grafeno. 2016, Centro de Investigación de Materiales Avanzados Unidad Monterrey: Apodoca N.L.

[13] Moliton, A. and J.-M. Nunzi, How to model the behaviour of organic photovoltaic cells. Polymer International, 2006. **55**(6): pp. 583–600.

[14] Brodie, B.C., On the atomic weight of graphite. Philosophical Transactions of the Royal Society of London, 1859. **149**: pp. 249–259.

[15] Staudenmaier, L., Verfahren zur darstellung der graphitsäure. Berichte der deutschen chemischen Gesellschaft, 1898. **31**(2): pp. 1481–1487.

[16] Shen, Y., et al., Evolution of the band-gap and optical properties of graphene oxide with controllable reduction level. Carbon, 2013. **62**: pp. 157–164.

[17] Paredes, J.I., et al., Graphene oxide dispersions in organic solvents. Langmuir, 2008. **24** (19): pp. 10560–10564.

[18] Qu, S., et al., Noncovalent functionalization of graphene attaching [6,6]-Phenyl-C61-butyric acid methyl ester (PCBM) and application as electron extraction layer of polymer solar cells. ACS Nano, 2013. **7**(5): pp. 4070–4081.

[19] Bai, S., et al., Reversible phase transfer of graphene oxide and its use in the synthesis of graphene-based hybrid materials. Carbon, 2011. **49**(13): pp. 4563–4570.

[20] Velasco-Soto, M. A., et al. Chapter 7 - Dispersion of Carbon Nanomaterials. pp 251 in Nanocolloids: A Meeting Point for Scientists and Technologists. Domínguez-Sánchez, M. and Rodriguez-Abreu, C. 2016. Amsterdam, Elsevier: 247–263.

[21] Rani, P. and V.K. Jindal, Designing band gap of graphene by B and N dopant atoms. RSC Advances, 2013. **3**(3): pp. 802–812.

[22] Novoselov, K.S., et al., Electric field effect in atomically thin carbon films. Science, 2004. **306**(5696): pp. 666–669.

[23] Zheng, F., et al., Charge transfer from poly(3-hexylthiophene) to graphene oxide and reduced graphene oxide. RSC Advances, 2015. **5**(109): pp. 89515–89520.

[24] Wang, Z., et al., Technology ready use of single layer graphene as a transparent electrode for hybrid photovoltaic devices. Physica E: Low-dimensional Systems and Nanostructures, 2011. **44**(2): pp. 521–524.

[25] Chen, S., et al., A graphene oxide/oxygen deficient molybdenum oxide nanosheet bilayer as a hole transport layer for efficient polymer solar cells. Journal of Materials Chemistry A, 2015. **3**(36): pp. 18380–18383.

[26] Jeon, Y.-J., et al., High-performance polymer solar cells with moderately reduced graphene oxide as an efficient hole transporting layer. Solar Energy Materials and Solar Cells, 2012. **105**: pp. 96–102.

[27] Robaeys, P., et al., Enhanced performance of polymer:fullerene bulk heterojunction solar cells upon graphene addition. Applied Physics Letters, 2014. **105**(8): p. 083306.

[28] Wang, J., et al., Composition and annealing effects in solution-processable functionalized graphene oxide/P3HT based solar cells. Synthetic Metals, 2010. **160**(23–24): pp. 2494–2500.

[29] Bernardi, M., et al., Nanocarbon-based photovoltaics. ACS Nano, 2012. **6**(10): pp. 8896–8903.

[30] Chuang, M.-K., et al., Gold nanoparticle-decorated graphene oxides for plasmonic-enhanced polymer photovoltaic devices. Nanoscale, 2014. **6**(3): pp. 1573–1579.

[31] Ou, C.F., The effect of graphene/Ag nanoparticles addition on the performances of organic solar cells. Journal of Materials Science and Chemical Engineering, 2015. **3**: pp. 30–35.

[32] Sutradhar, P. and M. Saha, Silver nanoparticles: synthesis and its nanocomposites for heterojunction polymer solar cells. The Journal of Physical Chemistry C, 2016. **120**(16): pp. 8941–8949.

3

Photoinduced Pseudospin Dynamical Effects in Graphene-Like Systems

Alexander López and Benjamin Santos

Abstract

In this chapter, we describe some of our recent results on the laser-induced manipulation of the energy band structure of graphene-like systems. We present numerical results on the quasi-energy spectrum as well as detailed calculations of semi-analytical approximations to other physical quantities of interest. The main message we would like to convey to the interested reader of the chapter is that by properly tuning the perturbation parameters of the radiation field one can control the size and shape of the photoinduced gaps. These in turn would allow the realization of new electronic phases on graphene and its related materials such as silicene.

Keywords: graphene, light-matter interaction, Floquet theory, Dirac fermions, Landau levels, pseudospin polarization

1. Introduction

Since its recent experimental realization [1], graphene has attracted a lot of research interest because of its remarkable transport properties. At low energy, single-layer graphene (SLG) has a *linear dispersion spectrum*, and charge carriers can be described as massless, chiral, Dirac fermions. Moreover, SLG is a zero-gap semiconductor, has a high mobility of charge carriers and features an unconventional half-integer quantum Hall effect which can be measured at room temperature [2]; therefore, it has been the focus of many experimental [3] and theoretical analyses [4, 5], and it is expected to have several potential applications in carbon-based spintronics. Thus, the study of spin-related transport phenomena is one of the most active research fields in graphene. One important breakthrough was the demonstration of a quantum spin Hall effect in

SLG [6]. This in turn relies on intrinsic spin-orbit coupling (SOC), which although weak compared to other energy scales in the problem [7] opens a gap in the energy spectrum, making SLG into a topological insulator. Another possibility for SOC is due to interactions with a substrate, the presence of electric fields or curvature of the SLG. Therefore, it is called extrinsic or Rashba spin-orbit coupling (RSOC). The RSOC is believed to be responsible for spin polarization [8], and spin relaxation phenomena in graphene [9]. In addition, in the presence of an external magnetic field B, Landau's quantization leads to a $\sqrt{B}$ energy relation which contrasts the linear in B dependence of conventional 2DEG, with parabolic energy bands.

Indeed, one interesting aspect of the new graphene-like materials is that they are predicted to support the so-called edge states which are related to the realization of new topological aspects of matter. As it has been acknowledged in the recent literature, since the discovery of the quantum Hall effect [10, 11], the quest for new topological states of matter has attracted a great deal of attention (for a recent review, see Ref. [12]). For instance, some proposals to topologically generate and characterize non-trivial phases in graphene have paved the road for research on how to implement models of non-trivial topological states in different physical systems ranging from semiconducting quantum wells [13], superconductors [14] and neutral [15] and cold atoms [16], just to mention a few.

In Ref. [13], a model for quantum spin Hall (QSH) state in telluride-based quantum wells was analysed. Yet, it was shown afterwards that this model really corresponds to a topological insulator state [17]. In general, a topological insulator is a quantum state of matter where a given material supports edge states that counter-propagate on its boundary. The existence of these edge states allows for non-dissipative transport, which is a most-wanted property in technological implementations, for instance, in quantum information and nanotechnological devices. In principle, the topological-insulating phase is protected by time-reversal symmetry. However, a time-reversal-symmetry-broken QSH state proposal was recently put forward by means of ferromagnetic leads attached to the sample [18].

Upon introduction of time-periodic dynamical modulation, for instance, in semiconducting quantum wells, with a zincblende structure, it has been recently shown that AC driving can induce a topological phase transition leading to the so-called Floquet topological insulator (FTI) phases [19]. This means that a non-trivial topological phase can be induced in a system that in equilibrium behaves trivially [20–22], that is, it does not possess non-dissipative or gapless edge states.

In this chapter, we show some of our results on the photoinduced effects on the Dirac fermions of graphene-like systems presenting two specific scenarios. In Section 2, we describe the Landau levels (LLs) in graphene when a laser is incident perpendicularly to the sample [23]. We show that an exact effective Floquet Hamiltonian can be found from which the dynamics is afterwards described [24–26]. Here, we show that the quasi-energy spectrum presents a level-dependent photoinduced gap and we also show that coherent Landau level states can be synchronized in order to produce Rabi oscillations and quantum revivals [27]. In Section 3, we describe how the energy spectrum and pseudospin polarization in monolayer silicene [28–33] can be manipulated beyond the so-called off-resonant regime [34, 35] when strong radiation effects are taken into account [36–38]. We find the explicit and realistic parameter regime for the realization of a single-valley polarized state in silicene. In Section 4, an outlook is given whereas in Section 5 we present some technical details of the calculations employed because although we know there are several works describing the dynamical aspects of periodically driven systems, we wanted to be self-contained. Another reason for delving into such details is that we think this explicit technical aspect could be useful for both students and researchers interested in the field.

2. Technical aspects of periodically driven systems

2.1. Floquet Fourier-mode approach

Before delving into the models of interest, let us present a summary of an important tool in the description of time-dependent Hamiltonian dynamics when the interaction term is periodic in time. These kinds of interactions are ubiquitous in physical systems ranging from cold atoms, cavity QED, superconducting interferometric devices, lasers, and so on. Let us then consider a generic time-dependent periodic Hamiltonian $H(t+T) = H(t)$. We write it as free Hamiltonian plus a time-dependent interaction

$$H(t) = H_0 + V(t). \tag{1}$$

The solution of the dynamics for the evolution operator

$$i\hbar\partial_t U(t) = H(t)U(t), \tag{2}$$

is formally given by

$$U(t) = \mathcal{T}\left[\exp\left(-\frac{i}{\hbar}\int_0^t H(t')dt'\right)\right], \tag{3}$$

with $\mathcal{T}$ the time-ordering operator. However, since the Hamiltonian $H(t)$ is periodic, one can resort to Floquet theorem [24, 25]. It asserts that the general solution to the dynamics (2) when $H(t)$ is periodic can be written as

$$U(t) = P(t)e^{-iH_F t/\hbar}, \tag{4}$$

with $P(t)$ periodic and H_F a constant matrix, respectively. Using Eq. (4), it is easy to verify the stroboscopic property

$$U(nT) = P(nT)e^{-inTH_F/\hbar} \tag{5}$$

$$U(nT) = [U(T)]^n, \tag{6}$$

where $U(T) = e^{-iH_F T/\hbar}$. Thus, although the time-dependent Hamiltonian $H(t)$ is periodic, the corresponding evolution operator $U(t)$ is not, that is, $U(t+T) \neq U(t)$. In fact, $U(T)$ carries non-trivial information on the dynamics of the periodic system. The eigenvalues of H_F give the quasi-energy spectrum of the periodically driven problem. In order to determine these quasi-energies, one standard approach (see below) consists of performing an expansion in the (infinite) eigenbasis of time periodic functions $\xi_n(t) = e^{in\omega t}$ (Fourier modes), where $\omega = 2\pi/T$. Then, in order to deal with the infinite eigenvalue problem one resorts to a truncation procedure in order to determine the Floquet exponents.

In the following, we would like to explicitly describe the Floquet-Fourier mode strategy. For this purpose, we assume that we can solve the dynamics of the free part H_0. Its eigenbasis is spanned by the spinors $|\phi_\alpha\rangle$, where α describes a set of quantum numbers. We now use the eigenstates $|\phi_\alpha\rangle$ as expansion basis for the eigenstates of the full Hamiltonian in Eq. (1).

In order to analyse the evolution equation,

$$i\hbar\partial_t|\Phi(t)\rangle = H(t)|\Phi(t)\rangle \tag{7}$$

we take advantage of the periodicity of the Hamiltonian so we can resort to Floquet's theorem [24, 25]. For this purpose, we define an auxiliary Hermitian Hamiltonian

$$\mathcal{H}(t) = H(t) - i\hbar\partial_t, \tag{8}$$

along with the so-called Floquet states

$$|\Psi_\alpha(t)\rangle = \exp\left(i\varepsilon_\alpha t/\hbar\right)|\Phi(t)\rangle \tag{9}$$

such that

$$\mathcal{H}(t)|\Psi_\alpha(t)\rangle = \varepsilon_\alpha|\Psi_\alpha(t)\rangle \tag{10}$$

which are periodic functions of time, $|\Psi_\alpha(t+T)\rangle = |\Psi_\alpha(t)\rangle$. In addition, the eigenvalues ε_α form the quasi-energy spectrum, and are the analogous of the quasi-momenta for Bloch electrons in a spatially periodic structure.

We can verify that the states

$$|\Psi_{\alpha n}(t)|\rangle = \exp\left(in\omega t\right)|\Psi_\alpha(t)\rangle \tag{11}$$

are also eigenstates of the Hamiltonian $\mathcal{H}(t)$ but with corresponding eigenvalues $\varepsilon_\alpha \rightarrow \varepsilon_\alpha + n\hbar\omega$. Thus, we can work in the *first Brillouin zone* $-\hbar\omega/2 \le \varepsilon_\alpha \le \hbar\omega/2$.

Using the periodic temporal basis $\xi_n(t) = \exp\left(in\omega t\right)$, which satisfies

$$\frac{1}{T}\int_0^T \xi_n^*(t)\xi_m(t) = \delta_{nm}, \tag{12}$$

we write the Fourier-mode expansion

$$\Phi_\alpha(t) = \exp\left(-i\varepsilon_\alpha t\right)\sum_{n=-\infty}^{n=\infty}|C_\alpha^{(n)}\rangle\xi_n(t), \tag{13}$$

Now, we use the expansion $|C_\alpha^{(n)}\rangle = \sum_\beta \Lambda_{\alpha\beta}^{(n)}|\phi_\beta\rangle$ such that Eq. (10) becomes

$$H(t)\sum_{n=-\infty}^{n=\infty}\sum_{\beta}\Lambda_{\alpha\beta}^{(n)}|\phi_\beta\rangle\xi_n(t)+ \\ \sum_{n=-\infty}^{n=\infty}\sum_{\beta}\Lambda_{\alpha\beta}^{(n)}|\phi_\beta\rangle\xi_n(t)(\varepsilon_\alpha - n\hbar\omega) = 0. \tag{14}$$

Multiplication by $\langle\phi_\gamma|\xi_m^*(t)$, then average over one temporal period, leads to

$$\sum_{n=-\infty}^{n=\infty}\sum_{\beta}[\langle\alpha|H^{(m-n)}|\beta\rangle - (\varepsilon_\alpha - n\hbar\omega)\delta_{nm}\delta_{\alpha\beta}]\Lambda_{\alpha\beta}^{(n)} = 0. \tag{15}$$

and we have used the simplifying notation $|\alpha\rangle \equiv |\phi_\alpha\rangle$, and $H^{(m-n)} = 1/T\int_0^T \xi_m^*(t)H(t)\xi_n(t)$.

Then, the quasi-energies ε_α are eigenvalues of the secular equation.

$$det|H_F - \varepsilon_\alpha| = 0 \tag{16}$$

where $\langle\langle\alpha m|H_F|n\beta\rangle\rangle = H_{\alpha\beta}^{(m-n)} + m\omega\delta_{nm}\delta_{\alpha\beta}$. Here, we simplify the notation to represent the direct product of orbital and periodic eigenfunctions $|\alpha m\rangle\rangle \equiv |\alpha\rangle \otimes \xi(t)$.

For example, for a two-level problem we could take the interaction in the form $V(t) \propto \vec{\sigma}\cdot\vec{A}(t)$. Then, if the vector potential given as $\vec{A}(t) = A(\cos\omega t, \sin\omega t)$, we find that H^{m-n} has non-vanishing elements only for $m - n = 0, \pm 1$ as it is described in the following section.

2.2. Off-resonant approximation

In some physical scenarios, the frequency of the radiation field is way much larger than any other energy scale in the problem. Within this regime, one can derive an approximate Floquet Hamiltonian that captures the essence of the photoninduced bandgaps. In order to describe this so-called off-resonant regime, let us begin by explicitly showing the effective Floquet Hamiltonian: For this purpose, let us start from the general periodic Floquet Hamiltonian given in Eq. (1)

$$H(t) = H_0 + V(t), \tag{17}$$

where H_0 is the static contribution and $V(t+T) = V(t)$ is the time-periodic interaction. Going to Fourier space, we get the general structure of the Floquet Hamiltonian in matrix form as

$$\begin{pmatrix} \ddots & & & & & & & \\ \cdots & V_{-2,-3} & H_0 - 2\Omega & V_{-2,-1} & V_{-2,0} & V_{-2,1} & V_{-2,2} & \cdots \\ \cdots & V_{-1,-3} & V_{-1,-2} & H_0 - \Omega & V_{-1,0} & V_{-1,1} & V_{-1,2} & \cdots \\ \cdots & V_{0,-3} & V_{0,-2} & V_{0,-1} & H_0 & V_{0,1} & V_{0,2} & \cdots \\ \cdots & V_{1,-3} & V_{1,-2} & V_{1,-1} & V_{1,0} & H_0 + \Omega & V_{1,2} & \cdots \\ & \vdots & \vdots & \vdots & V_{2,0} & V_{2,1} & H_0 + 2\Omega & \cdots \\ & & & & & & & \ddots \end{pmatrix}, \tag{18}$$

where the interaction submatrices are defined as

$$V_{nn'} = \frac{1}{T}\int_0^T dt V(t) e^{-i(n-n')\Omega t}. \tag{19}$$

For a monochromatic harmonic perturbation, this reduces to a block-tridiagonal matrix

$$\begin{pmatrix} \ddots & & & & & & & \\ \cdots & V_{-1} & H_{-2} & V_{+1} & 0 & 0 & 0 & \cdots \\ \cdots & 0 & V_{-1} & H_{-1} & V_{+1} & 0 & 0 & \cdots \\ \cdots & 0 & 0 & V_{-1} & H_0 & V_{+1} & 0 & \cdots \\ \cdots & 0 & 0 & 0 & V_{-1} & H_1 & V_{+1} & \cdots \\ & \vdots & \vdots & \vdots & 0 & V_{-1} & H_2 & \cdots \\ & & & & & & & \ddots \end{pmatrix}, \tag{20}$$

and to simplify the notation we have set $V_N = V_{n-m}$ and $H_j = H_0 + j\Omega$. If we set out the eigenstate for a given number of Fourier modes n, we will have

$$\Phi = \begin{pmatrix} \phi_{-n} \\ \phi_{-n+1} \\ \vdots \\ \phi_{-2} \\ \phi_{-1} \\ \phi_0 \\ \phi_1 \\ \phi_2 \\ \vdots \\ \phi_{n-1} \\ \phi_n \end{pmatrix}, \tag{21}$$

with each ϕ_n a vector of dimensionality determined by H_0. For instance, if we approximate the problem in such a way that we only consider one Fourier mode ($n = 1$), we have to solve the following system of coupled equations:

$$\begin{aligned} H_{-1}\phi_{-1} + V_{+1}\phi_0 &= E\phi_{-1} \\ V_{-1}\phi_{-1} + H_0\phi_0 + V_{+1}\phi_{+1} &= E\phi_0 \\ H_{+1}\phi_{+1} + V_{-1}\phi_0 &= E\phi_{+1}. \end{aligned} \tag{22}$$

From the first and last equations, we get

$$\begin{aligned} \phi_{-1} &= (E - H_{-1})^{-1} V_{+1}\phi_0 \\ \phi_{+1} &= (E - H_{+1})^{-1} V_{-1}\phi_0, \end{aligned} \tag{23}$$

such that we get an effective equation for ϕ_0

$$[V_{-1}(E - H_{-1})^{-1} V_{+1} + H_0 + V_{+1}(E - H_{+1})^{-1} V_{-1}]\phi_0 = E\phi_0 \tag{24}$$

which explicitly reads

$$[V_{-1}(E-H_0+\Omega)^{-1}V_{+1}+H_0+V_{+1}(E-H_0-\Omega)^{-1}V_{-1}]\phi_0=E\phi_0 \tag{25}$$

For $\Omega \gg ||H_0||$, that is, frequencies much larger than the typical energy scales of the static problem, we can approximate this as

$$\left(H_0+\frac{V_{-1}V_{+1}}{\Omega}-\frac{V_{+1}V_{-1}}{\Omega}\right)\phi_0\approx E\phi_0, \tag{26}$$

so we get the effective approximate Floquet Hamiltonian, valid for large frequencies

$$H_{\mathrm{F}}\approx H_0+\frac{[V_{-1},V_{+1}]}{\Omega}. \tag{27}$$

With a similar procedure, one can show that for $n=2$ one gets the approximate Floquet Hamiltonian

$$H_{\mathrm{F}}\approx H_0+\frac{[V_{-1},V_{+1}]}{\Omega}-\frac{1}{2\Omega}\frac{[V_{-1}^2,V_{+1}^2]}{\Omega^2}. \tag{28}$$

These results essentially imply that a photoinduced energy bandgap can be induced by means of the dressed Floquet states that emerge from the off-resonant condition. Needless to say that sometimes the photoinduced energy bandgaps can be tiny; yet, in some circumstances one only needs to make sure that there is a gap, however, small and the topological properties of the driven system can be qualitatively different as those of the undriven (gapless) system.

3. Photoinduced effects of Landau levels

In this section, we summarize our results reported in Ref. [23] where we theoretically analyse the dynamical manipulation of the LL structure of charge carriers on suspended monolayer graphene when a periodically driving radiation field is applied perpendicular to the sample. For this purpose, we focus on the low-energy properties of non-interacting spinless charge carriers in a suspended monolayer graphene subject to a perpendicular, uniform and constant magnetic field $\mathbf{B}=B\hat{z}$. The dynamics is governed by Dirac's Hamiltonian. In coordinate representation, it reads

$$\mathcal{H}_\eta(\mathbf{r})=v_F(\eta\pi_x\sigma_x+\pi_y\sigma_y), \tag{29}$$

where $v_F\sim 10^6 m/s$ is the Fermi velocity in graphene. In addition, the canonical momenta $\pi_j=p_j+eA_j$ $(j=x,y)$ contain the vector potential $(\nabla\times\mathbf{A}=\mathbf{B})$, $-e$ is the electronic charge $(e>0)$, and $\eta=\pm1$ describes the valley degree of freedom. Using the definition of the magnetic length $l_B^{-2}=eB/\hbar$ and the cyclotron energy $\hbar\omega_c=\sqrt{2}v_F\hbar/l_B$, the Hamiltonian Eq. (29) at each K (K′) Dirac point, which corresponds to $\eta=+1$ $(\eta=-1)$, can be written in the form

$$H_{+1} = \hbar\omega_c \begin{pmatrix} 0 & a \\ a^\dagger & 0 \end{pmatrix} \tag{30}$$

$$H_{-1} = -\hbar\omega_c \begin{pmatrix} 0 & a^\dagger \\ a & 0 \end{pmatrix} \tag{31}$$

where the annihilation and creation operators are defined by standard relations as

$$a = \frac{l_B(\pi_x - i\pi_y)}{\sqrt{2}} \qquad \text{and } a^\dagger = \frac{l_B(\pi_x + i\pi_y)}{\sqrt{2}}. \tag{32}$$

The eigenenergies of the Hamiltonian (29) are then

$$E_n^{s,\eta} = s\eta\sqrt{n}\hbar\omega_c \tag{33}$$

with $s = \pm 1$. Positive (negative) values of $s\eta$ represents the conduction (valence) band at each Dirac point. In addition, the integer quantum number $n = 0, 1, 2\ldots$ labels the Landau-level structure of monolayer graphene. Using the eigenstates $|n\rangle$ of the operator $a^\dagger a$, the corresponding eigenstates $|\varphi_n^{s,\eta}\rangle$ read

$$|\varphi_n^{s,+1}\rangle = \frac{1}{\sqrt{2}} \begin{pmatrix} s|n-1\rangle \\ |n\rangle \end{pmatrix} \tag{34}$$

$$|\varphi_n^{s,-1}\rangle = \frac{1}{\sqrt{2}} \begin{pmatrix} -s|n\rangle \\ |n-1\rangle \end{pmatrix} \tag{35}$$

for $n \neq 0$. The zero-energy eigenstate ($n = 0$) is given in each case by

$$|\varphi_0^{+1}\rangle = \begin{pmatrix} 0 \\ |0\rangle \end{pmatrix} \tag{36}$$

$$|\varphi_0^{-1}\rangle = \begin{pmatrix} |0\rangle \\ 0 \end{pmatrix}. \tag{37}$$

Due to time-reversal symmetry, we have $\mathcal{T}H_{+1}\mathcal{T} = H_{-1}$. Let us now consider the effect of intense circularly polarized terahertz electromagnetic radiation, incident perpendicularly to the sample. We assume that the beam radiation spot is large enough compared to the lattice spacing so we can neglect any spatial variation. According to the standard light-matter interaction formulation, the dynamical effects of a monochromatic radiation field incident perpendicular to the sample can be described by means of a time-dependent vector potential

$$\mathbf{A}(t) = \frac{\mathcal{E}}{\omega}(\cos\omega t, \delta\sin\omega t), \tag{38}$$

where $\boldsymbol{\mathcal{E}}$ and ω are, respectively, the amplitude and frequency of the electric field given in turn by the standard relation $\boldsymbol{\mathcal{E}}(t) = -\partial_t\mathbf{A}(t)$. In addition, we are using $\delta = +1$ ($\delta = -1$) for right (left) circular polarization. We are using circular polarization because it has been shown to provide the maximal photoinduced bandgap. Starting from the ordinary dipolar interaction term $-e\mathbf{p} \cdot \mathbf{A}(t)$, introduced to the tight-binding Hamiltonian via the Peierls substitution, we can evaluate the effects of the driving at each Dirac point as

$$V_\eta = ev_F[\eta\sigma_x A_x(t) + \delta\sigma_y A_y(t)], \tag{39}$$

which explicitly reads

$$V_\eta = \xi\eta(\sigma_x \cos\omega t + \eta\delta\sigma_y \sin\omega t). \tag{40}$$

with the effective coupling constant $\xi = ev_F\mathcal{E}/\omega$. This makes the total Hamiltonian

$$H_\eta(t) = H_\eta + V_\eta(t), \tag{41}$$

periodic in time $H_\eta(t+T) = H_\eta(t)$, with $T = 2\pi/\omega$ the period of oscillation of the driving field. Therefore, if we focus on the K Dirac point ($\eta = 1$), the physics at the K′ Dirac point ($\eta = -1$) can be easily found by the substitutions $\xi \to -\xi$ and $\omega \to -\omega$.

Thus, let us focus on the K point physics and afterwards, we can make the necessary substitutions. In order to simplify the notation, we set $H_{+1} = H_0$ and $V_{+1}(t) = V(t)$. Hence, defining rising σ_+ and lowering σ_- pseudospin operators by the standard formulas

$$\sigma_\pm = \frac{\sigma_x \pm i\sigma_y}{2}, \tag{42}$$

the time-dependent interaction potential can be rewritten as

$$V(t) = \xi(e^{-i\delta\omega t}\sigma_+ + e^{i\delta\omega t}\sigma_-), \tag{43}$$

Now, we invoke Floquet's theorem which states that the time evolution operator of the system induced by a periodic Hamiltonian can be written in the form [24]

$$U(t) = P(t)e^{-iH_F t/\hbar}, \tag{44}$$

with $P(t)$ a periodic unitary matrix and H_F a time-independent dynamical generator referred to as the Floquet Hamiltonian. The eigenvalues of the Floquet Hamiltonian H_F represent the quasi-energy spectrum of the periodically driven system. Typically, in order to solve for the quasi-energy spectrum, one can expand each term of the time-dependent Schrödinger equation in Fourier space and numerically solve an infinite eigenvalue problem. Instead, we will take a perturbative approach as discussed below.

Accordingly, for our problem we can find approximate solutions to the dynamics by modifying slightly the analytical strategy presented in Ref. [39]. Then, one finds that the excitation number operator, N_a, defined as

$$N_a = \left(a^\dagger a + \frac{1}{2}\right)\mathbb{1} + \frac{\sigma_z}{2}, \tag{45}$$

which commutes with the Hamiltonian H_0 and satisfies the eigenvalue equation

$$N_a|\varphi_n^s\rangle = n|\varphi_n^s\rangle. \tag{46}$$

N_a generates a time-dependent unitary transformation $|\Psi(t)\rangle = P(t)|\Phi(t)\rangle$ given as

$$P(t) = \exp\left(-iN_a\delta\omega t\right), \tag{47}$$

such that the time-dependent Schrödinger equation

$$i\hbar\partial_t|\Psi(t)\rangle = H(t)|\Psi(t)\rangle \tag{48}$$

can be transformed with a time-independent operator H_F governing the dynamics of the problem

$$i\hbar\partial_t|\Phi(t)\rangle = H_F|\Phi(t)\rangle, \tag{49}$$

where H_F and $|\Phi(t)\rangle$ are the Floquet Hamiltonian and Floquet eigenstate, respectively. Conducting the explicit calculation, H_F is found to be given by

$$H_F = H_0 - N_a\delta\hbar\omega + \xi\sigma_x, \tag{50}$$

where the effective coupling to the radiation field is $\xi = ev_F\mathcal{E}/\omega$. This Hamiltonian can be treated via non-degenerate perturbation theory to find the approximate quasi-energy spectrum (see Ref. [23] for details on the derivation), which read

$$\epsilon_m^s = s\sqrt{m(\hbar\omega_c)^2 + (m\xi)^2}, \,\mathrm{mod}\hbar\omega. \tag{51}$$

The associated mean energies are in turn found as

$$\bar{\epsilon}_m^s = \epsilon_m^s - \omega\frac{\partial\varepsilon_m^s}{\partial\omega}, \tag{52}$$

which are invariant under $\epsilon_m^s \to \epsilon_m^s + l\hbar\omega$, for l being an integer. Conducting the explicit calculation, the mean energies are found to be given by the expression

$$\bar{\epsilon}_m^s = s\left(\frac{m(\hbar\omega_c)^2 + 2m^2\xi^2}{\sqrt{m(\hbar\omega_c)^2 + m^2\xi^2}}\right). \tag{53}$$

As can be seen in **Figure 1**, these mean energies are plotted as function of the quantizing magnetic field B, for different values of the Landau-level index changing the effective coupling ξ. We notice that, at intermediate light-coupling strength, the energy resolution of these levels becomes much better and could experimentally be tested for not so large

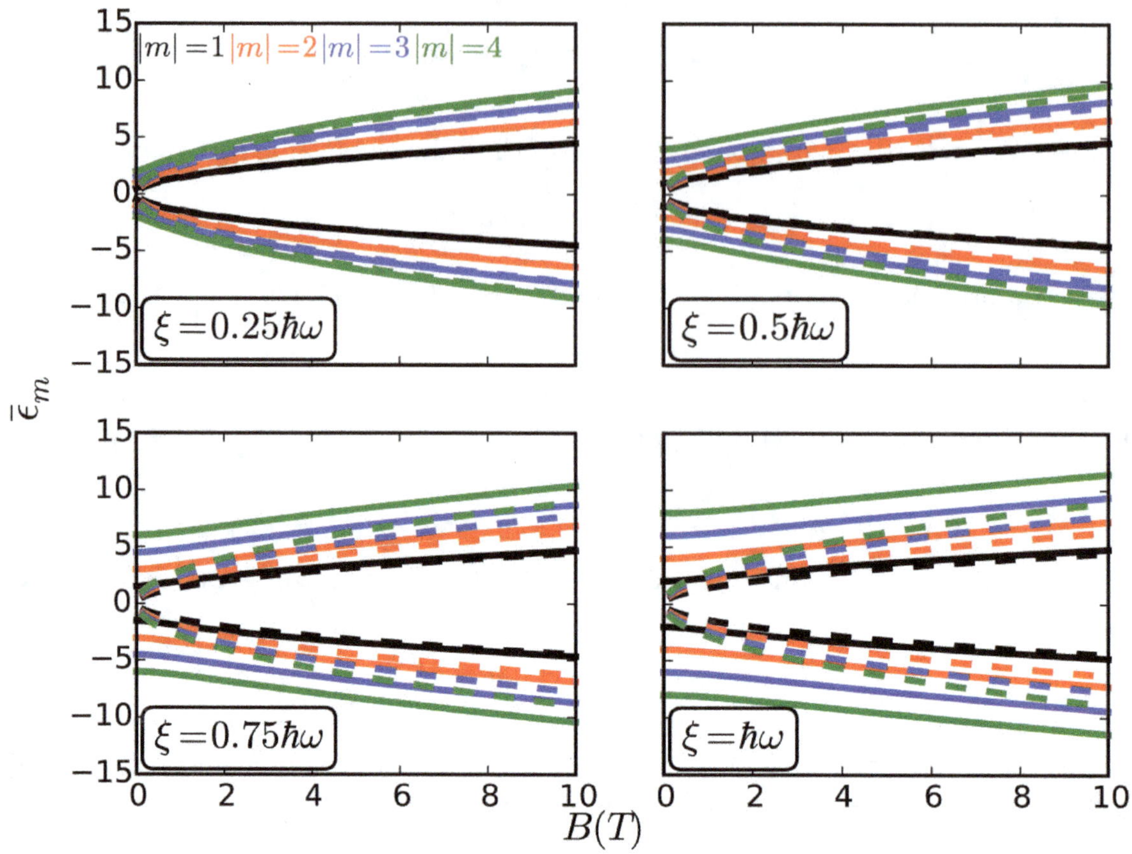

Figure 1. Approximate mean energies for the driven scenario (continuous lines) as function of the quantizing magnetic field B. The dotted lines represent the undriven spectra for the corresponding LL. We notice a level-dependent energy gapped that leads to non-trivial behaviour of physical quantities, as discussed below (see main text). Adapted from Ref. [23] with permission from APS.

quantizing magnetic fields B. Moreover, we find that to this order of approximation the LL becomes gapped, with the striking feature that the photoinduced gap is level-dependent. These gap openings appear except for the $m = 0$ level which, as discussed before, remains insensitive to the radiation field.

3.1. Discussion

So far, we have shown that upon introduction of a perpendicularly radiation field in the Terahertz frequency domain, the Landau-level-quantized scenario can be turned into a level-dependent-gapped system. As shown in reference, this energy bandgap effects can be traced via the oscillations of the pseudospin polarization as well as the temporal evolution of the autocorrelation function for an initially prepared coherent superposition of the static Landau-level configuration. Thus, due to the natural connection between coherent-state superpositions and applications in quantum optics, one could expect that the experimental consequences of such radiation effects on the Landau-level structure of graphene would have some applications within experimentally accessible parameter regimes in the quantum optics realm.

4. Irradiated silicene

As expected, the family of new two-dimensional materials with graphene-like properties has grown in the recent years. However, we show in this section that, although similar in lattice structure, both materials can be experimentally found to have different physical properties. In this section, we focus our attention on silicene which consists of a two-dimensional honeycomb lattice structure of silicon atoms analogous to that of graphene. Some works have reported the synthesis of silicene [30–32]. Silicene has a corrugated or buckled lattice structure that makes the silicon atoms in one sublattice to be perpendicularly displaced with respect to the other sublattice. For this reason, when a perpendicular electric field E_z is applied to silicene, the atoms belonging to each sublattice respond differently to E_z, giving rise to a staggered potential. Due to this peculiar pseudospin response to applied electric fields, and despite their similarities, the electronic properties of silicene are predicted to considerably differ from those of graphene.

In particular, since its intrinsic spin-orbit coupling is much larger than that of pristine graphene, an interesting interplay among intrinsic spin-orbit and electric field effects was predicted to appear because the bandgap can be electrically controlled. Moreover, the addition of an exchange potential term (which physically could represent the proximity effect due to coupling of ferromagnetic leads) allows for topological quantum phase transitions in the static regime [33]. Furthermore, in the presence of circularly polarized electromagnetic radiation, the realization of the so-called single Dirac cone phase in silicene has been recently proposed. At this topological phase, it is found that well-defined spin-polarized states are supported at every Dirac point. Moreover, within this configuration different spin components propagate in opposite directions giving rise to a pure spin current [34]. Yet, these photoinduced topological phase changes [20, 21] reported by Ezawa [34] were derived under the off-resonant assumption, that is, dynamical processes such that the frequency (coupling strength) of the radiation field is much larger (smaller) than any other energy scale in the problem. Under these assumptions, it is possible to derive an effective time-independent Floquet Hamiltonian [24, 25] with a tiny photoinduced bandgap correction that stems from virtual photon absorption and emission processes. Since the sign of the bandgap term (i.e., the effective bandgap) determines important topological properties of the material, it is vital both for potential practical implementations, for instance, in technological realizations of silicene-based devices, and from a fundamental point of view, to effectively achieve manipulation of this quantity.

In this section, we show via an exactly solvable model, where in order to detect physically relevant photoinduced effects in the energy band structure of silicene under strong circularly polarized electromagnetic radiation in the terahertz (frequency) domain one needs to go beyond the aforementioned off-resonant approximation. Indeed, we find that a zero momentum, the obtained exactly solvable time-dependent Hamiltonian, suggests the range of control parameters that physically might lead to experimentally feasible realization of new topological phases in silicene.

4.1. Model

Let us consider the Dirac cone approximation to describe the dynamics of non-interacting charge carriers in silicene subject to a perpendicular, uniform and constant electric field

$\mathbf{E} = E_z\hat{z}$. This is given by the 8×8 Hamiltonian [33] ($\hbar = e = 1$, with e being the electron's charge)

$$\begin{aligned}\mathcal{H}^\eta = v_F(k_x\sigma_x + \eta k_y\sigma_y) + \sigma_z(\eta s_z\lambda_{so} - lE_z) \\ +\eta\sigma_z h_{11} + h_{22}\end{aligned} \tag{54}$$

where $v_F = \frac{\sqrt{3}at_b}{2} \approx 8.1\times 10^5$ m/s is the Fermi velocity for charge carriers in silicene, with $a = 3.86$ Å the lattice constant and $t_b = 1.6$ eV the hopping parameter within a tight-binding formulation, whereas $l = 0.23$ Å measures half the separation among the two sublattice planes. In addition, $\eta = \pm 1$ describes the Dirac point, σ_i and s_i ($i = x, y, z$) are Pauli matrices describing pseudo- and real-spin degrees of freedom, respectively, whereas $\vec{k} = (k_x, \eta k_y)$ is the momentum measured from the Dirac point $\eta = \pm 1$. The parameter $\lambda_{so} = 3.9$ meV represents the strength of the intrinsic spin-orbit contribution. Moreover, the two contributions given by the terms

$$h_{11} = a\lambda_{R2}(k_y s_x - k_x s_y), \tag{55}$$

$$h_{22} = \lambda_{R1}(\eta\sigma_x s_y - \sigma_y s_x)/2, \tag{56}$$

describe the spin-orbit coupling associated to the next nearest neighbour hopping and nearest neighbour tight-binding formulation, respectively.

The term h_{11} has its origin in the buckled structure of silicone, whereas h_{22} is induced by the application of an external static electric field E_z. Using first-principle calculations, the authors of Ref. [40] found that $\lambda_{R2} = 0.7$ meV and typically h_{22} are of order μeV and thus much smaller than the other energy scales in the problem. Therefore, these two non-conserving contributions will be neglected in the following, although in the appendix we show that the largest contribution h_{11} can be easily incorporated in the solution to the dynamical evolution presented below. Yet, we have verified that our results do not qualitatively change by the introduction of these two small corrections.

Within the approximation $h_{22} = 0$, let us now consider the pseudospin dynamics under an *intense* radiation field represented by the time-dependent vector potential

$$\mathbf{A}(t) = A(\cos\Omega t, \sin\Omega t), \tag{57}$$

with $A = \mathcal{E}/\Omega$ and Ω its amplitude and frequency, respectively. It describes a monochromatic electromagnetic wave incident perpendicular to the sample. This vector potential can in turn be derived from the corresponding electric field by means of $\mathbf{E}(t) = -\partial_t\mathbf{A}(t)$, where $\mathcal{E}$ is the amplitude of the time-dependent electric field.

Using the standard minimal coupling prescription given as $\vec{k} \to \vec{k} + \vec{A}$, we get the dynamical generator

$$\begin{aligned}\mathcal{H}^\eta(\vec{k},t) = v_F(k_x\sigma_x + \eta k_y\sigma_y) + \sigma_z(\eta s_z\lambda_{so} - lE_z) \\ v_F A[e^{i\eta\Omega t}\sigma_- + e^{-i\eta\Omega t}\sigma_+].\end{aligned} \tag{58}$$

In the following, we explore the emerging photoinduced dynamical features at zero momentum since this scenario allows for an exact analytical solution to the dynamical evolution equations. Given the fact that we have an exact analytical solution, we can explore the low-, intermediate-, and strong-coupling regimes of the charge carriers in silicene under the radiation field. We then discuss this exact solution and argue the need to explore either the intermediate or strong light-matter-coupling regimes in order to obtain experimentally observable modifications in the physical properties of the system within the irradiation configuration.

4.2. Physics at $k = 0$

In this subsection, we explicitly analyse another exactly solvable model for a graphene-like system. Let us then focus on the zero-momentum scenario for which the extrinsic spin-orbit term h_{11} vanishes and the z-component of spin $s_z = \pm 1$ is a good quantum number. Therefore, the following analysis is independent of taking into account the aforementioned spin-orbit contribution. Setting for notational convenience $\alpha = v_F A$ and $V_z = lE_z$, the physics at zero momentum $\vec{k} = 0$ is described by the dynamical generator

$$\mathcal{H}^{\eta}(0,t) = (\eta s \lambda_{so} - V_z)\sigma_z + \alpha[e^{i\eta\Omega t}\sigma_- + e^{-i\eta\Omega t}\sigma_+]. \tag{59}$$

Using the unitary transformation

$$\mathcal{P}^{\eta}(t) = e^{-i\eta(\mathbb{1}+\sigma_z)\Omega t/2} \tag{60}$$

we get the effective time-independent Floquet Hamiltonian $\mathcal{H}_F(k=0) = (\mathcal{P}^{\eta})^{\dagger}(t)\mathcal{H}^{\eta}(0,t)\mathcal{P}^{\eta}(t)$ $-i(\mathcal{P}^{\dagger})^{\eta}(t)\dot{\mathcal{P}}^{\eta}(t)$

$$\mathcal{H}_F(k=0) = -\frac{\eta\Omega}{2}\mathbb{1} + \left[\eta\left(s_z\lambda_{so} - \frac{\Omega}{2}\right) - V_z\right]\sigma_z + \alpha\sigma_x. \tag{61}$$

Thus, the zero-momentum quasi-energy spectrum is given as

$$\varepsilon^{\eta}_{s\sigma}(k=0) = -\frac{\eta\Omega}{2} + \sigma\sqrt{\alpha^2 + (\Delta^{\eta}_s)^2}, \tag{62}$$

where $s, \sigma = \pm 1$ represent the real- and pseudospin degrees of freedom, respectively. In addition, we have defined the effective gap

$$\Delta^{\eta}_s = \eta\left(s\lambda_{so} - \frac{\Omega}{2}\right) - V_z. \tag{63}$$

On the other hand, the zero-momentum exact Floquet eigenstates are

$$|\psi^{\eta}_{s\sigma}(t)\rangle = \frac{e^{-i\varepsilon^{\eta}_{s\sigma}t}}{\sqrt{2\Gamma}}\begin{pmatrix} e^{-i\eta\Omega t}\sqrt{\Gamma + \sigma\Delta^{\eta}_s} \\ \sigma\sqrt{\Gamma - \sigma\Delta^{\eta}_s}\end{pmatrix}, \tag{64}$$

with $\Gamma = \sqrt{\alpha^2 + (\Delta^{\eta}_s)^2}$.

Some comments are in order at this point. The exact quasi-energy spectrum resembles the solution for the Rabi problem and one could expect that Rabi oscillations should appear in the dynamical evolution of this zero-momentum solution. Moreover, since we are interested in analysing the behaviour of a topological quantity not in terms of the momentum variable but assuming as a toy model that one explores the V_z parameter space along with the 'torus' for the time domain $0 \leq t \leq T$, one could evaluate the components of the pseudospin operator $\hat{\vec{S}}_\alpha$ defined within each subband as

$$\hat{\vec{S}}^\eta = \langle \psi^\eta_{s\sigma}(t) | \vec{s} | \psi^\eta_{s\sigma}(t) \rangle \tag{65}$$

After some algebra, we get

$$\hat{\vec{S}}^\eta = \eta(-\sin\beta \sin\Omega t, \sin\beta\cos\Omega t, -\cos\beta) \tag{66}$$

where

$$\tan\beta = \frac{2\alpha}{|\Delta^\eta_s|}. \tag{67}$$

Thus, within this limit, the average out-of-plane pseudospin polarization $\overline{S^z_\eta}$ is the only finite component and it reads

$$\overline{S^z_\eta} = -\frac{\eta|\Delta^\eta_s|}{\Gamma} \tag{68}$$

Thus, the main advantage of finding semi-analytical solutions to the dynamical evolution is that closed expressions for the physical quantities of interest can be found in such a way that one can provide further insight into the nature of the physical mechanisms involved and how their interplay leads to a given behaviour of the polarization, charge current and so on.

Let us now assume that the system is initially prepared in the arbitrary state

$$|\Phi(0)\rangle = \begin{pmatrix} \cos\frac{\theta}{2} e^{i\phi/2} \\ \sin\frac{\theta}{2} e^{-i\phi/2} \end{pmatrix}, \tag{69}$$

with $0 \leq \theta \leq \pi$ and $0 < \phi \leq 2\pi$ being spherical coordinates over the Bloch sphere. Thus, the evolution of the pseudospin polarization is given by the standard relation $\sigma_z(t) = \langle \Phi(0) | U^\dagger_F(t) \sigma_z U_F(t) | \Phi(0) \rangle$, with $U_F(t)$ being the unitary Floquet evolution operator $U_F(t) = \mathcal{P}^\eta(t) e^{-i\mathcal{H}_F t}$ (note that σ_z and $\mathcal{P}^\eta(t)$ commute with each other). The initial polarization in the state (65) is given by $\sigma_z(0) = \cos\theta$. After some algebra, we find

$$\sigma_z(t) = \frac{2\alpha}{\Gamma}\sin\theta\sin\Gamma t\left(\frac{\Delta^\eta_s}{\Gamma}\sin\Gamma t\cos\phi - \cos\Gamma t\sin\phi\right) + \cos\theta\left(1 - \frac{2\alpha^2}{\Gamma^2}\sin^2\Gamma t\right). \tag{70}$$

In addition, the one-period mean-value pseudospin polarization

$$\langle \sigma_z \rangle = \frac{1}{T} \int_0^T \sigma_z(t)dt, \tag{71}$$

with $T = 2\pi/\Omega$ being the period of oscillations of the driving field, gives then

$$\begin{aligned} \langle \sigma_z \rangle = \alpha \sin\theta \left[\frac{\Delta_s^\eta}{\Gamma^2} \cos\phi(1 - \mathrm{sinc}(2\Gamma T)) - T \sin\phi \mathrm{sinc}^2(\Gamma T) \right] \\ + \cos\theta \left[1 - \frac{\alpha^2}{\Gamma^2}(1 - \mathrm{sinc}(2\Gamma T)) \right], \end{aligned} \tag{72}$$

where $\mathrm{sinc}(x) = \frac{\sin(x)}{x}$.

In particular, for initial states that have zero polarization ($\theta = \pi/2$), we get the simplified expressions

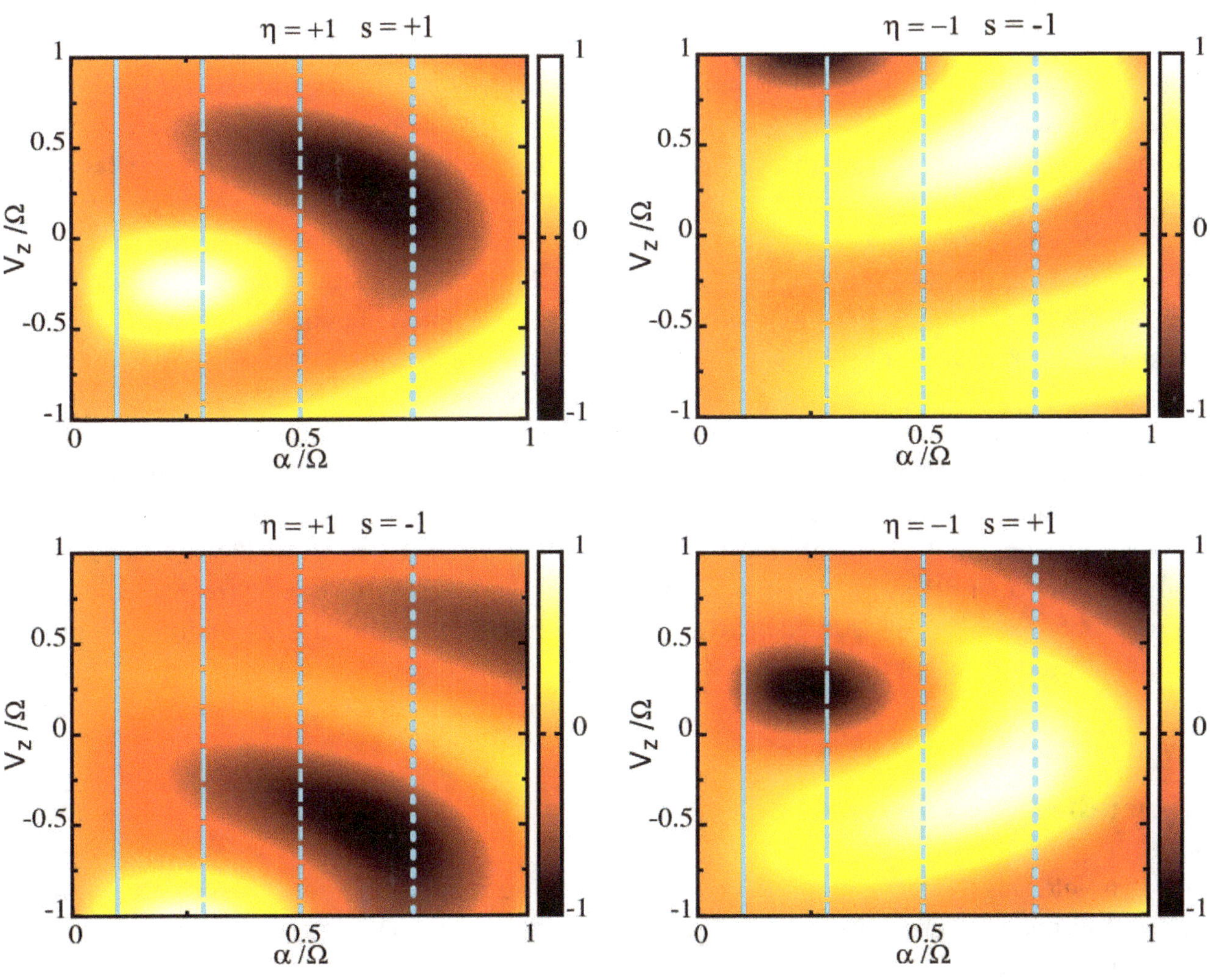

Figure 2. Zero-momentum pseudospin mean polarization for different combinations of the product $s\eta$. The vertical lines correspond to $\alpha = 0.1\Omega$ (continuous), $\alpha = 0.25\Omega$ (large dashed) $\alpha = 0.5\Omega$ (short dashed) and $\alpha = 0.75\Omega$ (large dots), respectively. See discussion in the main text. Adapted from Ref. [35] with permission from APS.

$$\langle \sigma_z \rangle = \alpha \left[\frac{\Delta_s^\eta}{\Gamma^2} \cos\phi (1 - \text{sinc}(2\Gamma T)) - T \sin\phi \text{sinc}^2(\Gamma T) \right]. \tag{73}$$

Setting the value $\phi = \pi/4$, we plot in **Figure 2** the mean pseudospin polarization for the different spin and valley $s\eta$ product combinations.

From this figure, we find that within the low coupling regime ($\alpha \leq 0.1\Omega$), it is in general not possible to induce appreciable changes of the pseudospin polarization and this is related to the fact that the quasi-energy behaviour is essentially controlled by the parameters V_z and λ_{so} which determine the gap behaviour in the static regime. On the other hand, for intermediate ($\alpha = 0.5\Omega$) and large ($\alpha = 0.75\Omega$) values of the coupling to the driving field, that is, beyond the off-resonant condition, effective pseudospin inversion is achievable and therefore a qualitatively different behaviour emerges within this coupling regime.

5. Conclusions

In this chapter, we have described some photoinduced consequences of using periodically driven interactions in graphene-like systems. In essence, these examples provide new insights on the non-trivial behaviour of systems taken out of equilibrium and given the fact that Floquet's theorem allows for a dynamical analysis in terms of an equivalent static description of the physics via the Floquet states, one can infer new physical aspects that can emerge on these systems when subject to such periodic interactions. We consider that these approaches do provide an arena for analysing some interesting phenomena beyond the static limit. The extension of the results presents more involved scenarios, where numerical tools can profit from the simple models described in this chapter and we hope the interested reader could profit from the material presented. One could further explore the dynamical features at finite momentum but refer the reader to Ref. [35] where it is proven that the intermediate light-matter-coupling regime is the physically correct scenario for describing the emergent topological features such as the realization of the single-valley polarized state in silicene. Needless to say that other quantities of interest such as charge and spin currents can be treated within the approximate scenarios discussed in the chapter, they go beyond the main interest of the chapter and are thus not further discussed here.

Appendix

For completeness, in this appendix we include one approximation strategy that can be useful in the semi-analytical treatment of periodically driven systems. This is called the rotating wave approximation concerns with the dynamical evolution of a periodically driven system when the parameters are closed to a resonance. These resonant effects have been shown to be relevant in several experimental situations such as within cavity QED as well as within the context of cold atoms, among others.

Approximate RWA dynamics

In some physical systems, the most important physical effects happen around the first resonance located at around $\omega_0 = \omega$, even a low coupling $\xi = 0.25\omega$. This motivates the search of an analytical solution by means of the rotating wave approximation (RWA) approximation. We now find an approximate analytical solution to the dynamics generated by

$$H(t) = \frac{\omega_0}{2}\sigma_z + A \sin \omega t \sigma_x. \tag{74}$$

Then, in the low coupling regime where the amplitude of the driving field satisfies $\frac{A}{\omega} \ll 1$, we can use the rotating wave approximation (RWA) which amounts to keep those contributions in the driving that are close to resonance, that is, $\omega \sim \omega_0$.

Changing to the interaction representation, the dynamics is dictated by $V_I(t) = e^{i\omega_0 t\sigma_z/2}V(t)$ $e^{-i\omega_0 t\sigma_z/2}$, where $V(t)$ is the time-dependent contribution to Eq. (74). Using the algebra of the Pauli matrices and the explicit time dependence of the driving field, we get

$$V_I(t) = A \sin \omega t(\cos \omega_0 t\sigma_x - \sin \omega_0 t\sigma_y) \tag{75}$$

At this point, we invoke the RWA to get the approximate interaction contribution

$$V_{RWA}(k, t) = i\frac{A}{2}(e^{-i\mu t}\sigma_+ - e^{i\mu t}\sigma_-), \tag{76}$$

where the frequency detuning that characterizes near-resonance contributions is defined as $\mu = \omega - \omega_0$. To get Eq. (76) from Eq. (75), we have assumed $\omega - \omega_0 \sim 0$ and thus have disregarded the quickly oscillating terms proportional to $\omega + \omega_0$ (which are known in the literature as the secular contributions).

Switching back to the Schrödinger representation, we get the total RWA Hamiltonian

$$H_{RWA}(t) = \frac{\omega_0}{2}\sigma_z + ig(e^{-i\omega t}\sigma_+ - e^{i\omega t}\sigma_-), \tag{77}$$

where we have introduced the effective coupling constant $g = \frac{A}{2}$ The Hamiltonian (77) is taking to a time-independent form H_F by means of a periodic unitary transformation $P(t) = e^{-i(\sigma_0+\sigma_z)\omega t/2}$ by means of the standard transformation rule $H_F = P^\dagger(t)H_{RWA}P(t) - iP^\dagger(t)\partial_t P(t)$. Explicitly, we have

$$H_F = -\frac{\omega}{2}\sigma_0 - \frac{\mu}{2}\sigma_z - g\,\sigma_y. \tag{78}$$

Defining the Rabi frequency $\Omega_R = \sqrt{\mu^2 + 4g^2}$, the approximate quasi-energies are found to be given as

$$\varepsilon_\alpha = -\frac{(\omega + \alpha\Omega_R)}{2} \tag{79}$$

where $\alpha = \pm 1$ is the energy subband index. The corresponding approximate Floquet eigenstates are

$$|\Phi_\alpha(k,t)\rangle = \frac{e^{-i\varepsilon_\alpha t}}{\sqrt{2\Omega_R}} \begin{pmatrix} e^{-i\omega t}\sqrt{\Omega_R - \alpha\mu} \\ i\alpha\sqrt{\Omega_R + \alpha\mu} \end{pmatrix}. \tag{80}$$

As before, once the Floquet quasi-energies and eigenstates are found, one can evaluate any physical quantity of interest. For instance, one can evaluate the spin (or any other related quantity that satisfies the Pauli matrices algebra) polarization. In addition, the evaluation of the charge as well as spin currents can be given analytical expressions within the RWA. The relevance of resonant phenomena in nature has already been highlighted in the introduction, but we would like to remember that within this scheme, photon-assisted transport is one of the many resonant processes that can be properly described via the RWA. However, we must mention that one of the drawbacks of the RWA is that it does not take into account a systematic shift of the resonant energies which occurs at higher coupling strengths to the driving field. These are the so-called Bloch-Siegert shifts which emerge from the neglected secular terms. Thus, in order to go to higher coupling strengths and preserve a more realistic physical picture, this secular term should be included. In this context, the Van Vleck quasi-degenerate perturbation theory as well as Magnus expansion approach can be useful semi-analytical tools and we refer the interested reader to Ref. [41] for a detailed treatment of the Magnus expansion approach and to Ref. [42] for a thorough description of the Van Vleck approach.

Author details

Alexander López[1*] and Benjamin Santos[2]

*Address all correspondence to: alexander.lopez@physik.uni-regensburg.de

1 Escuela Superior Politécnica del Litoral, ESPOL, Facultad de Ingeniería Mecánica y Ciencias de la Producción, Campus Gustavo Galindo Km 30.5 Vía Perimetral, Guayaquil, Ecuador

2 INRS-EMT, Université du Québec, Varennes, Québec, Canada

References

[1] Novoselov K. S. et al., Science **306**, 666 (2004).

[2] Novoselov K. S. et al., Nature **438**, 197 (2005).

[3] Geim A. K. and Novoselov K. S., Nat. Mater. **6**, 183 (2007).

[4] Castro Neto A. H. et al., Rev. Mod. Phys. **81**, 109 (2009).

[5] Das Sarma S. et al., Cond-mat.mes-hall, 1003.4701 (2010).

[6] Kane C. L. and Mele E. J., Phys. Rev. Lett. **95**, 226801 (2005), ibid, Phys. Rev. Lett. **95**, 146802 (2005).

[7] Huertas-Hernando D., Guinea F., and Brataas A., Phys. Rev. B **74**, 155426 (2006).

[8] Rashba E., Phys. Rev. B **79**, 161409(R) (2009).

[9] Ertler C. et al., Phys. Rev. B **80**, 041405(R) (2009).

[10] Ando T., Matsumoto Y. and Uemura Y., J. Phys. Soc. Jpn. **39**, 279 (1975).

[11] Klitzing K. v, Dorda G. and Pepper M., Phys. Rev. Lett. **45**, 494 (1980).

[12] Hasan M. Z. and Kane C. L., Rev. Mod. Phys. **82**, 3045 (2010).

[13] Bernevig B. A., Hughes T. L. and Zhang S. C., Science **314**, 1757 (2006).

[14] Fu L., Kane C. L. and Mele E. J., Phys. Rev. Lett. **98**, 106803 (2007).

[15] Goldman N. et al, Phys. Rev. Lett. **105**, 255302 (2010).

[16] Schneider U. et al., Nat. Phys. **8**, 213 (2012).

[17] König M. et al., Science **318**, 766 (2008).

[18] Yang Y. et al, Phys. Rev. Lett. **107**, 066602 (2011)

[19] Lindner N. H., Refael G. and Galinski V., Nat. Phys. **7**, 490 (2011).

[20] Kitagawa T. et al., Phys. Rev. B **82**, 235114 (2010).

[21] Kitagawa T. et al., Phys. Rev. B **84**, 235114 (2011).

[22] Cayssol J., Dóra B., Simon F., and Moessner R., Phys. Status Solidi RRL **7**, 101 (2013).

[23] López A., Di Teodoro A., Schliemann J., Berche B., Santos B., Phys. Rev. B **92**, 235411 (2015).

[24] Grifoni M. and Hänggi P., Phys. Rep. **304**, 229 (1998).

[25] Chu S.-I. and Telnov D. A., Phys. Rep. **390**, 1 (2004).

[26] Shirley J. H. Phys. Rev. **138**, B979 (1965).

[27] Krueckl V. and Kramer T., New J. Phys. **11**, 093010 (2009).

[28] Takeda K. and Shiraishi K., Phys. Rev. B **50**, 14916 (1994).

[29] Guzmán-Verri G. G. and Lew Yan Voon L. C., Phys. Rev. B **76**, 075131 (2007).

[30] Vogt P. et al., Phys. Rev. Lett. **108**, 155501 (2012).

[31] Fleurence A. et al., Phys. Rev. Lett. **108**, 245501 (2012).

[32] Chen L. et al., Phys. Rev. Lett. **109**, 056804 (2012).

[33] Ezawa M., Phys. Rev. Lett. **109**, 055502 (2012).

[34] Ezawa M., Phys. Rev. Lett. **110**, 026603 (2013).

[35] López A., Scholtz A., Schliemann J., and Santos B., Phys. Rev. B **91**, 125105 (2015).

[36] Oka T. and Aoki H., Phys. Rev. B **79**, 081406(R) (2009).

[37] Zhou Y. and Wu M. W., Phys. Rev. B **83**, 245436 (2011).

[38] Calvo H. L., Pastawski H. M., Roche S. and Foa Torres L. E. F., Appl. Phys. Lett. **98**, 232103 (2011).

[39] Lin Q.-g J. Phys. A: Math. Gen. **34**, 1903 (2001).

[40] Liu C. -C., Jiang H. and Yao Y., Phys. Rev. B **84**, 195430 (2011).

[41] Blanes S., Casas F., Oteo J. A. and Ros J., Phys. Rep. **470**, 151 (2009).

[42] Shavitt I. and Redmon L. T., J. Chem. Phys. **73**, 11 (1980).

4

Bilayer Graphene as the Material for Study of the Unconventional Fractional Quantum Hall Effect

Janusz Edward Jacak

Abstract

Fractional quantum Hall effect (FQHE) discovered experimentally in 1982 is still mysterious, not fully understood phenomenon. It fundaments are linked with a nontrivial topological effects in 2D space going beyond the standard description of FQHE with local quantum mechanics. The study of integer and fractional QHE in graphene might be helpful in resolution of this fundamental problem in many body quantum physics. FQHE has been observed both in monolayer and bilayer graphene with an exceptional accuracy due to advances in experimental techniques and purity of graphene samples. Recent experimental observations of FQHE in the bilayer graphene reveal different FQHE behavior than in the monolayer samples or in conventional semiconductor 2D materials. This unexpected phenomena related to Hall physics in the bilayer systems allows to better understand more than 30 years old puzzle of FQHE. In the chapter we will summarize the recent and controversial experimental observations of FQHE in bilayer graphene and describe the topology foundations which may explain the oddness of correlated multiparticle states in the bilayer system. These topological arguments shed also a new light on understanding of heuristic CF concept for FQHE and deeper the topological sense of the famous Laughlin function describing this strongly correlated state.

Keywords: monolayer graphene, bilayer graphene, FQHE, hierarchy of filling fractions, braid group for bilayer graphene, correlated state, commensurability

1. Introduction

Fractional quantum Hall effect (FQHE) discovered experimentally in 1982 [1] (Nobel Prize 1989 for Tsui and Störmer) is still a mysterious phenomenon robust against its complete understanding. The famous Laughlin function [2] proposed for theoretical description of this effect (Nobel Prize 1989 for Laughlin, together with Tsui and Störmer) was the illuminating concept towards explanation of the FQHE. This function was the next start-point for further attempts to understand and describe in detail the experimentally observed FQHE behavior with still growing number of new observations. Despite the breakthrough character of the Laughlin concept, the complete explanation of the FQHE has not been achieved as of yet. The basic problem is linked here with nontrivial topology-type effects in two-dimensional (2D)-charged systems exposed to strong perpendicular magnetic field going beyond the standard description of FQHE upon the local quantum mechanics but generating specific long-range correlations in the whole 2D system.

The Laughlin function has not been derived but it was rather proposed as variational trial multiparticle wavefunction and verified next by numerical exact diagonalization of the interaction of electrons. The archetype for the Laughlin function was the Slater determinant built of zeroth Landau level (LL) single-particle functions. The polynomial part of this Slater determinant has the form of the Vandermonde determinant polynomial and a heuristic change to the Jastrow polynomial results in the Laughlin function. This simple mathematical maneuver has, however, a pronounced physical consequence. The Laughlin function describes the strongly interacting particles, contrary to the Slater function and, moreover, pretty well describes multiparticle correlations of FQHE specific for 2D space topology. The ostensible simplicity of the Laughlin function triggered next efforts to describe FQHE in terms of quasiparticles common in solids, that is, in terms of effective single-particle objects dressed with interaction and defined as the poles of the single-particle-retarded Green function [3]. Nevertheless, in the case of the FQHE, the Coulomb interaction-induced mass operator is discontinuous in 2D at magnetic field presence, and this property precludes any definition of the quasiparticle as dressed electrons with the interaction.

Despite this inability to rigorously define quasiparticles in the FQHE system, a heuristic concept of single-particle picture for this system has been formulated in terms of the so-called composite fermions (CFs) [4]. CF is proposed as complex object consisting of the electron with attached local field flux (even the number of an auxiliary magnetic field flux quanta pinned to the electron). The origin of the auxiliary magnetic field as well as the manner of how the local flux is fixed to a moving electron is not explained, but the CF model works surprisingly well [4]. CFs defined as electrons with fictitious magnetic field flux quanta fixed to them gain the Laughlin function phase shift by the Aharonov-Bohm-type effect when composite particles interchange their position. Note, however, that the Aharonov-Bohm effect is of non-local character. The main advantage of CF effective single-particle picture resolves itself to the mapping of FQHE onto IQHE in the resultant magnetic field, that is, the true external magnetic field reduced by the averaged field of fluxes pinned to electrons. This mapping allows for the definition of the Jain-like hierarchy of FQHE in partial agreement with experimental data. In view of the above-mentioned discontinuity of the Coulomb interaction mass operator, CFs cannot be named as quasiparticles (moreover, an auxiliary magnetic field fluxes dressing electrons cannot be "produced" by the non-magnetic electric Coulomb repulsion of electrons). One can conclude that CFs describe phenomenologically a different and more fundamental property of the system in close relation to the specific topology of 2D charged system in magnetic field and connected with this topology specific long-range correlation induced by electron repulsion.

In order to decipher this enigmatic character of CFs and the topological correlations in FQHE, the algebraic-topology braid group-based approach has been developed and applied to FQHE [5]. Within this approach, the trajectories of interchanging positions of particles on the 2D plane in the system of N charged particles upon strong perpendicular magnetic field are considered including quantum indistinguishability of identical particles. The geometrically non-equivalent trajectories fall to disjoint classes of nonhomotopic trajectory classes (which cannot be continuously transformed between distinct classes), which form the algebraic group called the braid group. The corresponding quantum multiparticle correlated state is characterized by one-dimensional unitary representation (1DUR) of the braid group for a particular system. For particles in 3D, the braid group of the N particle system is always the N-element permutation group (regardless of a charge, interaction, or magnetic field presence). There exist only two 1DURs for any permutation group: $\sigma_j \to \{ {e^{i0} = 1 \atop e^{i\pi} = -1}$, where σ_j, $j = 1,\ldots,N$, are braids (multiparticle classical trajectories) for exchanges of positions of jth particle with (j+1) th particle, when other particles remain in rest. 1DUR = 1 defines bosons and 1DUR = –1 defines fermions in 3D multiparticle systems. For 2D multiparticle systems, the braid group is different than the permutation group [5–7] and 1DURs are different $\sigma_j \to e^{i\alpha}$, $\alpha \in (-\pi,\pi]$ and define 2D anyons (including 2D fermions for $\alpha = \pi$ and bosons for $\alpha = 0$).

Nevertheless, in 2D for charged-repulsing electrons in the presence of strong perpendicular magnetic field, the topological situation changes again. For magnetic field strong enough, the only possible on the plane cyclotron classical orbits may be too short to match neighboring particles uniformly distributed on the plane (with classical positions fixed by the repulsion of electrons). Too short braids must be thus removed from the braid group. The remaining braids form the subgroup of the former group called the cyclotronic braid subgroup. The cyclotronic subgroup collects only those braids which can match neighboring particles.

To clarify the structure of the cyclotronic braid subgroups, let us note that in the presence of strong magnetic field the 2D braids must be built from pieces of cyclotron orbits and this orbit must precisely fit to interparticle spacing. Otherwise, the definition of braids is precluded. Hence, the commensurability of planar braids with particle spacing is an unavoidable condition to define the braid group and establish the quantum statistics (by 1DURs of the braid group). The quantum statistics is a necessary prerequisite to multiparticle-correlated state. Therefore, the commensurability condition determines the magnetic field strength (or equivalently, Landau level-filling rate) at which correlated state can be formed. The archetype of the commensurability in 2D-charged N electron system is the exact fitting of cyclotron orbits of interacting 2D electrons to inter-electron spacing, which happens at completely filled lowest Landau level (LLL), that is, at filling factor $\nu = \frac{N}{N_0} = 1$, where $N_0 = \frac{BSe}{hc}$ is the LL degeneracy ($\frac{hc}{e}$ is the magnetic field quantum). In other words, the condition $\nu = 1$ is equivalent with the commensurability condition $\frac{S}{N} = \frac{S}{N_0} = \frac{hc}{eB}$, that is, the size of cyclotronic orbit in the LLL at $\nu = 1$

fits with particle spacing $\frac{S}{N}$ (S is the 2D plane size for the system; in the thermodynamic limit $\frac{S}{N}$ is constant even if S and N tend to infinity). For $\nu = 1$, the braid group can be thus established and the correlated state manifest itself as the IQHE.

For stronger magnetic fields, $\nu < 1$ and the commensurability fails. It means that ordinary cyclotron orbits $\frac{hc}{eB}$ are too short and the corresponding braids, σ_j, must be removed from the braid group as unavailable ones. In the braid group are, however, braids σ_j^q, which for q odd integer similarly as σ_j define exchanges of jth and (j+1)th particles. In distinction to σ_j, the braids σ_j^q realize exchange with additional $\frac{q-1}{2}$ loops [5]. This additional loops "take away" $\frac{q-1}{2}$ flux quanta. This is the origin of the auxiliary flux tubes pinned to CFs. In 2D, the external field flux per particle $\frac{BS}{N}$ is thus reduced by $\frac{q-1}{2}N$ similarly as in the CF concept. Braids σ_j^q define exchanges along multiloop cyclotron 2D orbits. Exclusively in 2D multiloop cyclotron orbits share the same external field flux per particle and per single loop falls only some fraction of $\frac{BS}{N}$ (noticeably, in contrary to 3D case, when each scroll of spiral adds a new field piercing by the field B). Hence, in 2D, the multiloop cyclotron orbits related to σ_j^q have larger size and σ_j^q can reach particles out of reach for single-loop braids σ_j. The braid subgroup generated by σ_j^q, $j = 1,\ldots,N$ is the proper braid group for $\nu < 1$ provided that the new commensurability condition holds: $\frac{S}{N} = \frac{qBSe}{hc}$, where the r.h.s of this condition expresses the q times larger range of multilooped cyclotron orbit in 2D. From this commensurability condition, it follows the relation: $\nu = \frac{N}{N_0} = \frac{1}{q}$ displaying the main line of LLL fillings for FQHE (described by the Laughlin function with qth order of the Jastrow polynomial). The braid group commensurability is thus the kernel of the Laughlin function derivation—any multiparticle ground-state wave function of the correlated 2D N-charged-interacting particles (electrons) must transform according to 1DUR of the related cyclotron braid group [6, 8, 9]. This feature together with requirement that the multiparticle function in the LLL must be holomorphic function of $z_1,\ldots,z_N$ ($z_j = x_j + iy_j$—complex coordinate on the plane of jth particle) —uniquely results in the form of the Jastrow polynomial $\prod_{i>j}^{N}(z_i - z_j)^q$ (multiplied by a factor independent of particle interchanges—in the Laughlin function $e^{-\sum_{i=1}^{N}|z_i|^2/4l_B^2}$ (l_B is the magnetic length). The latter term is common for all states from the LLL, thus the derivation of the Laughlin function resolves itself to the determination of the uniform with respect to all particles polynomial factor, which can be deduced according to 1DUR of related cyclotronic braid group as described above.

2. FQHE in graphene

Graphene is ideally two-dimensional but with some specific modification of LL structure due to pseudo-relativistic band structure of this gap-less semiconductor. Thus, examining FQHE in graphene is especially challenging. The specific for graphene electronic dynamics is referred to the Dirac point in all corners of the hexagonal first Brillouin zone. This specific pseudo-relativistic local band structure does not influence, however, the topology constraints imposed on trajectories and the FQHE has been observed in graphene. Easy methods for the manufacturing of monolayer and bilayer graphene ought to be experimentally noticeable. And indeed, several recent observations of FQHE in bilayer graphene [10–14] demonstrate strong distinctions in comparison to FQHE hierarchy in monolayer graphene [15–18].

The observed FQHE phenomena in bilayer graphene supply an opportunity to verify the cyclotron braid group commensurability approach because the topology of double sheet in bilayer graphene causes different commensurability conditions in comparison to the monolayer case. This will result in specific hierarchy of filling rates for FQHE in bilayer graphene not observed in monolayer graphene and in conventional semiconductor 2D systems.

In the present chapter, we summarize the recent and controversial experimental observations of FQHE in bilayer graphene and describe the topology foundations, which may explain the oddness of correlated multiparticle states for the bilayer system. The significance of such a discussion consists in experimental confirmation of new topological-based theory of FQHE going beyond the phenomenological model of CFs.

Advances in experiment allow now for measurements of longitudinal and transversal resistivity in Hall configurations in various graphene sample arrangements. The great progress in experiment has been achieved by the utilization of measurement technique in the so-called suspended graphene, where small scraping of the graphene sheet has been hanged in free air without any contact with substratum and thus without disturbing phonons in crystals and therefore with extremely high mobility triggering FQHE organization in relatively higher temperatures. It is possible to suspend both monolayer and bilayer sheets. Another experimental achievement is the utilization of homogeneous with graphene crystal substrate of boron nitride (BN) crystal similar as graphene hexagonal surface lattice geometry. The application of BN substrate to support graphene layer also allows for the observation of FQHE in higher temperatures because of reducing stress and perturbations caused by lattice mismatching. The new data for fractional quantum Hall effect both in suspended graphene scrapings [10–14, 17, 18] and in graphene samples on crystalline substrate of boron nitride [15, 16] have been recently collected.

The specific structure of Landau levels (LLs) in graphene leads to distinct than in conventional semiconductor 2DEG scheme for integer quantum Hall effect in graphene referred to as "relativistic" of its version [19]. The graphene planar sheet has a regular crystalline structure with two equivalent sublattices (two carbon atoms in the Bravais cell). This feature together with vanishing of the forbidden semiconductor gap on corners (two of them—non-equivalent) of hexagonal Brillouin zone, with cone-like shape of local valence and conduction bands meeting at these points (the so-called the Dirac points), results in the fourfold spin-valley degeneracy of LLs. Moreover, for Dirac points in graphene, the specific Berry phase-like shift for chiral 2D carriers additionally modifies the LL spectrum, which finally results in $\nu = 4(n + \frac{1}{2})$ series for fillings at which IQHE plateaus occur in graphene. These plateaus in the monolayer graphene occur at the centers of the consecutive LLs (not at completely filled LLs) [20]. In bilayer graphene the extra degeneracy of $n = 0$ and $n = 1$ LLs shifts IQHE plateau positions to edges of LLs, which in bilayer graphene are also fourfold-degenerated except for the eightfold-degenerated LLL [19, 20]. For both monolayer and bilayer graphene, FQHE hierarchy of fillings is also observed. In the case of monolayer graphene, the filling fractions for FQHE are observed in six first subbands of LLs with $n = 0$ and $n = 1$ in Refs. [15–18], which reproduce similar hierarchy known from observations in conventional semiconductors 2DEG. Especially interesting is, however, an observation of unusual even denominator fillings for FQHE in bilayer graphene including the most pronounced feature at $\nu = -\frac{1}{2}$ [10], which does not find counterpart in monolayer system. This state cannot be also explained upon the CF approach, as this approach suggests the Hall metal state at $\pm\frac{1}{2}$ filling ratio.

In the present paper, we propose the explanation of this exotic, beyond CF, state in bilayer graphene by the topological commensurability approach [6, 21 22]. In this braid group-based approach, we explain the structure of fractional fillings of LL subbands beyond the CF model and explain the reason for the failure of CF model in bilayer graphene. The topological braid group approach formerly developed in [23, 24] gives the hierarchy of FQHE in agreement with the available experimental data for monolayer and bilayer graphene in contrary to CF model. It is worth emphasizing the topological explanation of the even denominator-filling ratios for FQHE in bilayer graphene, including $\nu = -\frac{1}{2}$.

2.1. Commensurability condition

Let us briefly recall the concept of the commensurability of cyclotron trajectories with interparticle spacing in 2D-charged system at strong magnetic field within the braid group approach to multiparticle systems at magnetic field presence. It must be emphasized that for FQHE formation, the interaction of electrons is essential similarly as of any other correlated state. The strong Coulomb repulsion of electrons determines the steady uniform distribution of 2D electrons—in the form of triangle Wigner crystal lattice as the classical lowest energy state. Such a distribution of electrons rigidly fixed (in classical sense) by interaction is the starting point for quantization in terms of Feynman path integral including summation on braid group elements assigned with appropriate unitary weight supplied by 1DUR of the braid group [5]. The latter is necessary in the case of multiparticle system as to any path in the multiparticle configuration space an arbitrary closed loop from the braid group can be attached and the resulting paths with such loops fall into geometrically disjoint sectors (topologically non-equivalent) of the path space. This not-continuous structure of the path space, enumerated by the braid group, forces the additional sum over the braid group elements in the general path integration. Each component to this sum must enter the sum with some unitary factor (due to causality condition) and the collection of these factors for all braid group elements establishes its 1DUR. This explains the central role of 1DURs of the braid group for the definition of quantum statistics of multiparticle system. Each different 1DUR defines different sorts of quantum particles corresponding to the same classical ones [6, 9]. On the other hand, the collective behavior of a quantum multiparticle state must be assigned by the statistics phase shift acquired by the multiparticle wave function if one considers position exchanges of particle pairs [6, 9]. The exchange of particles is understood as the exchange of the position arguments of the multiparticle wave function, and these exchanges are defined by the braid group elements. Hence, the determination of the braid group for the multiparticle system is an unavoidable prerequisite for any correlated state.

The braid group is the first homotopy group of the multiparticle configuration space and describes all possible particle exchanges including quantum indistinguishability of identical particles. This group is denoted as $\pi_1(\Phi) = \pi_1((M^N - \Delta)/S_N)$, where M^N is the N-fold normal product of the manifold $M = R^2$ in the case of planar Hall system, Δ is the diagonal point set in this product (when coordinates of two or more particles coincide, subtracted in order to assure conservation of the number of particles), and S_N is the permutation group of N elements [7]. The quotient structure of the configuration space $\Phi = (M^N - \Delta)/S_N$ is caused by the indistinguishability of quantum identical particles. The braid group $\pi_1(\Phi)$ is a topological object collecting all classes of nonhomotopic trajectory loops in the configuration space Φ, where points which differ only by enumeration of particles are unified. The braid group does not reflect the dynamic details of the system but rather identifies only the topology restrictions imposed on interparticle trajectories conditioned by the geometry-type global features, like the dimension of the manifold M.

As mentioned above, the Coulomb repulsion of electrons on the plane is a central prerequisite for the braid group definition at magnetic field presence, because the interparticle separation rigidly fixed by the Coulomb repulsion (in the classical equilibrium state) must interfere with the planar cyclotron orbits, discriminating in this way possible correlation types, by the commensurability condition.

For the same classical particles, the quantumly different particles (as, e.g., bosons and fermions in 3D or anyons in 2D) can be defined as assigned by distinct 1DURs of the related braid group. At the magnetic field presence for 2D-charged particles (as in quantum Hall configuration), there are, however, also other important consequences of 2D topology beyond solely 1DURs of the braid group. For the magnetic field strong enough that the classical cyclotron orbits are shorter in comparison to the interparticle separation on the plane, the braid group must be modified itself. The classical trajectories of charged particles obligatory defined by cyclotron orbits at the presence of a magnetic field allow for braid exchanges only when the size of the cyclotron orbit fits accurately to the separation between particles, as is illustrated in **Figure 1** (the separation between particles on the plane is fixed by the Coulomb interaction). It

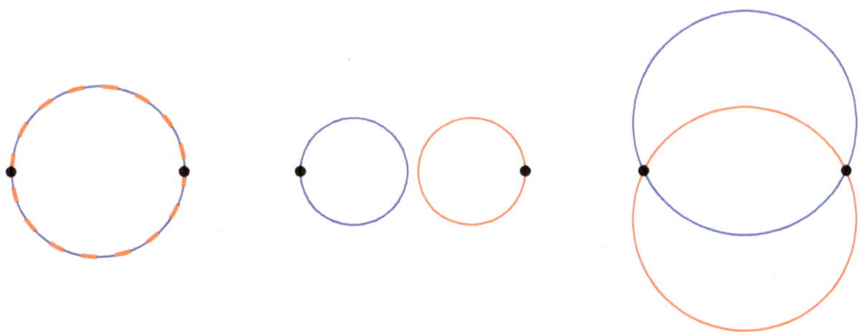

Figure 1. Pictorial illustration of the commensurability (left) between the cyclotron orbits (schematically drawn as circular ones, although at the interaction presence they might be deformed) and the interparticle spacing. This commensurability (also needed at any deformation of the cyclotron orbit by the interaction) satisfies the topology requirements unavoidably needed for the braid interchange implementation in the uniformly distributed particles on the plane with (rigidly fixed particle separation by the electric repulsion). In the case of stronger magnetic fields resulting in shorter cyclotron orbits, particles cannot be matched (center). Similarly for larger cyclotron orbits (at the lower magnitude of the magnetic field), interchanges also cannot be arranged as the interparticle spacing cannot be conserved during exchanges (right).

is easy to notice that the geometrical planar commensurability between the cyclotron orbit and the interparticle spacing is an unavoidable requirement needed for the definition (implementation) of the generators σ_j of the braid group. The generators σ_j describe exchanges of jth and (j+1)th particles. Otherwise, when the cyclotron orbits are incommensurate with the interparticle spacing on the plane, then cyclotron orbits cannot match neighboring particles, and the classical trajectories describing generators σ_j do not exist in this case—thus the braid group generators σ_j cannot be implemented. It means that the quantum statistics cannot be defined in this case and any correlated multiparticle quantum state cannot be organized. The possibility to implement the braid group selects thus the magnetic field strengths (or equivalently, filling rates of the LLL) at which the correlated state (FQHE) can be organized. If the braid group cannot be implemented, none of the correlated state is possible at corresponding filling factor. In this way, one can determine the hierarchy of filling rates for FQHE by analyzing the corresponding braid groups. Remarkably, the result fully agrees with all current experimental data for FQHE in any observable Hall systems.

Though the generators σ_j of the braid group cannot be implemented when the cyclotron orbits are shorter than interparticle separation, there exist other braids, which in 2D do fit to interparticle separations. Exclusively in 2D, the multilooped cyclotron orbits have larger size, which allows the braids σ_j^q (q odd positive integer) to fit to interparticle separation. This opportunity happens, however, at some "magic" fractional fillings of the LLL—the same ones at which FQHE is observed. The reason for the enhancement of the size of planar multilooped orbits (and the braid σ_j^q) is linked with the same surface spanned by 2D orbits regardless of its multilooped character (in opposition to 3D case when each additional loop of the spiral adds an additional surface portion spanned by this loop—in 2D spiral does not exist, however). The external field flux passing through the 2D multilooped orbit must be thus divided among all loops. Hence, the portion of the flux per each loop diminishes, which subsequently results in loop size growth. This is illustrated in **Figure 2**.

In **Figure 2** (left), the scheme of the cyclotron orbit at magnetic field B is shown as accommodated to the quantum of the magnetic field flux, that is, $BA = \frac{hc}{e}$. This is the definition of the cyclotron orbit size A. For interacting 2D electrons distributed in the Wigner crystal form, the cyclotron orbits are not circular but their surface A fits exactly to the interparticle separation in the case of the completely filled LLL, $\frac{S}{N}$, S—the sample area, N—the number of particles. If only single-looped orbits are considered, then at, for example, q = 3-times larger field, $3B$, the cyclotron orbit accommodated again to the flux quantum is too short in comparison to the

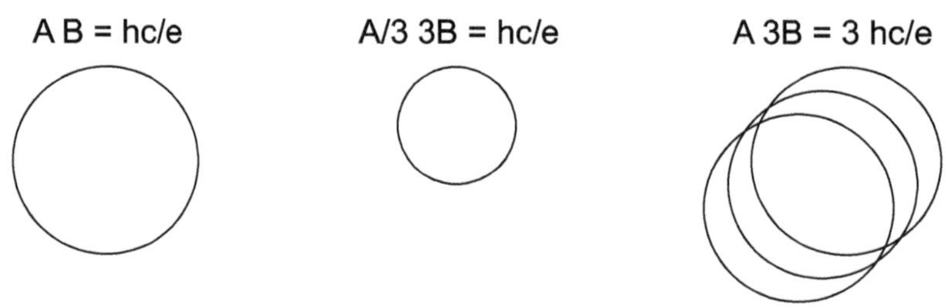

Figure 2. Schematic illustration of cyclotron orbit enhancement in 2D due to multiloop trajectory structure (third dimension added for visual clarity).

interparticle separation $\frac{S}{N}$—as is sketched in the central panel of **Figure 2**. However, in the case when three-loop orbits are considered (note that such orbits are present in the braid group generated by σ_j generators), then exclusively in 2D space, the external flux $3BA$ passing through this orbit must be divided among three loops. In the case of the uniform division (when all loops have the same surface A) per each loop falls a BA fraction of the total $3BA$ flux. Therefore, one can notice that each loop in this situation, as usual accommodated to the flux quantum $\frac{hc}{e}$, has the orbit with the surface A (the same surface A as for the single-looped orbit at three times weaker magnetic field). Three loops result together in the total flux $3BA$ per particle, as needed for considered stronger field $3B$—cf. **Figure 2** (right). It means that the three-loop orbits surprisingly fit to the interparticle separation $\frac{S}{N} = A$.

Because the braid group generator must be defined by half of the cyclotron orbit (cf. **Figure 3**), thus the braid with one additional loop corresponds to the cyclotron orbits with three loops—such a generator has the form σ_j^3. The group generated by σ_j^3, $j = 1,\ldots,N-1$ (new elementary braid exchanges) is obviously the subgroup of the original braid group because its generators σ_i^3 are built from original group generators σ_j. This subgroup is called the cyclotron braid subgroup. It is clear that 1DURs of this subgroup define statistics of 2D-charged particles at sufficiently strong magnetic fields, that is, at fields corresponding to the fractional fillings of the LLL provided that the commensurability condition is fulfilled (in the presented example, for $\nu = \frac{1}{3}$). The generalization to more loops attached to the braid generator one by one results

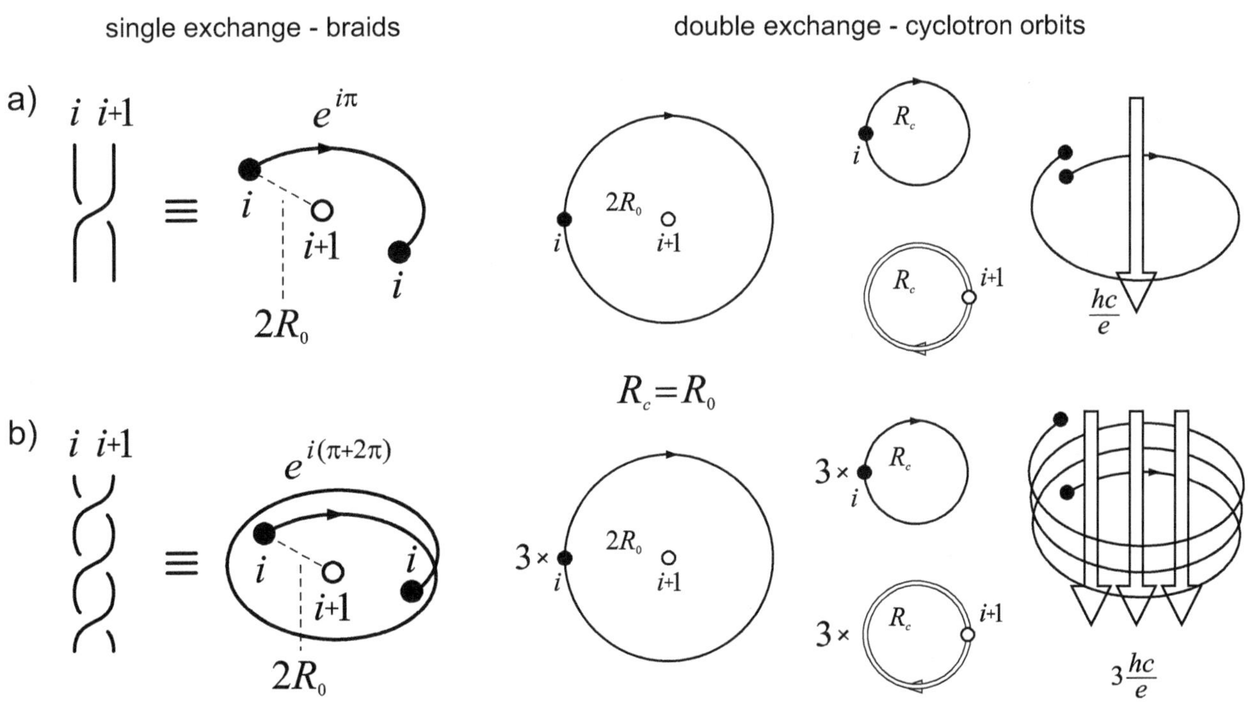

Figure 3. (a) A geometrical presentation of the braid generator σ_i—it corresponds to an ordinary single exchange of neighboring particles, ith and (i+1)th (left); the cyclotron orbit which realizes this braid exchange at magnetic field presence at $\nu = 1$—drawn in the relative view (right), it is visible that the closed cyclotron orbit corresponds to the double exchange of particles matched by single-looped cyclotron trajectories for $\nu = 1$, as indicated in (a) by $R_c = R_o$. (b) The three-looped braid generator σ_i^3 must be considered for $\nu = \frac{1}{3}$ to match neighboring particles at three times stronger magnetic field; the additional loop is needed in this case to enhance an effective cyclotron radius, and again in (b) $R_c = R_o$ (R_c—cyclotron radius, $2R_o$—particle separation, i.e., $\pi R_c^2 = \frac{hc}{eB}$, $\pi R_o^2 = \frac{S}{N}$).

in double increase of loop number in multilooped cyclotron orbits and thereby in fractions $\nu = \frac{1}{q}$, q – odd integer. This approach can be easy generalized to higher LLs also in exact agreement with the experimental observations [22].

This generalization to higher LLs resolves itself to the observation that for the filling fractions corresponding to the commensurability condition written as $x\frac{S}{N} = A$ with x – positive integer, the cyclotron orbits also fit to equidistantly separated particles, though not to just any but rather to every xth particle. Thus, the commensurability condition $x\frac{S}{N} = \frac{S}{N/x}$ also allows for the definition of the generators σ_j in the form of ordinary single-looped braids (linking every xth particles, however). This happens for the completely filled higher LLs. The resulting statistics (expressed by 1DUR of these braid groups with longer braids for $x > 1$) is the same as for IQHE (the longer braids are also single-looped as for IQHE at $\nu = 1$). Remarkably, such an opportunity for commensurability happens in higher LLs not only when they are completely filled but also at some fractional fillings, which is illustrated in Ref. [22] in satisfactory good correspondence with the experimental observations available now up to the third LL for 2DEG in conventional semiconductor Hall systems [25–27]. It must be emphasized, however, that the described above situation of too large cyclotron orbits in comparison to the particle separation (and fitting to every xth particle with $x > 1$) may happen exclusively in such LL subbands for which the commensurability condition $x\frac{S}{N} = A$ (x – integer) is admitted to be fulfilled. Apparently, such a situation may happen only for $n \geq 1$ where n is the LL number. Only in the LLs with $n \geq 1$ (i.e., for relatively weaker magnetic field in comparison to the LLL, $n = 0$) are cyclotron orbits of sufficiently large size to match every second, third particle, and so on. This is because the size of the cyclotron orbits in nth LL grows proportionally to the factor $2n$+1 present in Landau kinetical energy in higher LLs, that is, the cyclotron orbit size attains the value $A = \frac{(2n+1)hc}{eB}$ for particles with kinetic energy $(2n+1)\frac{eB}{2mc}$ in nth LL. Moreover, for $n \geq 1$ too short cyclotron orbits may occur as well as in the LLL, but in higher LLs rather close the subbands edges, that is, for sufficiently small density of particles and thus larger separation which can commensurate with longer cyclotron orbits at $n \geq 1$. Let us emphasize, however, that the quantization of the transverse resistance R_{xy} related to any fractional-filling rates ν (also in higher LLs) is always as for ordinary FQHE prerequisite from the LLL, that is, equal to $\frac{h}{e^2\nu}$, but the correlations expressed by the exponent in the Jastrow polynomial vary displaying single-looped or multilooped braid exchanges, both accessible in higher LLs in the sharp opposition to the LLL (where only multilooped orbits were possible for fractional fillings $\nu < 1$).

2.2. Hierarchy for FQHE in graphene

Due to the specific band structure in graphene with Dirac points on the corners of the hexagonal Brillouin zone [19], the LL spectrum is not equidistant as for ordinary 2DEG but is proportional to $\sqrt{n}$ instead of n (n is the number of the LL) [28]. This specific "relativistic" form of LL energy emerges from the linearity in the momentum of the local Hamiltonian to the vicinity of Dirac points—this Hamiltonian determines carrier quantization in graphene. Note, however, that the degeneracy of each LL subband in graphene is the same one as in the conventional 2DEG and equals $\frac{BS}{hc/e}$. Nevertheless, the number of subbands per each LL in graphene is different from the conventional semiconductor case and equals 4 in graphene, which corresponds to the Zeeman spin splitting and to the valley splitting (absent in conventional semiconductors) due to two inequivalent Dirac points mixed with two sublattices in crystal lattice of the graphene sheet [28]. The Zeeman splitting in graphene is small [19], and the valley splitting is small as well [28], thus the fourfold approximate spin-valley additional degeneracy may be assumed. It results in the following form of IQHE hierarchy in monolayer graphene $\nu = 4(n + \frac{1}{2})$ in agreement with experimental observations [19]. The remarkable difference with respect to the ordinary 2DEG resoles itself to the factor 2 instead of 4, due to only spin degeneracy in conventional 2DEG, and the presence of the shift by 2. This shift is due to the Berry phase in monolayer graphene and sharing of the LLL states between particles and holes from the conduction and valence band at the zero-energy level [28]. Hence, the bottom of the LLL is shifted by 2 (in terms of the filling factor) upward. Conventionally, filling rates for holes from the valence band are assigned as negative numbers mirror reflection of those positive numbers denoting filling rates for electrons from the conduction band. An additional opportunity in graphene, beyond the ability of conventional 2DEG, is the possibility to control a transition between particles and holes in graphene by the shift of the Fermi level passing the Dirac point. Experimentally, it is realized by the application of a lateral relatively small voltage. The above-mentioned Berry phase contribution to the Landau energy spectrum in graphene manifests itself by an additional π phase shift born by the chiral valley pseudospin [28].

For the bilayer graphene, the situation slightly differs [28, 29]. Due to an interlayer hopping, the Hamiltonian for bilayer graphene attains back the quadratic form with respect to the momentum. Hence, the LL spectrum in bilayer graphene resembles that one for the ordinary 2DEG but with four subbands for each LL level except for the LLL, which has eightfold degeneration [28, 29]. As usual in graphene, the division of the LLL subbands equally among particles and holes causes the bottom for uniformly charged carriers (electrons or holes) to be placed in the center of the eightfold-degenerated LLL (let us remind that in the monolayer graphene, the LLL was only fourfold degenerated; the additional degeneracy in the bilayer graphene is caused by the inclusion of states with $n = 0$ and $n = 1$ to the LLL in the bilayer graphene in opposition to the monolayer one). In the bilayer graphene, the Berry phase shift for chiral particles is also different in comparison to the monolayer one and equals 2π [28]. Hence, the consecutive plateaus of IQHE are located in bilayer graphene at integer-filling rates, whereas in monolayer graphene, they were located at half-fillings of the LLs [28, 29].

2.3. FQHE hierarchy in monolayer graphene

For the magnetic field strong enough that $\nu \in (0, 1)$ and for Fermi level shifted (by a lateral voltage) to the conduction band, we deal with fractionally filled first conduction subband of the particle LLL $n = 0, 2\uparrow$ (in this numeration of subbands 2 marks the valley pseudospin orientation and the arrow $\uparrow$ marks the orientation of the ordinary spin). For $N < N_0$ the filling rate, $\nu = N/N_0$, is fractional (the degeneracy N_0 of each subband is $N_0 = \frac{BS}{hc/e}$)

An important property follows from the fact that cyclotron orbits in graphene must be accommodated to the bare kinetic energy $T = \hbar\omega_c(n + \frac{1}{2})$ with $\omega_c = \frac{eB}{mc}$, similarly in the conventional semiconductor 2DEG (as in non-interacting 2D gas), despite the different "relativistic" version of Landau levels energy in graphene. This "relativistic" oddness is caused, however, by peculiar crystal field features in graphene which do not change the topology of braids. Thus, the bare kinetic part of Landau energy is not modified by the crystal field and the dimension of braid cyclotron orbits repeat the corresponding orbit size from the non-interacting gas. Therefore, for graphene the cyclotron orbit structure is governed by ordinary Landau level restrictions despite a specific band structure with Dirac points, being the result of the crystal field in graphene and not of kinetic energy origin. The difference between the conventional 2DEG system and graphene will be thus related with the different number of LL subbands in graphene in comparison to the conventional 2DEG. The uniform shift in filling factors will be caused also by the Barry phase shift, nonzero for chiral valley pseudospin in graphene [28].

Therefore, the cyclotron orbit size in the subband $n = 0,2\uparrow$ is equal to $\frac{hc/e}{B} = \frac{S}{N_0}$ (S is the sample surface). Because this orbit size is lower than the interparticle spacing expressed by $\frac{S}{N}$ (as $N < N_0$), the multilooped braids are needed to match neighboring particles. The commensurability condition reads here as $q\frac{S}{N_0} = \frac{S}{N}$, which gives $\nu = \frac{N}{N_0} = \frac{1}{q}$, ($q$ – odd integer [21]). For the local band holes in this subband, the particle-hole symmetric-filling rates $\nu = 1 - \frac{1}{q}$ are expected.

One can generalize this fundamental fractional series by an assumption that in the case when the portion of the residual flux per last loop must be different than the portions per former loops, the last loop of the multilooped cyclotron orbit may be commensurate with every lth particle separation, whereas the former loops take away an integer number of flux quanta (are commensurate with neighboring particles). For $l = 2,3,\ldots$, the last loop reaches every lth particles (next neighbors). In this manner, we obtain the hierarchy of fillings for FQHE in this LLL subband in the following form: $\nu = \frac{l}{l(q-1)\pm1}$, $\nu = 1 - \frac{l}{l(q-1)\pm1}$, where $l = 1,2,\ldots$ and minus in the denominator corresponds to the possibility of the reverse eight-figure orientation of the last loop with respect to the antecedent loops in the multilooped orbit. Additionally, we notice that the Hall metal states can be achieved in the limit $l \to \infty$ in the above formula, which corresponds to the situation when the residual flux passing through the last loop tends to zero. This means that the last loop can reach in such a case the infinitely distant particles as for fermions at the absence of the magnetic field, which is referred to the case of the Hall metal archetype for $\nu = \frac{1}{2}$ in the conventional 2DEG. In the limit right arrow ∞, we arrive thus with the hierarchy for the Hall metal states in the form: $\nu = \frac{1}{q-1}$, $\nu = 1 - \frac{1}{q-1}$.

One can observe also that other variants of commensurability may be concerned with multilooped orbits. Namely, each loop of the multilooped structure may be in principle accommodated to particle separation in a different and mutually independent manner matching nearest and next neighbors in various schemes. One of such a possibility may correspond with the situation when in q-looped orbit $q - 1$ loops are accommodated to every xth particle ($x = 1,2,3,\ldots$), whereas the last one fits to every lth particle separation. The similar commensurability scheme is observed in ordinary 2DEG Hall systems within the LLL for exotic fractions, for example, $\nu = \frac{4}{11}, \frac{5}{13}, \frac{3}{8}, \frac{3}{10}$ (beyond the CF hierarchy). It is noticeable, however, that a similar series of exotic FQHE-filling fractions are not observed in the LLL in graphene up to date, though it is observed in the first LL in monolayer graphene and also in bilayer graphene (as will be analyzed below).

In order to include the Berry phase anomaly in graphene, the overall shift of ν by –2 must be performed (for monolayer case), but we confine our notation to the net-filling fractions, neglecting this uniform shift.

With lowering of the magnetic field, one can achieve the completely filled subband $n = 0,2\uparrow$, which corresponds to $\nu = 1$ and the IQHE state. For still lower magnetic fields, the last subband of LLL $n = 0, 2\downarrow$ is gradually filled by electrons. In this subband, the cyclotron orbit size $\frac{S}{N_0}$ is still lower than the interparticle separation $\frac{S}{N-N_0}$ (because $N - N_0 < N_0$) similarly in the antecedent subband. Therefore, the multilooped structure of orbits must be repeated here from the previous subband. The FQHE hierarchy is exactly the same as for antecedent subband, only shifted ahead by 1. After complete filling of this subband, the LLL is completely filled, that is, there are all four subbands of the LLL filled. This gives the IQHE according to its main-line hierarchy $\nu = 4(n + \frac{1}{2})$.

In an analogous way, one can consider fillings of the following LLs. The nearest one corresponds to $n = 1$. This level has four subbands of electron type. Let us emphasize, however, that the bare kinetic energy in this level (in all its subbands) is equal to $\frac{3\hbar\omega_c}{2}$ and the related cyclotron orbit size is $\frac{3hc}{eB} = \frac{3S}{N_0}$ (larger in comparison to the LLL). When $N \in (2N_0, 3N_0]$, then electrons fill the $n = 1, 1\uparrow$ subband. In this subband, the cyclotron orbits are of size $\frac{3S}{N_0}$ and it must be accommodated with interparticle separation between electrons in this subband, $\frac{S}{N-2N_0}$. For small amount of electrons in the subband, we thus can meet with the multilooped orbits if single-looped orbits are too short, $\frac{3S}{N_0} < \frac{S}{N-2N_0}$. The q-looped orbits satisfy the commensurability condition $q\frac{3S}{N_0} = \frac{S}{N-2N_0}$, q – odd integer, which defines the main series for FQHE (multiloop) in this subband, $\nu = 2 + \frac{1}{3q}$. This main line can be supplemented to the complete related hierarchy, $\nu = 2 + \frac{l}{l3(q-1)\pm 1}$, $\nu = 3 - \frac{l}{l3(q-1)\pm 1}$ $l = i/3, i = 1, 2, \ldots$, with the Hall metal hierarchy in the limit $l \to \infty$, as described in the case of the LLL. These series of FQHE filling rates are located closer to the subband edges in comparison to the FQHE rates in the LLL. Simultaneously, in the central part of this higher LL subband the new type of commensurability is possible, not accessible in the LLL. This new commensurability occurs when $\frac{3S}{N_0} = \frac{xS}{N-2N_0}$ and $x = 1,2,3$ corresponding to fitting of the single-looped orbits (large enough in this subband) with every xth particle (x order next-nearest neighbors). From this new commensurability opportunity, one finds fractions $\nu = \frac{7}{3}, \frac{8}{3}, 3$ corresponding to single-looped cyclotron orbits similar as for IQHE. Thus, for $\nu = \frac{7}{3}, \frac{8}{3}$ we deal with the FQHE (single loop). This is a new Hall feature manifesting itself only in higher LLs, where cyclotron orbits may be larger than the interparticle separation and orbits can reach next-nearest neighbors.

Finally, let us note that at the special case of the commensurability, $\frac{3S}{N_0} = \frac{1.5S}{N-2N_0}$, one can meet at $\nu = \frac{5}{2}$. This commensurability concerns rather the paired particles and not the single ones. The pairing does not change the cyclotron radius but reduces twice the carrier number $\frac{N-2N_0}{2}$, which gives the above commensurability for pairs at $\nu = \frac{5}{2}$. Hence, at this filling rate, one can expect a manifestation of the IQHE but for paired electrons (the considered correlation corresponds to p-like pairing due to the spin polarization in this subband).

An interesting new possibility for commensurability happens next for q-looped orbits. As mentioned previously, the size of particular loops in the multilooped structure may be in general accommodated to the interparticle spacing in an independent manner resulting in the abundance of possible new filling rates. The loop size may be accommodated to xth-order next-nearest neighbors independently. In particular, it results in the additional hierarchy in all subbands of the first LL, $\nu = 2(3,4,5) + \frac{xl}{l3(q-1)\pm 1}$, $\nu = 3(4,5,6) - \frac{xl}{l3(q-1)\pm 1}$, which for $q = 3$, $x = 2,3$, $l = i/3$, $i = 1,2,3$ reproduces $\nu = \frac{7}{3}, \frac{8}{3}, \frac{12}{5}, \frac{13}{5}, \frac{17}{7}, \frac{18}{7}, \frac{22}{9}, \frac{23}{9}, \frac{10}{3}, \frac{11}{3}, \frac{17}{5}, \frac{18}{5}, \frac{24}{7}, \frac{25}{7}, \frac{13}{3}, \frac{14}{3}, \frac{22}{5}, \frac{23}{5}$. This opportunity for FQHE pretty well agrees with the recent observations of FQHE in three first subbands of the $n = 1$ LL in monolayer graphene at ultra-low temperatures [16].

The similar scheme of commensurability may be applied to the following subbands. For the subband $n = 1, 1\downarrow$, the cyclotron orbit size is again equal to $\frac{3hc}{eB} = \frac{3S}{N_0}$, whereas the interparticle spacing is governed by plaque $\frac{S}{N-3N_0}$, for $N \in (3N_0, 4N_0)$. The basic commensurability condition has thus the form: $\frac{q3S}{N_0} = \frac{S}{N-3N_0}$ which gives the main series for FQHE (multiloop), $\nu = 3 + \frac{1}{3q}$. This series can be developed to the full hierarchy via more complicated commensurability opportunities, as described above. In this way, the condition $\frac{3S}{N_0} = \frac{xS}{N-3N_0}$ with $x = 1,2,3$ gives fractions with single-looped correlations of FQHE (single-loop)-type for $\nu = \frac{10}{3}, \frac{11}{3}$ and fully developed IQHE for $\nu = 4$. A possible paired state can be realized in this subband at $\nu = \frac{7}{2}$.

2.4. FQHE hierarchy in bilayer graphene

In bilayer graphene, the topology of braid trajectories considerably changes in comparison to the monolayer system. The bilayer graphene is not strictly two-dimensional and the bilayer graphene opens new possibility for the topology of trajectories. Two sheets of the graphene plane lie in close distance and electrons can hop between these two sheets. We deal here with electrons residing on a two-sheet structure instead of the single sheet as it was the case for the monolayer graphene.

All the above-described requirements to fulfill commensurability condition in order to define the related braid group are binding also for the bilayer graphene, with a single exception. This exception is linked with the fact that *double-looped* cyclotron orbits may have in bilayer graphene the same size as the single-looped orbit, that is, the one additional loop may not enhance the cyclotron radius. This is in opposition to the monolayer case, where the additional loops always enhance cyclotron radius. This difference in bilayer graphene follows, however, from the fact that the second loop may be located in the opposite sheet of bilayer graphene with respect to the first one and the external field passing through such a double-looped orbit is twice larger than the flux passing through the single-looped orbit (and the double-looped orbit ideally planar as well). When the two loops are located in the distinct sheet of the bilayer structure, then each loop has its own and separate surface, as usual in 3D (here reduced, however, to double-sheet structure). This is in sharp distinction from the multilooped (double-looped, in particular) cyclotron orbit located in the purely 2D plane (like monolayer graphene). More generally, in the bilayer system loops of the multilooped orbit may be located partly in both 2D sheets. To account this effect in topological terms adjusted to braid trajectories and commensurability requirements, one must neglect the contribution of the one loop in the multilooped structure when the total flux of the external field is divided for fractions per all loops. This one loop may capture its own additional flux, whereas the remaining loops must share the same flux similarly as passing through any cyclotron orbit in monolayer case.

Removing the single loop must be done independently from how the loops are distributed among two sheets. When we consider a selected loop located in the opposite sheet with respect to the antecedent loop, the next loops must fill both sheets of bilayer structure as additional ones regardless their specific distribution—thus these all loops take part in dividing the external field flux exactly in the same manner as in monolayer case provided that the selected loop is omitted together with the flux passing this loop. This trick reduces the bilayer system to monolayer one in braid-loop topology sense. Thus, we can denote the commensurability condition in the bilayer graphene for the usual case of too short single-looped cyclotron orbits in the following form (as an example, for the subband $n = 0, 2\uparrow$ of the LLL—i.e., the first particle-type subband of the LLL),

$$\begin{array}{l} (p-1)\frac{hc}{eB} = \frac{S}{N}, \\ \nu = \frac{N}{N_0} = \frac{1}{p-1} = \frac{1}{2}, \frac{1}{4}, \frac{1}{6}, \ldots, \end{array} \tag{1}$$

here N is the total number of particles counted in both graphene sheets, $N_0 = \frac{BSe}{hc}$ is the degeneracy also for both sheets together, S is the surface of the sample (the surface of the single sheet), p is an odd integer (it must be odd in order to assure that half of the cyclotron orbit defines the braid, similarly as in monolayer case).

The factor $p - 1$ in l.h.s. of Eq. (1) is caused by the fact that in an enhancement of the effective cyclotron orbit size participates only orbits from the ideally 2D sheet of bilayer graphene (no matter where the doubling loops are located) with exception of a single loop which may be located in opposite sheet to the former loop, and thus must be avoided in order to reduce the bilayer system to an effective monolayer one.

The contribution of this sole loop is independent from other loops and must be omitted. After avoidance of this single loop, the next loops must duplicate the former ones without any rise of the surface per total cyclotron orbit no matter in which way loops are distributed among both sheets (in both sheets, the remaining loops will duplicate loops already present there). Thus, only $p - 1$ loops take part in the enhancement of the effective p-looped cyclotron orbit in bilayer graphene.

It must be emphasized that for multilooped orbits in bilayer graphene the total number of loops still is p (despite avoiding one loop at the formulation of the commensurability condition)—therefore, the generators of the corresponding cyclotron subgroup are of the form σ_j^p, which results in the standard Laughlin correlations with the Jastrow polynomial with p exponent. However, due to the specific for bilayer structure commensurability (1) the resulting main line of filling fractions is $\nu = \frac{1}{p-1}$ (p-odd) in the first particle-type subband of the LLL, that is, in the subband $n = 0, 2\uparrow$. The even denominator in this main series for the FQHE hierarchy for bilayer graphene pretty well coincides with the experimental observations [10].

For holes in this subband (holes corresponding to empty states in the almost filled subband of particle type), we can write, $\nu = 1 - \frac{1}{p-1}$. The generalization to the full hierarchy of FQHE in this subband attains thus the form, $\nu = \frac{l}{l(p-2)\pm 1}$, $\nu = 1 - \frac{l}{l(p-2)\pm 1}$, where $l > 1$ corresponds to some filling factor for other correlated Hall state, similarly as discussed in the monolayer case. In the following subbands of the LLL, $n = 0, 2\downarrow$ (assuming that this subband succeeds the former one), the hierarchy is repeated in the same form only uniformly shifted ahead by one (because the commensurability conditions are similar for all subbands with the same n due to the same size of the cyclotron orbits).

The considerable novelty occurs, however, in the next two subbands of the LLL, $n = 1, 2\uparrow$ and $n = 1, 2\downarrow$. Because of the larger size of cyclotron orbits for n = 1, the FQHE main series in the first of these subbands of the LLL, $n = 1, 2\uparrow$, attains the form,

$$\begin{aligned} &\tfrac{3hc}{eB} = \tfrac{3S}{N_0} < \tfrac{S}{N-2N_0}, \\ &(p-1)3\tfrac{hc}{eB} = (p-1)\tfrac{3S}{N_0} = \tfrac{S}{N-2N_0}, \\ &\nu = \tfrac{N}{N_0} = 2 + \tfrac{1}{3(p-1)} = 2 + \tfrac{1}{6}, 2 + \tfrac{1}{12}, 2 + \tfrac{1}{18}, \ldots, \end{aligned} \tag{2}$$

The generalization of this main series for holes in the subband and to the full FQHE hierarchy in this subband looks as follows: for subband holes, $\nu = 3 - \frac{1}{3(p-1)}$ and for the full FQHE hierarchy in this subband, $\nu = 2 + \frac{l}{l3(p-2)\pm 1}$, $\nu = 3 - \frac{l}{l3(p-2)\pm 1}$, $l = \frac{i}{3}$, $i = 1, 2, 3, \ldots$ (the Hall metal hierarchy may be obtained in the limit $l \to \infty$).

In the subband $n = 1, 2\uparrow$ of the LLL, the new commensurability opportunity occurs (a similar one as that which occurred in the first LL of the monolayer graphene): $\frac{3}{N_0} = \frac{x}{N-2N_0}$ for $x = 1, 2, 3$. It gives filling ratios $\nu = \frac{7}{3}, \frac{8}{3}, 3$. These rates are related with single-looped cyclotron trajectories, thus with single-loop correlations similar as for IQHE (though the first two for not integer-filling rates). This is typical for LLs with $n \geq 1$ Hall features we called as FQHE (single loop). Similarly as in the monolayer case one can consider paired state for x = 1.5 in the above formula, which corresponds to the perfect commensurability of cyclotron orbits of electron pairs with the separation of these pairs at electron-filling rate $\nu = \frac{5}{2}$.

Filling of the last subband $n = 1, 2\downarrow$ in the LLL in bilayer graphene undergoes the similar constraints as for all subbands with n = 1, the cyclotron orbits have the same size. Thus, the FQHE hierarchy is only shifted by 1 from the antecedent subband. The situation changes, however, in the next LL (the first one beyond the LLL). The cyclotron orbits are determined here by the bare kinetic energy for n = 2, which gives the orbit size: $\frac{5hc}{eB} = \frac{5S}{N_0}$. The similar analysis as in the LLL gives in the subband $n = 2, 1\uparrow$ the main series and the full hierarchy for FQHE (multiloop) $\nu = 4 + \frac{1}{5(p-1)}$, $\nu = 4 + \frac{l}{l5(p-2)\pm 1}$, $l = \frac{i}{5}$, $i > 1$, respectively (for subband holes, the substitution of 4+ by 5− in both the above formulae is needed). As previously, the limit $l \to \infty$ determines the Hall metal hierarchy. The next feature in this subband is the presence of four (instead of two) satellite FQHE (single-loop) states symmetrically located around the central paired state. In the subband $n = 2, 1, \uparrow$, these satellite states occur at $\nu = \frac{21}{5}, \frac{22}{5}, \frac{23}{5}, \frac{24}{5}$ and the central paired state at $\nu = \frac{9}{2}$. This similar hierarchy scheme is repeated in all four subbands of the first LL with n – 2.

The evolution of the fractional-filling hierarchy of subsequent LLs is pictorially presented in **Figures 4** and **5** for the monolayer and bilayer graphene, correspondingly. It is also summarized in **Tables 1** and **2**.

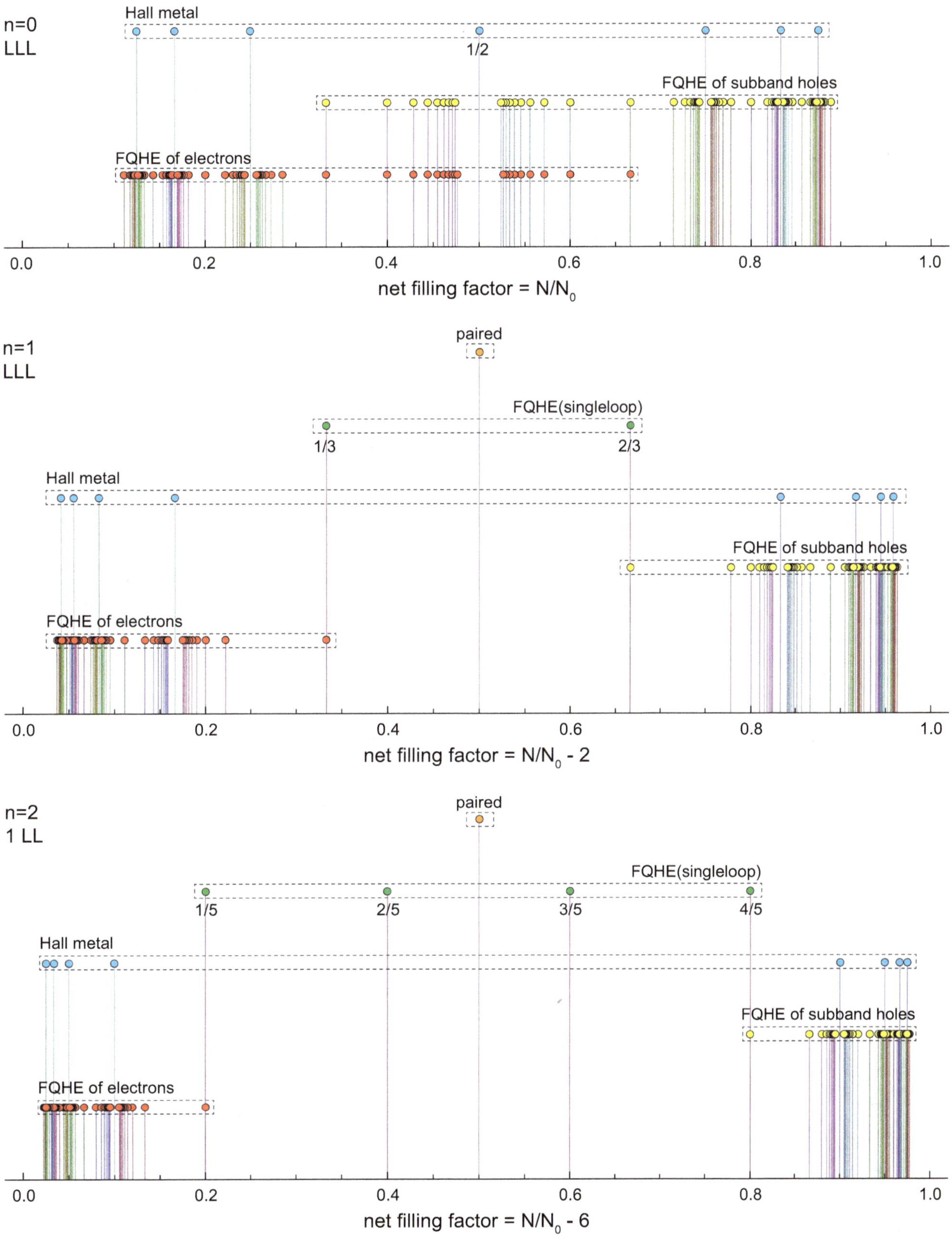

Figure 4. Illustration of the FQHE hierarchy in three first LLs of the monolayer graphene; for each LL the first particle subband is illustrated, the next subbands in each LL repeat the hierarchy from the first one. Different types of ordering are indicated with spikes of various heights. Series for ordinary FQHE (multiloop), FQHE (single-loop), Hall metal, and paired state are displayed acc. to the hierarchy described in **Table 1** with $q = 3$–9, 1=1–10; only a few selected ratios from these series are explicitly written out.

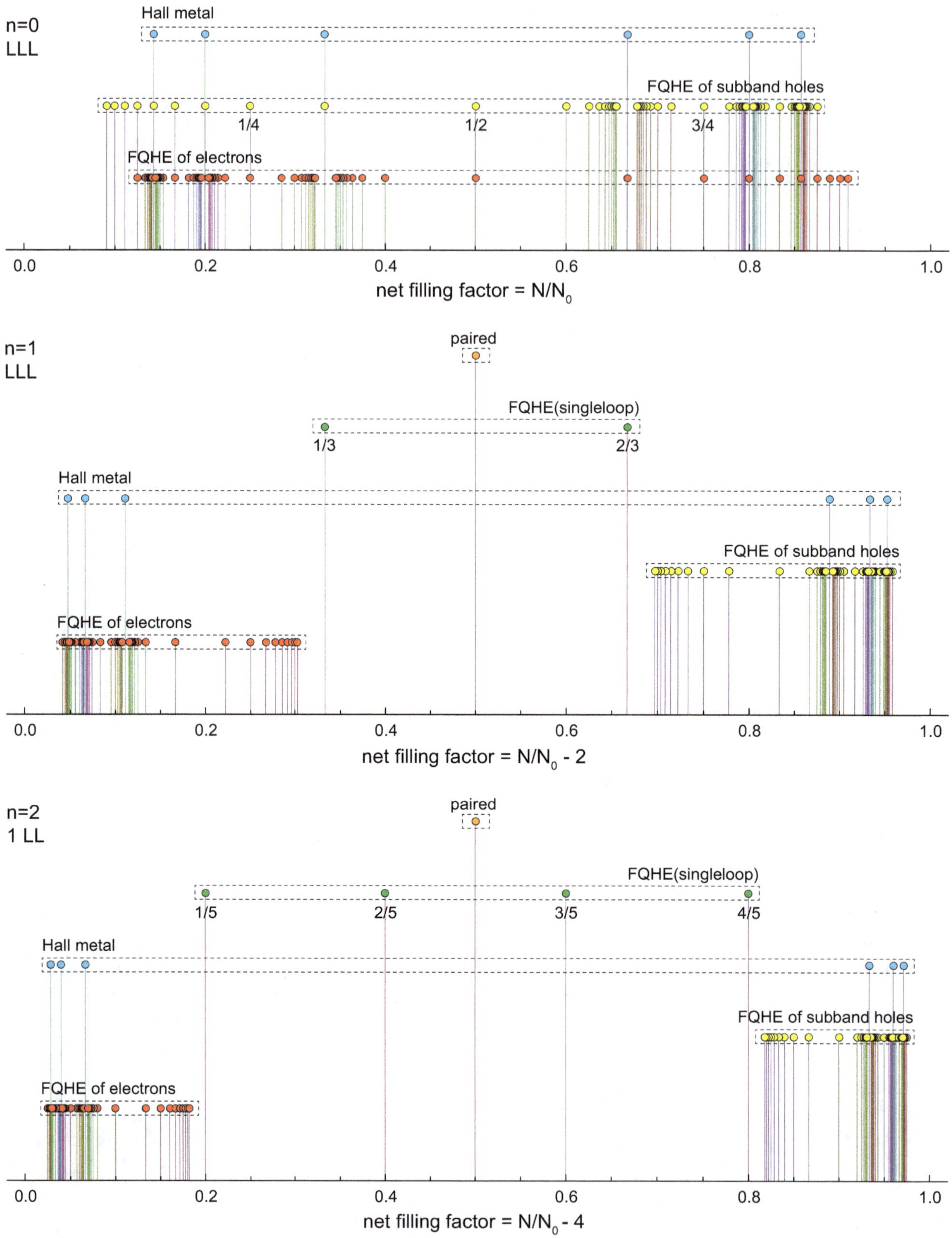

Figure 5. FQHE hierarchy in two first LLs of the bilayer graphene; for the LLL two subbands with $n = 0$ and $n = 1$ are included. Different types of braid correlations are indicated with spikes with various heights. Hierarchy series are displayed acc. to the formulae presented in **Table 2** with $q = 3 - 9$, $i = 1 - 10$; limited number of ratios from these series are explicitly written out.

LL subb.	FQHE (single-loop), paired, IQHE	FQHE (multiloop) (q – odd, $l = \frac{i}{2n+1}, i = 1, 2, 3, \ldots$)	Hall metal
$n = 0, 2\uparrow$	1	$\frac{1}{q}, 1 - \frac{1}{q}, \frac{l}{l(q-1)\pm 1}, 1 - \frac{l}{l(q-1)\pm 1}$	$\frac{1}{q-1}, 1 - \frac{1}{q-1}$
$n = 1, 1\uparrow$	$\frac{7}{3}, \frac{8}{3}$, ($\frac{5}{2}$ paired), 2, 3	$2 + \frac{1}{3q}, 3 - \frac{1}{3q}, 2 + \frac{l}{3l(q-1)\pm 1}, 3 - \frac{l}{3l(q-1)\pm 1}$	$2 + \frac{1}{3(q-1)}, 3 - \frac{1}{3(q-1)}$
$n = 2, 1\uparrow$	$\frac{31}{5}, \frac{32}{5}, \frac{33}{5}, \frac{34}{5}$ ($\frac{13}{2}$ paired), 6, 7	$6 + \frac{1}{5q}, 7 - \frac{1}{5q}, 6 + \frac{l}{5l(q-1)\pm 1}, 7 - \frac{l}{5l(q-1)\pm 1}$	$6 + \frac{1}{5(q-1)}, 7 - \frac{1}{5(q-1)}$

Table 1. LL-filling factors for FQHE determined by commensurability condition (paired indicates condensate of electron pairs), for the first particle subband in each of the three first LLs ($n = 0, 1, 2$) for the monolayer graphene.

LL subb.	FQHE (single-loop), paired, IQHE	FQHE (multiloop) (q – odd, $l = \frac{i}{2n+1}$, $i = 1, 2, 3, \ldots$)	Hall metal
$n = 0, 2\uparrow$ (LLL)	1	$\frac{1}{(q-1)}, 1 - \frac{1}{(q-1)}, \frac{l}{l(q-2)\pm 1}, 1 - \frac{l}{l(q-2)\pm 1}$	$\frac{1}{q-2}, 1 - \frac{1}{q-2}$
$n = 1, 2\uparrow$ (LLL)	$\frac{7}{3}, \frac{8}{3}$, ($\frac{5}{2}$ paired), 2, 3	$2 + \frac{1}{3(q-1)}, 3 - \frac{1}{3(q-1)}, 2 + \frac{l}{3l(q-2)\pm 1}, 3 - \frac{l}{3l(q-2)\pm 1}$	$2 + \frac{1}{3(q-2)}, 3 - \frac{1}{3(q-2)}$
$n = 2, 1\uparrow$	$\frac{21}{5}, \frac{22}{5}, \frac{23}{5}, \frac{24}{5}$, ($\frac{9}{2}$ paired), 4, 5	$4 + \frac{1}{5(q-1)}, 4 + \frac{l}{5l(q-2)\pm 1}, 5 - \frac{1}{5(q-1)}, 5 - \frac{l}{5l(q-2)\pm 1}$	$4 + \frac{1}{5(q-2)}, 5 - \frac{1}{5(q-2)}$

Table 2. FQHE hierarchy found by commensurability braid group approach (paired marks condensate of electron pairs), in the first subbands of two first LLs ($n = 0, 1$ correspond to the LLL whereas $n = 2$ for the first LL beyond the LLL) in the bilayer graphene.

2.5. Specific to bilayer graphene FQHE hierarchy change caused by the type of the LLL degeneracy lifting

For bilayer graphene the degeneration of n = 0 and n =1 states results in eightfold degeneracy of the LLL, doubling fourfold spin-valley degeneracy of the LLL in comparison to the monolayer case. The degeneracy is not exact and for enhancing magnetic field amplitude both the Zeeman splitting and the valley splitting grow. Stress, deformation, and structure imperfections also cause lifting of the valley degeneracy. Moreover, the Coulomb interaction causes mixing of n = 0,1 states lifting their degeneracy within the LLL. Of particular importance is, however, an order of subbands after the degeneracy lifting allowing for mutually inverted ordering of LLL subbands with distinct $n = 0, 1$. The orders $n = 0, 1$ or $n = 1, 0$ lead to distinct FQHE-filling hierarchy. In the case when the LLL subbands with n = 1 is filled earlier than the n = 0 subband, one arrives with the following hierarchy for the first subband $n = 1, 2\uparrow$: multilooped orbits for $\nu = \frac{l}{l3(p-2)\pm 1}$, $\nu = 1 - \frac{l}{l3(p-2)\pm 1}$, single-looped orbits for $\nu = \frac{1}{3}, \frac{2}{3}$, and a paired state for $\nu = \frac{1}{2}$. For the next subband, $n = 0, 2\uparrow$, we get the hierarchy in the form: multilooped orbits for $\nu = 1 + \frac{l}{l(p-2)\pm 1}$, $\nu = 2 - \frac{l}{l(p-2)\pm 1}$ and no single-looped orbits. The comparison of reverted orderings of two first LLL subbands is summarized in **Table 3**.

One can consider also the situation in the LLL of bilayer graphene, when the degeneracy of n = 0,1 states is lifted in such a way that both levels cross at certain filling factor $\nu^* < 1$ (cf. Ref. [30], where mixing between n = 0,1 states has been analyzed numerically on small models on torus or sphere). Let us assume for an example, that the n = 1 subband

LL subb.	FQHE (single-loop), paired, IQHE	FQHE (multiloop) (q – odd, $l = \frac{i}{2n+1}$, $i = 1, 2, 3, \ldots$)	Hall metal
$n = 0, 2, \uparrow$	1	$\frac{1}{(q-1)}, 1 - \frac{1}{(q-1)}, \frac{l}{l(q-2) \pm 1}, 1 - \frac{l}{l(q-2) \pm 1}$	$\frac{1}{q-2}, 1 - \frac{1}{q-2}$
$n = 1, 2\uparrow$	$\frac{4}{3}, \frac{5}{3}$, ($\frac{3}{2}$ paired), $1, 2$	$1 + \frac{1}{3(q-1)}, 2 - \frac{1}{3(q-1)}, 1 + \frac{l}{3l(q-2) \pm 1}, 2 + \frac{l}{3l(q-2) \pm 1}$	$1 + \frac{1}{3(q-2)}, 2 - \frac{1}{3(q-2)}$
$n = 1, 2\uparrow$	$\frac{1}{3}, \frac{2}{3}$, ($\frac{1}{2}$ paired), 1	$\frac{1}{3(q-1)}, 1 - \frac{1}{3(q-1)}, \frac{l}{3l(q-2) \pm 1}, 1 - \frac{l}{3l(q-2) \pm 1}$	$\frac{1}{3(q-2)}, 1 - \frac{1}{3(q-2)}$
$n = 0, 2\uparrow$	1,2	$1 + \frac{1}{q-1}, 1 + \frac{l}{l(q-1) \pm 1}, 2 - \frac{1}{q-1}, 2 - \frac{l}{l(q-2) \pm 1}$	$1 + \frac{1}{q-2}, 2 - \frac{1}{q-2}$

Table 3. Comparison of filling hierarchy in the LLL level in the bilayer graphene for two mutually inverted successions of two lowest subbands: $n = 0, 2\uparrow$ $n = 1, 2\uparrow$ (upper) and $n = 1, 2\uparrow$ $n = 0, 2\uparrow$ (lower).

($n = 1, 2\uparrow$) is energetically favorable up to some filling fraction ν^*. At this filling rate, the subband $n = 1, 2\uparrow$ crosses with the subband $n = 0, 2\uparrow$ and the latter starts to be more convenient for $1 + \nu^* > \nu > \nu^*$. The hierarchy of fractional fillings corresponding to such a situation looks like for ordinary filling of the subband $n = 1, 2\uparrow$, though with an insertion of $n = 0, 2\uparrow$ subband hierarchy. Depending on the value of ν^*, the various patterns are possible by a combination of hierarchy patterns as illustrated in **Table 3**.

3. Comparison with experiment

The filling factor in graphene (monolayer and bilayer) can be changed both by the external magnetic field strength variation and by the particle concentration variation via shifting of the Fermi level near the Dirac point by the application of a lateral voltage of relatively small magnitude (typically 10–60 V). Due to spin-valley degeneracy and Berry phase contribution related with the chiral valley pseudospin, the IQHE is observed in monolayer graphene for fillings $\nu = 4(n + \frac{1}{2}) = 2, 6, 10, 14, \ldots$ for particles from the conduction band and for the mirror-negative fillings for holes from the valence band. However, FQHE was not detected in graphene samples deposited on a substrate of SiO_2, despite using very strong magnetic fields (up to 45 T) [20]. Instead of FQHE features in this configuration in so strong magnetic fields, the emergence of additional plateaus of IQHE has been observed for the fillings $\nu = 0, \pm 1, \pm 4$. This additional IQHE manifestation is linked with spin-valley degeneracy lifting due to an increase of the mass of Dirac fermions at stronger magnetic fields [20]. After mastering the technology of the so-called suspended ultra-small graphene scrapings with extreme purity and high mobility of carriers above 200,000 $cm^2V^{-1}s^{-1}$, it was possible to observe FQHE in graphene at net fillings ν = 1/3 and –1/3 (the latter for holes in valence band) [31, 32]. Both these papers report the observation of FQHE for not extremely strong magnetic fields: in the paper [31], at the field of 14 T, for electron concentration of $10^{11}/cm^2$ and in the paper [32] at the field of 2 T, but for a concentration smaller by one order of magnitude ($10^{10}/cm^2$ and the mobility of 200,000 $cm^2V^{-1}s^{-1}$).

FQHE in suspended graphene is observed at relatively high temperatures around 10 K [33], and even higher (up to 20 K) [34], which seems to be explained by the relative strengthening of the electric interaction due to the absence, for suspended samples, of a dielectric substrate (with the dielectric constant in case of SiO_2, ~3.9), and on the other hand, with very high cyclotron energy in graphene (i.e., large energy gap between incompressible states).

The nowadays experimental data of FQHE in graphene [31–35] are consistent with the predictions of the braid group commensurability approach. In the case of graphene, the specific band structure with pseudo-relativistic conical Dirac bands leads to equal division of the LLL between the valence band of holes and the conduction band of electrons, which itself via an anomalous "relativistic" IQHE [20, 36, 37]. Applying lateral gate voltage (within the range ca. 5–60 V [10, 15, 31]) allows for the control of the density of carriers at a constant magnetic field, which is exceptionally valuable from an experimental point of view. For relatively small densities of carriers, the cyclotron orbits will be too short for braid exchanges of particles (leading to FQHE (multilooped)) at a relatively low magnetic field—the weaker for smaller concentrations is). The experimental observations exactly support this prediction [31, 32]. For low concentration, while closing on the Dirac point, one may expect, on the other hand, too strong which would correspond to an emergence of the insulating state near the Dirac point in a sufficiently large magnetic field. This prediction also agrees with the experiment [38]. In the case of the hexagonal structure of graphene, electron (or hole) Wigner crystallization [39] may exhibit interference between the triangular crystal sublattices, and including such resonance between two sublattices may cause blurring of the sharp transition to the insulator state, which is also consistent with observations.

The rapid progress in the recent experiment allowed for the precise observation of FQHE in graphene on the crystal substrate of boron nitride (*BN*) in large magnetic fields of the order of 40 T (remarkably, FQHE features were noticed in this case up to $\nu = 4$) [15, 16]. This achievement is caused by avoiding crystalline disorder on the frontier of graphene layer with the substrate. For BN with similar graphene hexagonal surface lattice, the disorder perturbations are minimal which guarantee high mobility of carriers in graphene triggering FQHE organization.

The mobility of carriers in graphene is lower than in traditional 2DEG, but taking into account that the carrier concentration in graphene can be lower in comparison to semiconductor heterostructure [40, 41], the corresponding mean free path in both cases well exceeds the sample dimension (of μm order, as the mobility is proportional to the concentration and to the mean free path of carriers).

It is worth noting that energy gaps protecting incompressible FQHE states in graphene are larger than in traditional semiconductor materials, reaching the high magnitude of the order of 10 K, which is linked with Dirac massless character of carriers. In conventional semiconductor heterostructure, the corresponding gaps are much lower and the observed FQHE is much more fragile in comparison to graphene.

The recent development in experiments with monolayer graphene on BN substrate [15, 16] and with suspended small sheets [17, 18] allowed for the observation of more and more Hall features referred to FQHE in subsequent subbands of two first LLs. While the sequence of fillings in the lowest subband of the LLL fits well to CF predictions (including CFs with two- and four-flux quanta attached), an explanation of the FQHE-filling structure in next subbands strongly deviates from this simple picture. CF theory fails also in all subbands of the first LL

[16–18]. Various scenarios of breaking of the approximate SU(4) spin-valley symmetry in graphene do not solve this problematic situation despite many theoretical attempts, which evidences insufficiency of the CF model in this case.

For much more effective understanding of FQHE in graphene, there occurs the braid group base commensurability approach. The hierarchy for FQHE predicted in this way is consistent with all experimental data known as of yet—the corresponding filling fractions can be reproduced by the topology braid group method (cf. **Table 1**). From this comparison, it is also clearly visible why the CFs are efficient only in the LLL and fails in many other situations.

Partial efficiency of CF model in the LLL is linked with the fact that exclusively in the LLL cyclotron orbits are always shorter than the interparticle spacing and additional loops are necessary. These loops can be simulated by fictitious field flux quanta attached to CFs. In the case, however, when the more complicated commensurability condition supports particular FQHE states in the LLL or in higher LLs the CF model fails. The braid group approach reproduces all features described correctly by CF model and moreover explains details inaccessible for CF approach. The usefulness of the CF model is in particular strongly limited in higher LLs because in this level the simple multilooped commensurability is needed only close to the subband edges, whereas the central regions of all subbands in higher LL are of single-looped type commensurability with next-nearest neighbors beyond the CF concept. For example, in the first LL on monolayer graphene the doublets of fillings are as follows: $(\frac{7}{3},\frac{8}{3})$, $(\frac{10}{3},\frac{11}{3})$, $(\frac{13}{3},\frac{14}{3})$, $(\frac{16}{3},\frac{17}{3})$, corresponding to single-looped commensurability condition, not related to CFs. These doublets are visible in experiment [15–18]. The number of centrally located filling rates for FQHE (single loop) grows next with the LL number as $2n$. The repeating doublet of filling ratios for $n = 1$ is noticeable in very accurate measurements in suspended samples [17, 18] besides those on the BN substrate [15, 16]. The observation [16] that the stability of FQHE(single-loop) states is of similar order as of IQHE states and higher in comparison to FQHE (multiloop) state features as is visible in **Figure 6** is worth noting. This might be associated with stronger correlations related to the single-looped braids as for IQHE states.

The commensurability braid group approach successfully reproduces all the positions of observed features in two lowest LLs of the monolayer graphene. The elongate plateaus at edges of subbands with IQHE in center seem to embrace also minimums related to FQHE(multiloop) rates which are located closely to edges – such location makes them out of the experimental resolution. In higher LL, the new features were observed in between the abovementioned doublets but with not developed vanishing longitudinal resistivity as at other FQHE states. It suggests that not all particles participate in correlated state, but might correspond to multilooped correlation commensurate to every second or every third particles. And indeed, these new features in the first LL of monolayer graphene recently reported [16] at $\nu = \frac{7}{3},\frac{8}{3},\frac{12}{5},\frac{13}{5},\frac{17}{7},\frac{18}{7},\frac{22}{9},\frac{23}{9}$, $\frac{10}{3},\frac{11}{3},\frac{17}{5},\frac{18}{5},\frac{24}{7},\frac{25}{7},\frac{13}{3},\frac{14}{3},\frac{22}{5},\frac{23}{5}$ are reproduced one to one by the commensurability series $\nu = 2(3,4) + \frac{xl}{l3(q-1)\pm 1}$ with $q = 3$, $x = 2, 3$, $l = \frac{i}{3}$, $i = 1, 2, 3$ as shown in **Figure 6**. One can notice that the FQHE-filling rates $\frac{7}{3},\frac{8}{3},\frac{10}{3},\frac{11}{3},\frac{13}{3},\frac{14}{3}$ are repeated in this hierarchy. It is likely that the single-looped correlated states at the same filling rates are more stable than the multilooped ones which seem to be consistent with the experimental data presented in **Figure 6** and in the upper panel of **Figure 7**.

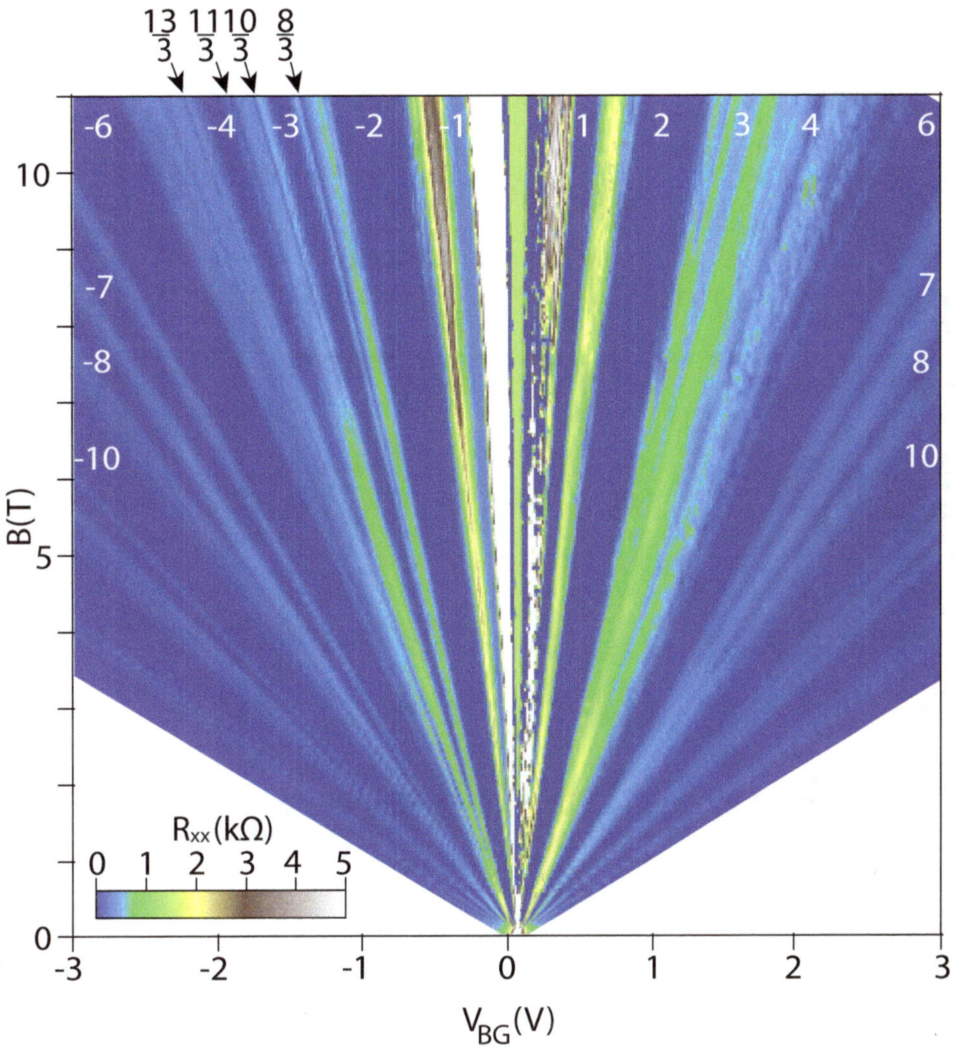

Figure 6. Fan diagram for $\rho_{xx}(\nu, B)$ for graphene up to 11 T (after [16]).

The next evidence supporting the correctness and usefulness of the commensurability condition is the consistence of the related predictions with the experimental observations in bilayer graphene. The remarkable feature observed in bilayer graphene is the occurrence of FQHE in the LLL in bilayer graphene at fractions with even denominators [10] which is completely beyond the CF model explanation ability. The commensurability braid group approach for bilayer graphene reproduces, however, perfectly all the experimentally observed FQHE hierarchy, including the controversial state at $\nu = -\frac{1}{2}$—cf. **Figure 8**, the illustration in **Figure 5**, and in **Table 2**.

Note finally that the FQHE hierarchy in bilayer systems with the characteristic even denominators manifested itself previously also in bilayer 2DEG of conventional Hall setups—the existence of FQHE state at $\nu = \frac{1}{2}$ state has been discovered there [42, 43], which also is consistent with the commensurability braid group predictions.

The interesting observation in bilayer graphene is reported by the comparison of FQHE measurement in bilayer graphene in suspended samples and in samples on the BN substrate

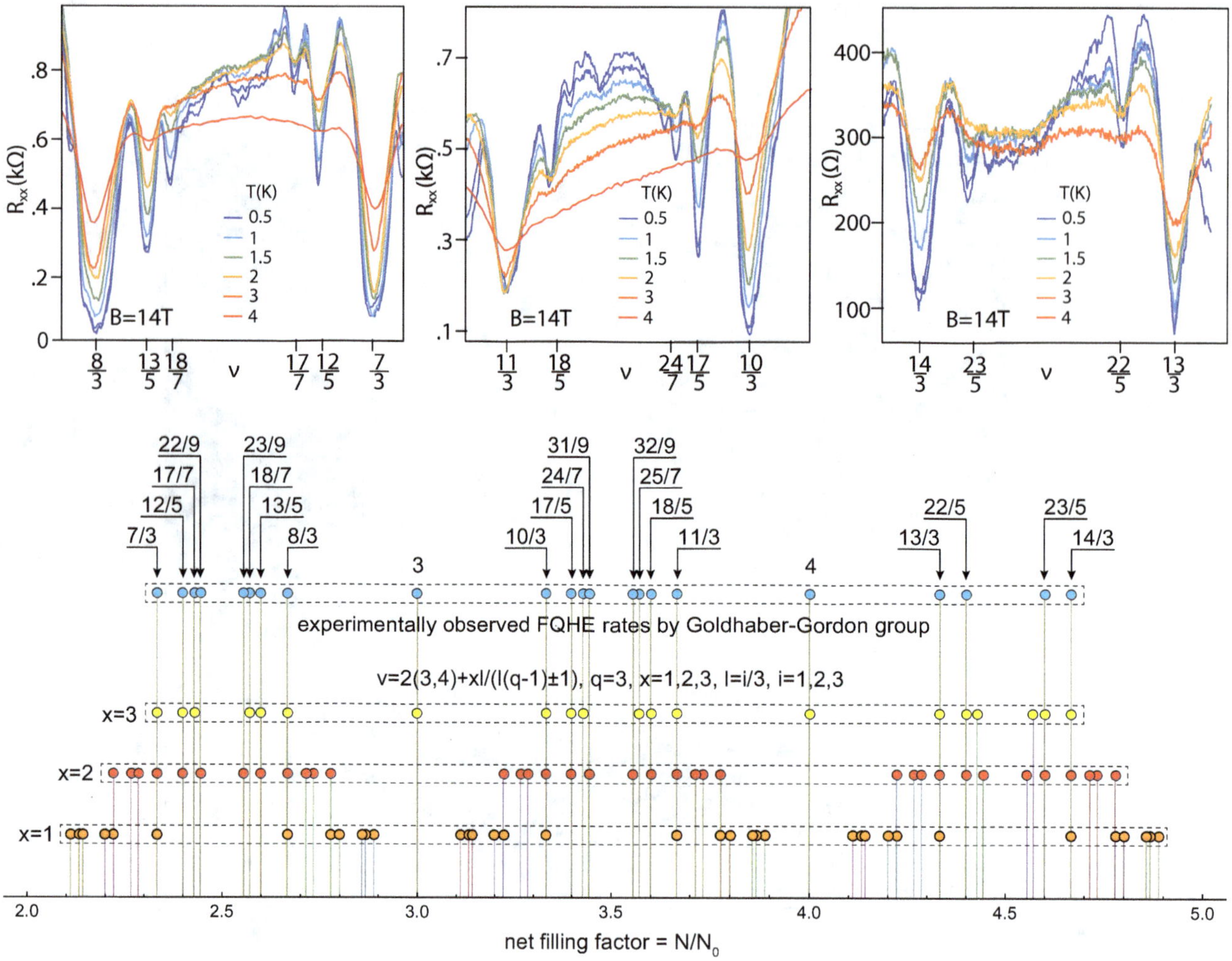

Figure 7. Not fully developed FQHE states with residual longitudinal resistance as corresponding to correlation of every second or every third particles at fractional rates reproduced by the commensurability series $\nu = 2(3,4) + \frac{xl}{l3(q-1)\pm1}$ with $q = 3$, $x = 2, 3$, $l = \frac{i}{3}$, $i = 1, 2, 3$ (upper panels visualize longitudinal resistivity measurements after [16]).

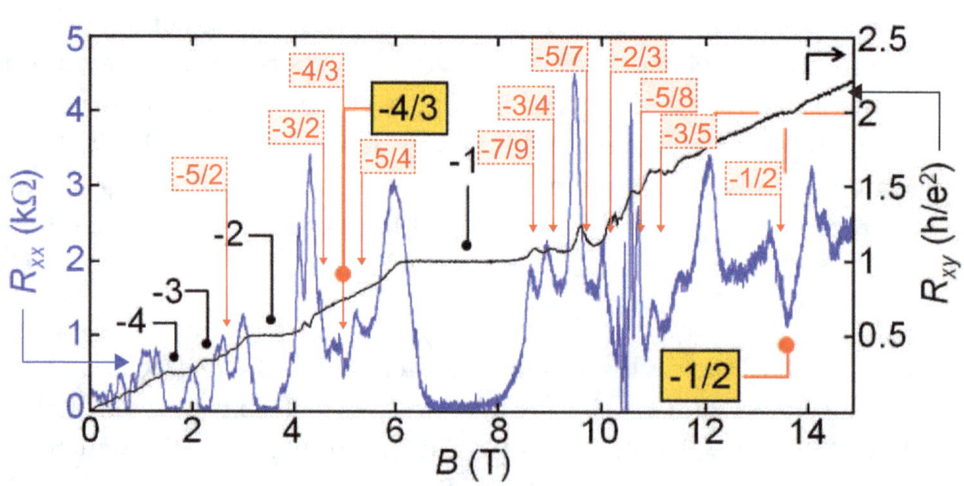

Figure 8. Observation of FQHE at T = 0.25 K in bilayer suspended graphene. (a) magneto-resistance R_{xx} and R_{xy} at the lateral voltage −27 V (after [10]). In dotted frames, fitting with the hierarchy given in **Table 2** is added.

[10–14]. Surprisingly, the FQHE state at $\nu = -\frac{1}{2}$ observed in bilayer structure on BN substrate disappears in the suspended sample of bilayer graphene. We propose to explain this behavior by our commensurability braid group approach, noting that the occurrence of the FQHE state depends of the order of the LLL subband degeneracy lifting, as shown in **Table 3**. We suppose thus that the external conditions related to the presence of the BN substrate can reverse the ordering of the LLL subbands resulting in the occurrence of the FQHE state at $\nu = -\frac{1}{2}$, which is, however, no case for the suspended sample with opposite lowest subband ordering.

4. Conclusion

The commensurability of cyclotron orbits building cyclotron braids with interparticle spacing in homogeneous 2D-charged systems in magnetic field is formulated in order to verify the possibility of arrangement of correlated multiparticle Hall states and decipher the "magic" hierarchy of filling ratios for FQHE. By application of the commensurability braid group approach to study the statistics of interacting many particle systems, the FQHE hierarchy of filling rates in graphene has been determined. The evolution of this hierarchy with growing number of the LL has been deciphered in consistence with experimental observations. The new opportunities for commensurability in higher LLs were established leading to different than ordinary FQHE (multiloop)-correlated states and the related fractional fillings of LLs have been identified beginning from the first LL. This new states have been named as FQHE (single loop) because the related correlations are described by single-looped braids. Both the monolayer and bilayer graphene have been analyzed and the essential difference of related hierarchy structures has been identified and described in detail. The even denominator main line of the fractional-filling hierarchy in the bilayer graphene is found in satisfactory consistence with the experimental observations. The presented hierarchy for the monolayer and bilayer graphene found the confirmation in available experimental data for graphene on BN substrate as well as for suspended samples including bilayer graphene for fillings up to sixth spin-valley subband. The impressive success of the commensurability braid group approach is the explanation of the disappearance of the FQHE at $\nu = -\frac{1}{2}$ in bilayer graphene in suspended sample, whereas this state has been formerly observed in bilayer graphene on the BN substrate. We link this phenomenon with the inversion of the ordering of the LLL lowest subbands caused by an influence of the crystalline BN substrate.

It is worth noting that the hierarchy observed in bilayer graphene is out of reach of the conventional CF theory.

Acknowledgements

This work is supported by the NCN projects P.2011/02/A/ST3/00116 and P.2016/21/D/ST3/00958.

Author details

Janusz Edward Jacak

Address all correspondence to: janusz.jacak@pwr.edu.pl

Department of Quantum Technologies, Faculty of Fundamental Problems of Technology, Wroclaw University of Science and Technology, Wroclaw, Poland

References

[1] Tsui DC, Störmer HL, Gossard AC. Two-dimensional magnetotransport in the extreme quantum limit. Phys Rev Lett. 1982;48:1559.

[2] Laughlin RB. Anomalous quantum Hall effect: an incompressible quantum fluid with fractionally charged excitations. Phys Rev Lett. 1983;50:1395.

[3] Abrikosov AA, Gorkov LP, Dzialoshinskii IE. Methods of Quantum Field Theory in Statistical Physics. Dover: Dover Publ. Inc.; 1975.

[4] Jain JK. Composite Fermions. Cambridge: Cambridge UP; 2007.

[5] Jacak J, Gonczarek R, Jacak L, Jóźwiak I. Application of Braid Groups in 2D Hall System Physics: Composite Fermion Structure. Singapore: WorldScientific; 2012.

[6] Wu YS. General theory for quantum statistics in two dimensions. Phys Rev Lett. 1984;52:2103.

[7] Birman JS. Braids, Links and Mapping Class Groups. Princeton: Princeton UP; 1974.

[8] Imbo TD, Imbo CS, Sudarshan CS. Identical particles, exotic statistics and braid groups. Phys Lett B. 1990;234:103.

[9] Laidlaw MG, DeWitt CM. Feynman functional integrals for systems of indistinguishable particles. Phys Rev D. 1971;3:1375.

[10] Ki DK, Falko VI, Abanin DA, Morpurgo A. Observation of even denominator fractional quantum Hall effect in suspended bilayer graphene. Nano Lett. 2014;14:2135.

[11] Diankov G, Liang CT, Amet F, Gallagher P, Lee M, Bestwick AJ, et al. Robust fractional quantum Hall effect in the N=2 Landau level in bilayer graphene. Nature Comm. 2016;7:13908.

[12] Kou A, Feldman BE, Levin AJ, Halperin BI, Watanabe K, Taniguchi T, et al. Electron-hole asymmetric integer and fractional quantum Hall effect in bilayer graphene. Science. 2014;345(6192):55–57.

[13] Maher P, Wang L, Gao Y, Forsythe C, Taniguchi T, Watanabe K, et al. Tunable fractional quantum Hall phases in bilayer graphene. Science. 2014;345(6192):61–64.

[14] Kim Y, Lee DS, Jung S, SkÃąkalovÃą V, Taniguchi T, Watanabe K, et al. Fractional Quantum Hall states in bilayer graphene probed by transconductance fluctuations. Nano Lett. 2015;15(11):7445–7451.

[15] Dean CR, Young AF, Cadden-Zimansky P, Wang L, Ren H, Watanabe K, et al. Multicomponent fractional quantum Hall effect in graphene. Nature Physics. 2011;7:693.

[16] Amet F, Bestwick AJ, Williams JR, Balicas L, Watanabe K, Taniguchi T, et al. Composite fermions and broken symmetries in graphene. Nature Comm. 2015;6:5838.

[17] Feldman BE, Krauss B, Smet JH, Yacoby A. Unconventional sequence of fractional quantum Hall states in suspended graphene. Science. 2012;337:1196.

[18] Feldman BE, Levin AJ, Krauss B, Abanin DA, Halperin BI, Smet JH, et al. Fractional quantum Hall phase transitions and four-flux states in graphene. Phys Rev Lett. 2013;111:076802.

[19] Castro Neto AH, Guinea F, Peres NMR, Novoselov KS, Geim AK. The electronic properties of graphene. Rev Mod Phys. 2009;81:109.

[20] Zhang Y, Jiang Z, Small JP, Purewal MS, Tan YW, Fazlollahi M, et al. Landau-level splitting in graphene in high magnetic fields. Phys Rev Lett. 2006;96:136806.

[21] Jacak J, Jacak L. Recovery of Laughlin correlations with cyclotron braids. Europhysics Lett. 2010;92:60002.

[22] Jacak J, Jacak L. The commensurability condition and fractional quantum Hall effect hierarchy in higher Landau levels. JETP Lett. 2015;102:19–25.

[23] Łydżba P, Jacak L, Jacak J. Hierarchy of fillings for the FQHE in monolayer graphene. Sci Rep (Nature). 2015;5:14287.

[24] Jacak J, Jacak L. Difference in hierarchy of FQHE between monolayer and bilayer graphene. Phys Lett A. 2015;379:2130.

[25] Eisenstein JP, Lilly MP, Cooper KB, Pfeiffer LN, West KW. New physics in high Landau levels. Physica E. 2000;6:29.

[26] Dolev M, Gross Y, Sabo R, Gurman I, Heiblum M, Umansky V, et al. Characterizing neutral modes of fractional states in the second Landau level. Phys Rev Lett. 2011;107:036805.

[27] Willett RL. The quantum Hall effect at 5/2 filling factor. Rep Prog Phys. 2013;76:076501.

[28] Goerbig MO. Electronic properties of graphene in a strong magnetic field. Rev Mod Phys. 2011;83:1193.

[29] McCann E, Falko VI. Landau-level degeneracy and quantum Hall effect in a graphite bilayer. Phys Rev Lett. 2006;96:086805.

[30] Papić Z, Abanin DA. Topological phases in the zeroth Landau level of bilayer graphene. Phys Rev Lett. 2014;112:046602.

[31] Du X, Skachko I, Duerr F, Luican A, Andrei EY. Fractional quantum Hall effect and insulating phase of Dirac electrons in graphene. Nature. 2009;462:192.

[32] Bolotin KI, Ghahari F, Shulman MD, Störmer HL, Kim P. Observation of the fractional quantum Hall effect in graphene. Nature. 2009;462:196.

[33] Abanin DA, Skachko I, Du X, Andrei EY, Levitov LS. Fractional quantum Hall effect in suspended graphene: transport coefficients and electron interaction strength. Phys Rev B. 2010;81:115410.

[34] Skachko I, Du X, Duerr F, Luican A, Abanin DA, Levitov LS, et al. Fractional quantum Hall effect in suspended graphene probed with two-terminal measurements. Phil Trans R Soc A. 2010;368:5403.

[35] Papić Z, Goerbig MO, Regnault N. Atypical fractional quantum Hall effect in graphene at fillig factor 1/3. Phys Rev Lett. 2010;105:176802.

[36] Novoselov KS, Geim AK, Morozov SV, Jiang D, Katsnelson MI, Grigorieva IV, et al. Two-dimensional gas of massles Dirac fermions in graphene. Nature. 2005;438:197.

[37] Zhang Y, Tan YW, Störmer HL, Kim F. Experimental observation of the quantum Hall effect and Berry's phase in graphene. Nature. 2005;438:201.

[38] Yannouleas C, Romanovsky I, Landman U. Edge and bulk components of lowest-Landau-level orbitals, correlated fractional quantum Hall effect incompressible states, and insulating behavior of finite graphene samples. Phys Rev B. 2010;82:125419.

[39] Dahal HP, Joglekar YN, Bedell KS, Balatsky AV. Absence of Wigner crystallization in graphene. Phys Rev B. 2006;74:233405.

[40] Pfeiffer L, West KW. The role of MBE in recent quantum Hall effect physics discoveries. Physica E. 2003;20:57.

[41] Bolotin KI, Sikes KJ, Jiang Z, Klima M, Fudenberg G, Hone J, et al. Ultrahigh electron mobility in suspended graphene. Solid State Comm. 2008;146:351.

[42] Suen YW, Engel LW, Santos MB, Shayegan M, Tsui DC. Observation of a ν=1/2 fractional quantum Hall state in a double-layer electron system. Phys Rev Lett. 1992;68:1379.

[43] Eisenstein JP, Boebinger GS, Pfeiffer LN, West KW, He S. New fractional quantum Hall state in double-layer two-dimensional electron systems. Phys Rev Lett. 1992;68:1383.

5

Controlled Functionalization of Graphene Layers

Maurizio Galimberti, Vincenzina Barbera and

Annalisa Sironi

Abstract

Controlled functionalization of graphene layers is one of the most important research objectives in the material chemistry. A well established procedure is the oxidation with strong acids and oxidizing agents often in harsh and dangerous reaction conditions giving products of unknown precise structure. In this chapter, the controlled functionalization of graphene layers with a derivative of serinol is presented, avoiding toxic reagents and dangerous reaction conditions. The derivative is the bio-based serinol pyrrole, obtained through the neat reaction of serinol with 2,5-hexanedione; the graphitic substrate was high surface area graphite (HSAG) with high-shape anisotropy. The functionalization reaction, characterized by a 85% atomic efficiency (water is the only by-product), evolved with high yields leading to functionalized graphene layers through the controlled introduction of oxygen and nitrogen-containing functional groups. Sustainable processes were adopted, such as ball milling and heating. The mechanism pathway, the characterization of HSAG and reaction products through a wide range of analytical methods, some successful applications of the adducts are discussed in this chapter. The functionalization left the bulk crystalline structure of the layers substantially unaltered. Stable dispersions in water and eco-friendly solvents were prepared.

Keywords: graphene, functionalization, serinol derivative

1. Introduction

Graphene [1–3] is the thinnest material on Earth and is commonly considered as a single layer of graphite. It has a two-dimensional conjugated structure, good thermal stability, very high aspect ratio and specific surface area and has, as a consequence, outstanding electronic, thermal and mechanical properties. At low temperatures and high magnetic fields, quantum Hall effect has been observed in graphene layers for both electrons and holes [4, 5]. The in-plane thermal conductivity of graphene is among the highest recorded for known materials, about 2000–4000 W m^{-1} K^{-1} at room temperature [6, 7]. A graphene sheet has theoretical elastic modulus of over 1 TPa and Young modulus of about 1060 MPa [8, 9].

In the light of such properties, impressive research activity is currently carried out for applying graphene in high-performance materials [10], in fields such as nanoelectronics [3], energy storage and energy conversion [11]. Graphene composites allow contributing the exceptional properties of graphene to the macroscopic scale. In particular, light, flexible, robust and conductive graphene papers find a broad spectrum of applications such as electrochemical energy storage devices [12], catalyst supports and fuel cells [13], sensors and actuators [14], chemical filters and membranes [15] and structural composites [16].

Graphene and derivatives are also finding increasing applications in the field of catalysis [17, 18]. Indeed, carbocatalysis is largely used for promoting synthesis and transformation of organic or inorganic substrates: carbons favour reduction, oxidation and bond-forming reactions [19].

All the applications mentioned above become successful if graphene or graphitic aggregates made by only few layers of graphene can be used. Large research efforts are thus made in order to prepare graphene and few-layer graphene [3, 10, 20–23].

Moreover, high interest is on functionalization of graphene layers [24–33]. Graphene oxide is considered a stable carbon framework to be functionalized [24] and, as it will be discussed in the next paragraph, is the product of the first step of the oxidation-reduction process aimed at preparing graphene starting from graphite. It is thus the subject of much of the research on functionalization [24–26].

In our group, research was performed with the following objectives:

(i) To prepare graphene and few-layer graphene through a simple, environmentally friendly, sustainable and economically viable method that could be applied for large-scale development

(ii) To introduce functional groups on graphene layers, containing in particular heteroatoms such as oxygen and nitrogen

(iii) To preserve the ideal structure of graphene, in graphene and functionalized graphene layers.

In a nutshell, our main goal was the controlled functionalization of graphene layers.

The present chapter summarizes the results of such a research.

2. Preparation of graphene and functionalized graphene: prior art

In the light of what is reported in Section 1, it can be easily understood that primary objectives of the research in the field of graphene are as follows:

(i) To prepare graphene or graphitic nanofillers made by only few layers of graphene, through a simple method, suitable for large-scale production

(ii) To preserve the ideal graphene structure.

Nowadays, single-layer or few-layer graphene are obtained through bottom-up and top-down approaches [3, 10, 20–23], such as epitaxial growth of graphene films, micromechanical cleavage and dilution in appropriate solvents.

Best practice, particularly in view of large scale applications, is considered the oxidation of graphite or graphitic nanofiller to graphite oxide (well known as GO), followed by thermal or chemical reduction. However, such a pathway is affected by several flaws, in the oxidation and reduction steps and also in consideration of the features of reduced GO.

Harsh and even dangerous reaction conditions are required for the oxidation of pristine graphitic material. First papers on carbon oxidation date back to the nineteenth century. Brodie reported the use of fuming HNO_3 and $KClO_3$ as intercalant and oxidant [34], in the frame of a multicycle process, that produced toxic and explosive gases such as NO_2/N_2O_4 and ClO_2, respectively. Indeed, explosions have been documented [35, 36]. Towards the end of the century, Staudenmeier used a blend of H_2SO_4 and HNO_3 (2/1) with $KClO_3$ as oxidant, in a one-step process that still produced explosive ClO_2 [37]. These methods of the eighteenth century cannot be taken into consideration for large-scale production of GO. Towards the 1960s of the last century, Hummers reported the use of $H_2SO_4/NaNO_3$ and $KMnO_4$ to intercalate and oxidize graphite [38]. The Hummers' method is considered promising for large-scale production of GO as $KMnO_4$ is very efficient and reaction takes only few hours, explosive ClO_2 is not formed and the replacement of HNO_3 with $NaNO_3$ eliminates acidic smokes. However, NO_2/N_2O_4 are still formed and reaction products such as manganese ions can be hardly removed, only by washing with acids such as HCl which, in turn, remains strongly absorbed on the graphitic substrate. Finally, only partial oxidation is obtained [39–41]. Research has been thus dedicated to improve the Hummers' method [40–43]. $NaNO_3$ has been removed by increasing the amount of $KMnO_4$ and by using a H_2SO_4/H_3PO_4 mixture [40]. $NaNO_3$ has been simply removed, without observing negative effects on the reaction [41]. However, only an incomplete conversion of graphite to GO was obtained. The pre-oxidation of graphite with P_2O_5 and $K_2S_2O_8$ in H_2SO_4 adds a further step [44]. The complete conversion to GO has been reported [42] by using graphite flakes with sizes in the range of 3–20 μm. In another variation of the Hummers' method, the concentration of $NaNO_3$ and $KMnO_4$ and the residence times have been modified [43]. GO can be also prepared by solvent-free mechanochemical oxidation of graphite [45]. However, such a method can be hardly employed for large-scale production.

The Hummers' method based on the use of $NaNO_3$, applied by one of the authors of the present chapter is fully described in Ref. [46].

It is evident that the objective of developing a really simple method, suitable for large-scale production, has not been achieved yet.

The reduction step is crucial in order to obtain graphene. Well-known methods are based on the use of hydrazine [47–50] or hydrogen plasma [49]. To avoid reaggregation of graphitic flakes, a stabilizing agent should be added [9, 51]. Hydrazine is well known as a toxic reagent. Hence, eco-friendly methods have been attempted. Quick deoxygenation of graphite oxide assisted by a base (NaOH, 0.1 M) has been performed at moderate temperatures (80°C) [52]. Ascorbic acid has been used as well [53]. Thermal [54] and flash [55] reductions have been reported.

However, it is widely acknowledged that reduction is still incomplete and that the ideal graphene structure is neither preserved nor restored [20, 56]. Hence, also this objective is not achieved yet.

However, the pathway that could lead to graphene through graphite oxidation gives the opportunity of preparing GO that could be a suitable building block for further reactions [24–26] as well as for catalytic applications [17, 18]. Moreover, functionalized single-layer graphene sheets can be prepared by splitting graphite oxide [57]. The structure of GO has been investigated for decades, but it is substantially still unknown [20, 21, 58], and this makes GO not the ideal building block for further reactions. It was reported [59] that hydroxyl and epoxide groups are on the surface of basal planes and carbonyl and carboxyl groups are on the edges. Moreover, it is widely acknowledged that oxidation leads to extensive disruption of sp^2 hybridization of graphene layers, that is, graphene properties are drastically damaged and GO loses the benefits of a graphitic structure.

This brief sum up allows to comment that important research efforts are required in order to achieve the objective of preparing graphene and few-layer graphene (optionally containing functional groups) preserving the ideal graphene structure and through a simple, eco-friendly, economically viable and scalable method.

3. Serinol derivative for graphene functionalization

3.1. Serinol derivative: 2-(2,5-dimethyl-1H-pyrrol-1-yl)-1,3-propanediol

To pursue the (i)–(iii) objectives reported in Introduction, a serinol derivative was used: 2-(2,5-dimethyl-1H-pyrrol-1-yl)-1,3-propanediol, in the text referred to as serinol pyrrole (SP). Chemical structure of serinol pyrrole is shown in **Figure 1**.

In this paragraph, synthesis of serinol pyrrole and motivation for using such a molecule in combination with a high surface area nanosized graphite (HSAG) is discussed.

Serinol is 2-amino-1,3-propane diol that can be directly obtained from renewable sources [60] and is produced at the industrial scale from glycerol. Glycerol is a non-toxic and biodegradable raw material, cheap and easily available as it is the main co-product of biodiesel production.

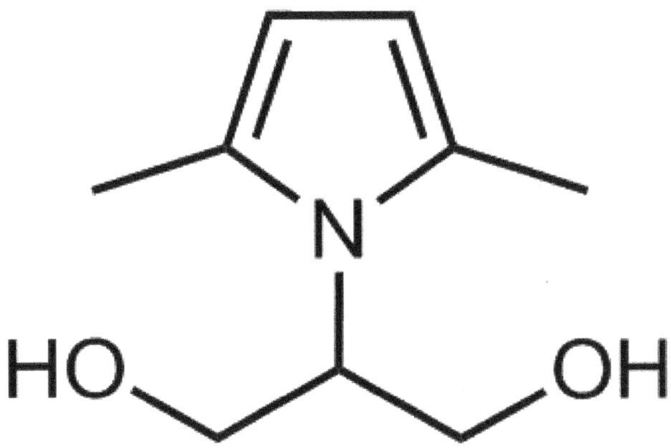

Figure 1. 2-(2,5-dimethyl-1*H*-pyrrol-1-yl)-1,3-propanediol (serinol pyrrole, SP).

Glycerol can be thus considered as a waste of an important industrial process. To use and to give added value to such a waste is particularly meaningful, as glycerol can be the building block for a C3 platform alternative to the oil-based one. It is well known that large research efforts are made nowadays to replace oil with biomass as the source for preparing chemicals and materials. First and second generation biomasses are identified. The first generation includes biomass that has impact on the food chain. For example, first-generation bioethanol is produced from sucrose, oligosaccharides and starch, whereas lignocellulose materials are used for the second-generation bioethanol. Biodiesel was produced from oil crops, whereas nowadays it is increasingly obtained from algae. In order to reduce the use of biomasses that could be employed for food production, the exploitation of wastes is encouraged. The European Parliament has made clear statements about the use of wastes and residues in place of original biomass: 'The recent experience of the development of certain renewable energy sources, particularly biofuels from food and feed crops such as cereals, oilseeds and sugar, has stimulated concern that new biorefinery processes must as far as possible be based on non-competing wastes and residues to minimize impacts on food availability and prices' [61]. More than 1 million tons of glycerol are produced per year, with a price well below 1 Euro/kg, and many routes are available for the selective conversion of glycerol to value-added products [62–65]. Serinol, produced via its reductive amination, presents the precious tool of chemoselectivity between amino and hydroxyl groups which allows in developing innovative synthetic strategies. Indeed, objective of our research was to exploit the chemoselectivity of serinol, transforming the amino moiety into a group suitable for interaction with sp^2 carbon allotropes and, in particular, with graphene layers.

3.2. Synthesis of serinol derivative, 2-(2,5-dimethyl-1H-pyrrol-1-yl)-1,3-propanediol

3.2.1. Neat synthesis

As it is shown in **Scheme 1**, Paal-Knorr reaction [66, 67] was carried out between 2-amino-1,3-propane diol (serinol, S) and 2,5-hexanedione (HD), transforming the amino group into a pyrrole ring and thus giving rise to 2-(2,5-dimethyl-1H-pyrrol-1-yl)-1,3-propanediol (SP) [68–71].

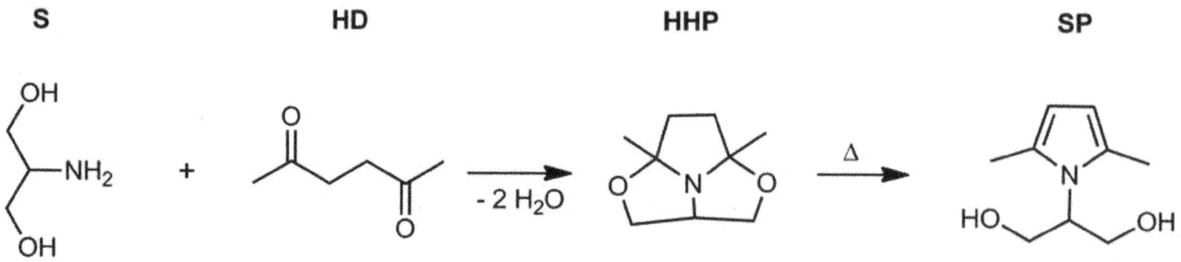

Scheme 1. Neat reaction between serinol and 2,5-hexanedione for preparing serinol pyrrole. S and HD in equimolar amount.

It is definitely worth observing that the reaction that leads to SP is characterized by very high atom economy, 82.5%, with H_2O as the only by-product. Neat reaction was carried out, in the absence of solvents and catalysts [68], with temperatures ranging from 130 to 180°C and reaction times ranging from 30 to 120 min. In Section 8, an example characterized by very high reaction yield to SP is reported: 96%. In consideration of this yield and of the atom economy, the atom efficiency is 85%.

Scheme 1 reveals a two-step synthesis. The first step is the reaction between equimolar amounts of S and HD and leads to the preparation of a tricyclic compound: 4a,6a-dimethyl-hexahydro-1,4-dioxa-6b-azacyclopenta[cd]pentalene (HHP). The reaction was performed at room temperature, with almost complete conversion and selectivity, the only by-product being water (as mentioned above). In the second step of the reaction, the tricyclic compound was then isomerized to the aromatic SP, with high conversion and selectivity, by simply increasing the reaction temperature. Indeed, Nuclear Magnetic Resonance (NMR) spectra of the products of reaction steps 1 and 2 did not reveal the presence of any chemical compounds other than HHP and SP, respectively. The Paal-Knorr reaction between S and HD had been already reported [72]. Reagents were refluxed in toluene in the presence of acidic substances such as glacial acetic acid (in large amount) and *p*-toluenesulfonic acid (in catalytic amount). The product mixture contained SP in low amount (yield was about 13% mol), the tricyclic compound (HHP) and polymeric materials, whose relative amount was enhanced by acids. The neat two-step synthesis described here surprisingly led to SP with high atom efficiency. As published by some of the authors [69], this result could be ascribed, at least in part, to absence of acids and mild reaction conditions. Such experimental frame favoured, in the first reaction step, hemiaminal formation and polycyclization, preventing aromatization to SP and formation of polymeric species. Temperature increase allowed then the establishment of the thermodynamic control of the reaction.

3.2.2. *Synthesis on high surface area nanographite (HSAG)*

Synthesis of serinol pyrrole was as well performed on high surface area nanosized graphite, HSAG. Characteristics of HSAG are discussed in Section 4.1. The simple reaction process is shown in **Scheme 2**.

Details are in Section 8. Nuclear Magnetic Resonance (NMR) analysis of the solution of the reaction product in D_2O revealed the presence of serinol pyrrole only. This result indicates that the reaction between S and HD occurs with high yields also by supporting the ingredients on a graphitic substrate.

$$\text{HSAG + S + HD} \xrightarrow[\text{flask, slow rotation}]{\text{120 min at 150°C}} \xrightarrow[\text{to room temperature}]{\text{Mixture cooled}} \text{HSAG + SP}$$

Scheme 2. Reaction between serinol (S) and 2,5-hexanedione (HD) on high surface area graphite (HSAG), for the selective preparation of serinol pyrrole (SP). S and HD are in equimolar amount.

3.3. Why to use serinol pyrrole for the functionalization of graphene layers?

SP is a serinol derivative, hence it comes from a molecule that is naturally occurring whether it comes from natural sources or from a waste such as glycerol. In the light of that, SP can be considered as a bio-sourced molecule. Its preparation appears to have also another good feature in view of a sustainable process, that is, the high atom efficiency.

By observing the chemical structure of serinol pyrrole in **Figure 1**, it can be commented that SP is a *Janus* molecule [70], that is, a molecule with two faces. Such a definition, that comes from the portray of a Roman god, was first used to indicate micro- and nanoparticles with at least two physically or chemically different surfaces [73, 74] and, subsequently, to describe molecules with two moieties (one hydrophobic and one hydrophilic) such as colloids [75] and block copolymers [76]. A *Janus* molecule is known to have a dual reactivity.

What are the two moieties in SP? One is represented by the pyrrole ring. The Paal-Knorr reaction changes the hybridization of the sp^3 nitrogen atom of the amino group and leads to the formation of sp^2 atoms in the aromatic pyrrole ring, which could give rise to π-π stacking with aromatic compounds such as graphene layers and hence to a stable interaction; moreover, the pyrrole ring could interact with lipophilic substances. Conversely, the moiety that contains hydroxy groups is hydrophilic and can promote the interaction with polar environments. Moreover, hydroxy groups can be reactive functional groups for the preparation of step growth polymers [70, 72].

Thanks to its hydrophilic moiety, SP could then promote the dispersion of graphitic aggregates in polar solvents, such as water. Exfoliation of such aggregates into few or single graphene layers could be thus promoted. As mentioned in the introduction, dilution in appropriate solvents is a method for the preparation of graphene and few-layer graphene. However, in most cases, organic (mainly aromatic) substances have to be used [22]. The structure of the graphitic substrate is expected to remain substantially unaltered by the interaction with SP: this would lead to introduce functional groups with heteroatoms such as oxygen and nitrogen on graphene layers that would maintain their pristine structure. It is evident that the success of such a design essentially depends on the strength of the interaction between SP and the graphene layers and also on the amount of SP which should be used. Moreover, the choice of the graphitic substrate for the preparation of adducts with SP appears to be of great importance.

4. Adducts of graphene layers with serinol pyrrole

In this paragraph, characteristics of HSAG (selected as the graphitic substrate) preparation and characterization of HSAG-SP adducts are discussed.

4.1. Graphitic substrate: high surface area nanosized graphite

High surface area nanosized graphite was selected for the preparation of adducts with SP. Characterization of HSAG has been already reported [57, 70, 77]. Some relevant data are shown in **Table 1**, in comparison with those of other sp^2 carbon allotropes: multiwalled carbon nanotubes (CNT) and a furnace carbon black, CBN326.

HSAG has a high surface area and is able to establish extended interactions with a polymer matrix, as shown by the diisobutyl phthalate (DBP) absorption number. By comparing HSAG and CNT data in **Table 1**, a correlation cannot be seen between surface area and DBP absorption number. As commented in previous publications [77], this finding could be explained considering that graphene layers in HSAG are stacked in crystalline domains: layers are more accessible to small nitrogen molecules used in Brunauer Emmet Teller (BET) analysis than to bulky phthalates. However, the peak shape analysis of wide angle X-ray diffraction (WAXD) pattern (shown in **Figure 2**, with the peak assignment), performed by applying the Scherrer equation to (002) reflection [56], revealed quite a low number of graphene layers stacked in a crystalline domain (about 35). Moreover, the analysis of (100) and (110) reflections, typical of the crystalline order inside the layer, allowed to estimate the shape anisotropy, defined as the ratio between the crystallites dimensions in directions parallel and orthogonal to structural layers [56]. HSAG was found to have the largest shape anisotropy, when compared with CBN326, expanded graphites and coke and calcinated petroleum cokes. (Transmission electron microscopy (TEM) analyses revealed that HSAG layers had an average size of about 300 nm [70].

Characterization of HSAG was also performed via Raman spectroscopy [70], a crucial technique for the study of carbonaceous materials [79–83]. Raman spectrum of HSAG is reported in **Figure 2** below in the text. In the Raman spectrum, two lines, named D and G, reveal the presence of graphitic sp^2-phase: they are located at 1350 cm^{-1} and 1590 cm^{-1}, respectively. Bulk crystalline graphite (graphene) gives rise to G peak, whereas the D peak occurs in the presence of either structural defects or confinement (e.g. by edges) of the graphitic layers [78, 79, 82–84]. Carbon atoms with different hybridization and grafted functional groups are indeed structural defects. Graphene layers have finite dimensions and irregular boundaries whose relative importance depends on the size of the layer. These boundaries contribute to

Carbon allotrope	Surface area[a] (m^2/g)	DBP absorption number (mL/100 g)[b]	Number of stacked layers[c]
HSAG	330	162	35
CNT[d]	200	316	10
CB N326[e]	77	85	5

[a]From BET measurements (see Ref. [78]).
[b]mL of absorbed DBP/100 g of CB (see Ref. [78]).
[c]Estimated from WAXD pattern (see Ref. [56]).
[d]Baytubes C150 P from Bayer Material Science (see Ref. [78]).
[e]From Cabot.

Table 1. Surface area, DBP absorption number and number of graphene layers stacked in crystalline domain for sp^2 carbon allotropes.

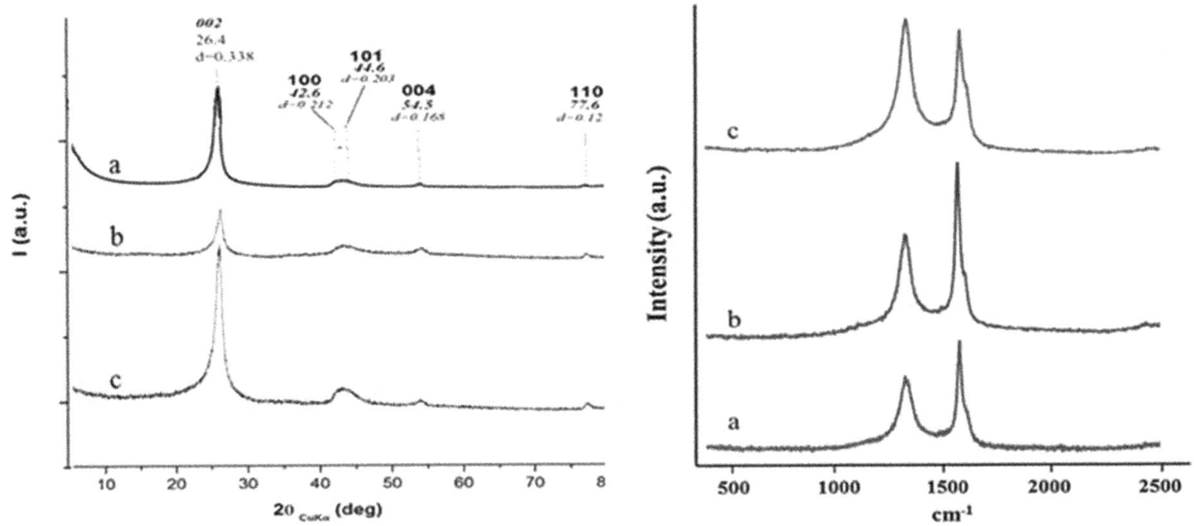

Figure 2. WAXD patterns (on the left) and Raman spectra (on the right) of HSAG (a), HSAG-SP-M (b) and HSAG-SP-T (c).

the D band. In the Raman spectrum of HSAG in **Figure 2**, both D and G bands are present, with similar intensities. To justify the intensity of the D band, some of the authors reported an interpretation [70] based on the existence of a confined crown region close to the edge, affected by confinement effects and electronically perturbed. This interpretation takes into account the small size of the HSAG graphitic layers and the obtainment of HSAG through ball milling. It was demonstrated [85] that the intensity of the D band increases in a graphitic sample by reducing the size of the layers by progressive ball milling.

HSAG appears thus an ideal substrate for the preparation of adducts with SP, thanks to the following features: high surface area, low number of stacked layers, high crystalline order inside the layers and remarkable presence of edges, that could favour interactions or even reactions with a suitable molecule.

4.2. Adducts of SP with HSAG: preparation and characterization

Adducts were prepared as illustrated in **Scheme 3**. In a nutshell, HSAG and SP were mixed in acetone, the mixture was sonicated for few minutes, solvent was removed and either mechanical or thermal energy was given to the solid HSAG-SP mixture, via ball milling or simply heating. The following adducts were prepared: HSAG-SP-M from mechanical treatment and HSAG-SP-T from thermal treatment. Typical examples of the preparation of mechanical and thermal adducts are in Section 8.

The strength of interaction between HSAG and SP was investigated. HSAG-SP adducts, upon extraction (see Section 8) were weighed and analysed through thermogravimetric analysis (TGA). As reported in a previous publication [70], the amount of SP in the adduct was estimated on the basis of the mass loss in the temperature range from 150 to 500°C. Data are collected in **Table 2**. The SP: HSAG ratio was estimated as molar ratio (the moles of HSAG are the moles of benzene rings), while the yield of functionalization was calculated by applying the following equation:

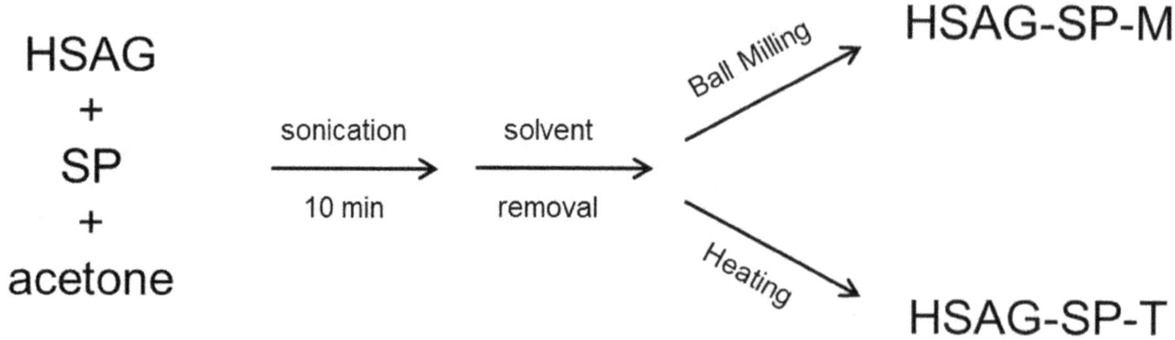

Scheme 3. Block diagram for the preparation of HSAG-SP adducts, by using either mechanical or thermal energy (for details, see Section 8).

Method	SP/HSAG molar ratio	Yield (%)
Mechanical	0.100	97.3
Thermal	0.100	91.2[b]
Thermal	0.050	79.9
Thermal	0.010	63.6
Thermal	0.005	59.5

[a]Evaluated from TGA, with the equation reported in the text.
[b]Variation coefficient, calculated for a population of 10 samples, was 1.7%.

Table 2. Preparation of HSAG-SP adducts: yield of functionalization[a].

$$\text{Yield} = \frac{\text{SP in (HSAG SP adduct)}_{\text{after the extraction}}}{\text{SP in (HSAG SP adduct)}_{\text{before the reaction}}} \times 100 \tag{1}$$

Very high yield of functionalization (larger than 90%) was achieved, both through mechanical and thermal treatments. In particular, thermal treatment with 0.1 as the SP:HSAG molar ratio gave a yield of 99%. This indicates that SP is able to establish strong interactions with HSAG. It is worth noting that the yield of functionalization decreased with the starting SP:HSAG molar ratio.

WAXD and Raman analyses were used to study the organization at the solid state of HSAG-SP adducts (details are in Ref. [71]). X-ray diffraction patterns and Raman spectra of HSAG (a), HSAG-SP-M (b) and HSAG-SP-T (c) are shown in **Figure 2**.

As regards WAXD patterns, in HSAG-SP-M and HSAG-SP-T, (002) reflection remains at the same 2θ value as in pristine HSAG, indicating that SP was not intercalated in the graphitic interlayer space. The number of layers stacked in a crystalline domain, calculated by applying the Scherrer equation to (002) reflection, as explained above, was found to decrease from HSAG (35) to HSAG-SP-T (29) to HSAG-SP-M (24). Peak shape analysis was as well performed on (100) and (110) reflections, analysing pristine HSAG and the adducts: the correlation length was found in a range from 26.5 to 28 nm. Hence, the in-plane order of HSAG was not substantially altered by the reaction with SP, even via ball milling.

Considering Raman characterization, D band is present in the spectrum of pristine HSAG and its relative intensity with respect to G band does not substantially change in HSAG-SP-T, after the thermal treatment. This finding suggests that the treatment with SP does not appreciably modify the structure of HSAG. On the contrary, the intensity of D band increases in HSAG-SP-M, after the mechanical treatment. In the light of what observed from WAXD analysis, the increase of disorder in the milled sample could be attributed to electronic perturbation of the confined crown region close to the edge, in line with what reported in the literature [85] about the effect of milling on graphitic samples. Different types of disorder can occur at the very edges: grafting of molecules and loss of sp^2 hybridization are among them.

The chemical nature of HSAG-SP adduct was investigated by means of FTIR analysis (experimental details are in reference [71]). The IR spectra of HSAG, HSAG-SP-T, HSAG-SP-M and SP, in the region 4000–700 cm^{-1} are reported in **Figure 3**.

The spectrum of HSAG (a) is characterized by the peak near 1590 cm^{-1} which can be assigned to the absorption of E_{1u} IR active mode of collective C=C stretching vibration of graphite/graphene materials. The increasing background towards high wavenumbers is due to diffusion/reflection phenomena of the IR beam passing through HSAG microparticles.

In the SP spectrum (d), the broad band at 3370 cm^{-1} can be reasonably attributed to hydrogen-bonded OH groups. The pyrrole ring is evidenced by the collective vibration mode of C=C/C-C stretching, located at about: 1530, 1395, 1490 and 802 cm^{-1}. Such bands can be taken as fingerprint of SP.

Spectra of extracted HSAG-SP-T and HSAG-SP-M adducts (**Figure 3b** and **c**) show bands that cannot be attributed to HSAG. Such bands are located as follows: at about 2900 cm^{-1}, in the region of sp^3 C-H stretching; at 1590 and 1470 cm^{-1}, in the region of C-C stretching of aromatic rings; at 1383 cm^{-1}, in the region of vibrations of diols and at 802 cm^{-1}, where vibration of the alkenyl groups absorbs. C-H stretching can be due to methyl groups of the pyrrole ring and to methylenes of the serinol moiety. Diols come from serinol as well. Aromatic C-C stretching is due to the presence of pyrrole ring. Moreover, in both the adducts' spectra, there are bands that are not present in either HSAG or SP spectra. Such bands are located at 1742 and 1662 cm^{-1}. Absorbances in this region are usually due to the presence of carbonyl groups. In particular, these new bands can be assigned to an aldehydic group. As reported in Section 8, the preparation of adducts is performed at high temperatures and in the presence of air. Oxidation process of the pyrrole ring could be thus hypothesized. Moreover, the efficient synthesis of graphene sheets using pyrrole as a reducing agent has been recently reported [86]: pyrrole oxidation was obtained by mixing graphene oxide with pyrrole at 95°C for 12 h.

It is worth observing that the relative intensity of the peak at 802 cm^{-1} decreases in the spectra of the adducts, in particular in the spectrum of HSAG-SP-M adduct. This peak is due to the C=C stretching of the pyrrole ring and such a reduction might lead to hypothesize that C=C bonds are involved in a chemical reaction. Working hypothesis for such a process could be the Diels-Alder reaction with the graphitic substrate. It is known that graphite is able to react with dienes and dienophiles through Diels-Alder reaction [27]. The presence of electron

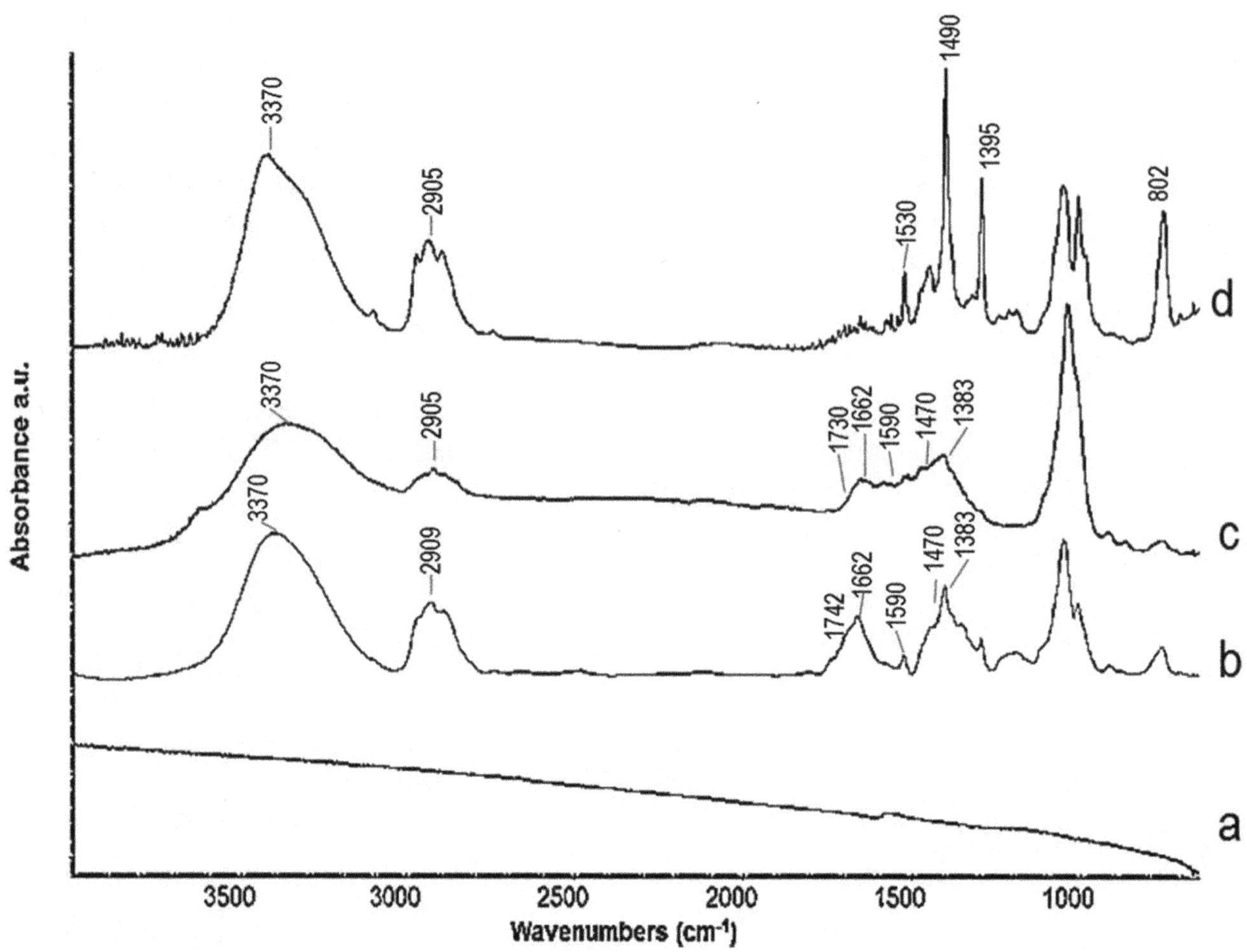

Figure 3. FTIR spectra of HSAG (a), HSAG-SP-T adduct (b), HSAG-SP-M adduct, (c) and SP (d).

withdrawing groups in alpha position with respect to the C=C bond of the pyrrole ring, as it could occur if the methyl group was oxidized to aldehyde, would favour the cycloaddition reaction.

The mechanical adduct HSAG-SP-M, with 13% by mass of SP (determined from TGA), was characterized by means of high-resolution transmission electron microscopy (HRTEM). As it has been reported [71], stable water suspensions (1 mg/mL) were prepared and samples were isolated from supernatant suspensions, after centrifugation for 10 min at 2000 rpm and for 5 and 60 min at 9000 rpm. HRTEM micrographs at lower and higher magnifications are shown in **Figure 4**.

Micrograph in **Figure 4a** reveals that the lateral size of HSAG-SP-M adduct is of few hundreds nanometers, that means of the same order of magnitude of pristine HSAG. Two comments can be made: the milling step does not appreciably affect the structure of the graphitic layers and SP promotes the dispersion in water of graphene layers with pretty large lateral size. Stacks of graphene layers (indicated in the boxes), disposed perpendicularly to the beam, are in **Figure 4b** and **c**. These figures show the most abundant stacks in populations that contained little larger of lower number of layers. In **Figure 4b**, stacks isolated after centrifugation for 10 min at 2000 rpm are of about 3.5–4.2 nm: they are made by about 10–12 stacked graphene layers. Stack in **Figure 4c**, isolated after centrifugation for 5 min at 9000 rpm, is of about 2.8 nm and is made by about eight graphene layers.

Stacks of graphene layers can be fractionated by means of centrifugation; such fractionation appears to be prevailingly due to the number of stacks, rather than to their lateral size. Thanks to the functionalization with SP, nano-stacks of HSAG can be isolated, with a number of graphene layers which depends on the adopted experimental conditions.

4.3. Adducts of SP with other graphitic substrates

Other synthetic graphites were used for preparing adducts with SP. Their main characteristics are shown in **Table 4** in Section 8. Adducts were prepared by giving thermal energy to the graphite / SP mixture and functionalization yield was estimated by means of TGA. Values are shown in **Table 3**.

These findings indicate that the formation of graphite-SP adducts is not due to the serendipitous combination of SP with HSAG, but to the affinity and/or reactivity of the pyrrole compound with the graphitic substrate. More in particular, it is worth commenting that, in spite of the different surface area, by using 0.1 as the molar ratio between SP and the graphitic substrate, essentially the same functionalization yield was obtained.

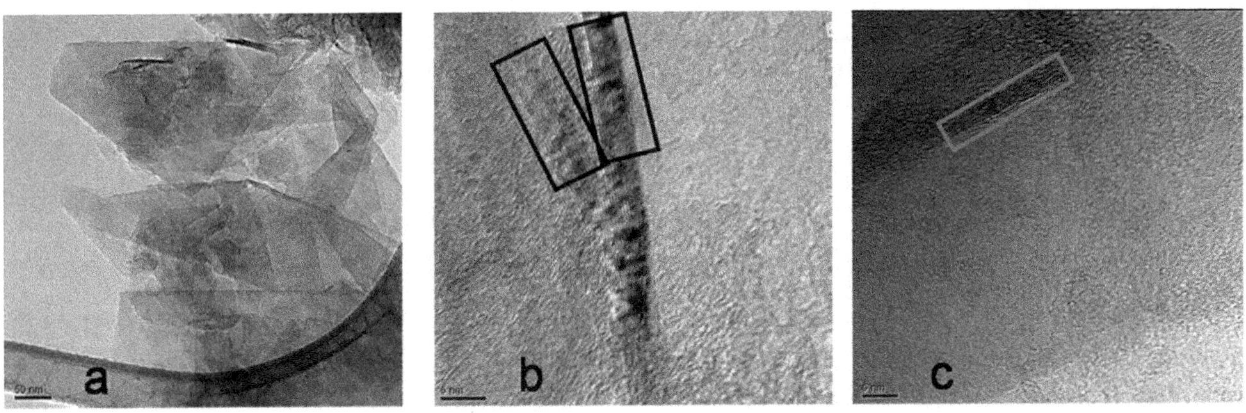

Figure 4. Micrographs of HSAG-SP-M adduct isolated from supernatant solutions, after centrifugation for 10 min at 2000 rpm (a, b) and 5 min at 9000 rpm (c). Micrographs are low magnification bright field TEM, target: 50 nm (a) and HRTEM image, target: 5 nm (b, c).

Type of graphite[a]	Nano 24	HSAG	Timrex SFG6	3807
Functionalization yield (%)	91	99	92	95

[a]Characteristics are collected in **Table 4** in Section 8.

Table 3. Graphite-SP thermal adducts: functionalization yield.

Code	Grade	C (%)	Ash (%)	Moisture (%)	Surface area (m^2/g)	Density (g/cm^3)
8427[a]	HSAG	99.8	n. a.	n. a.	330.3	n. a.
Nano 24[a]	HSAG	99.7	0.33	1.93	353.2	n. a.
3807[a]	Surface-enhanced flake graphite	98.9	n. a.	0.36	17.2	2.26
Timrex SFG6[b]	Expanded graphite	99.0	0.07	0.10	17.0	0.07

[a]Supplier: Asbury Carbon.
[b]Supplier: Imerys.

Table 4. Synthetic graphites used for the preparation of adducts with serinol pyrrole.

4.4. Adducts of SP with HSAG: what are they?

Results discussed in previous paragraphs allow commenting as follows. Serinol pyrrole and a graphitic substrate can establish a strong interaction and form very stable adducts. The functionalization of the graphitic substrate with SP leaves the bulk structure of graphene layers substantially unaltered.

The origin of such strength and stability could be attributed to the π-π interaction of the aromatic moieties, the pyrrole ring in SP and the C6 rings in the graphene layers. However, experimental indications suggest the possibility that SP is able to give rise to a chemical reaction with the aromatic layers of the substrate, more exactly to a cycloaddition reaction. More experimental data would be required to support this hypothesis. At present, rather than stretching too far intriguing inferences, it is worth examining applications that HSAG-SP adducts allow to pursue.

5. Adducts of SP and SP-based polymers with CB and CNT

This chapter is focused on the functionalization of graphene layers with a serinol derivative containing a pyrrole ring. In Section 4.3, it has been reported that the formation of stable graphite-SP adducts does not occur only by using a high-surface area nanosized graphite, such as HSAG:SP, but is indeed able to interact/react with different types of graphites.

Objective of the present paragraph is to demonstrate that SP is able to interact also with sp^2 carbon allotropes other than graphites. Adducts were formed by SP with carbon allotropes such as carbon black (CB N326) [87, 88] and multiwalled carbon nanotubes. Moreover, adducts with CNT were formed by polyurethane [69, 89] and polyethers [71, 89] oligomers containing SP as a comonomer.

CB-SP adduct [88] was prepared by using thermal energy, in the absence of solvents or catalysts, as described in Section 8. IR spectrum of the adduct, upon extraction, revealed the presence of peaks characteristic of SP. Water suspensions of CB-SP adducts were prepared and were first sonicated for 10 min and then centrifuged at 2000 rpm for 5 min. UV-Vis measurements were taken after each step of the procedure and related spectra are shown in **Figure 5a**.

UV-Vis absorption remained substantially the same after centrifugation. Stability of CB-SP and CB water suspensions were compared after centrifugation: suspensions are in the vials shown in **Figure 5b**. These results indicate that the treatment with SP confers a hydrophilic nature to CB, such as to allow the obtainment of stable water suspensions.

Polyurethanes (PU) oligomers (with molar mass up to about 11 × 10^3 g/mol) were prepared through the solvent-free polymerization of serinol pyrrole and 1,6-hexamethylene diisocyanate [69]. Adducts of such oligomers with multiwalled CNT were prepared. Suspensions in acetone of CNT-PU adducts (prepared by sonication, with 1 mg/mL as the concentration and 46% by mass of PU in the adduct) were stable even after centrifugation. HRTEM analysis of CNT-PU adduct revealed prevailingly disentangled CNTs, with intact skeleton, decorated by PU oligomers, tightly adhered to CNT surface. Micrograph of a CNT-PU adduct is shown in **Figure 6a**. Extraction tests were performed at room temperature, with ethyl acetate as the solvent, on adducts of CNT with PU oligomers with or without SP (in the latter case, 2,2-dimethyl-1,3-propanediol was the comonomer): mass loss was about 1 and 23% for the respective adducts. It is well known [90] that polyurethanes are very effective CNT modifiers, because they are able to establish very stable interaction, thanks to the π-π interaction between the carbonyl groups and the aromatic rings. It is thus worth mentioning that SP is able to enhance such interaction, revealing a sort of synergistic effect between different π systems.

Polyether (PE) oligomers (molar mass up to about 2600 g/mol) were prepared from the reaction of serinol pyrrole with 1,6-dibromo-hexane and their adducts with CNT were prepared by sonication (in acetone) [71]. Suspensions (1 mg/mL) were stable over months, even after centrifugation. HRTEM analysis revealed prevailingly disentangled CNTs, with intact skeleton, with PE oligomers wrapped on the CNT surface. Micrograph of a CNT-PE adduct is shown in **Figure 6b**. Extraction tests at room temperature with ethyl acetate were performed on CNT adducts with PE oligomers with or without SP as comonomer. Mass loss was thus about 24% for the adducts with SP based PE and 98% for the ones with a pluronic surfactant without SP.

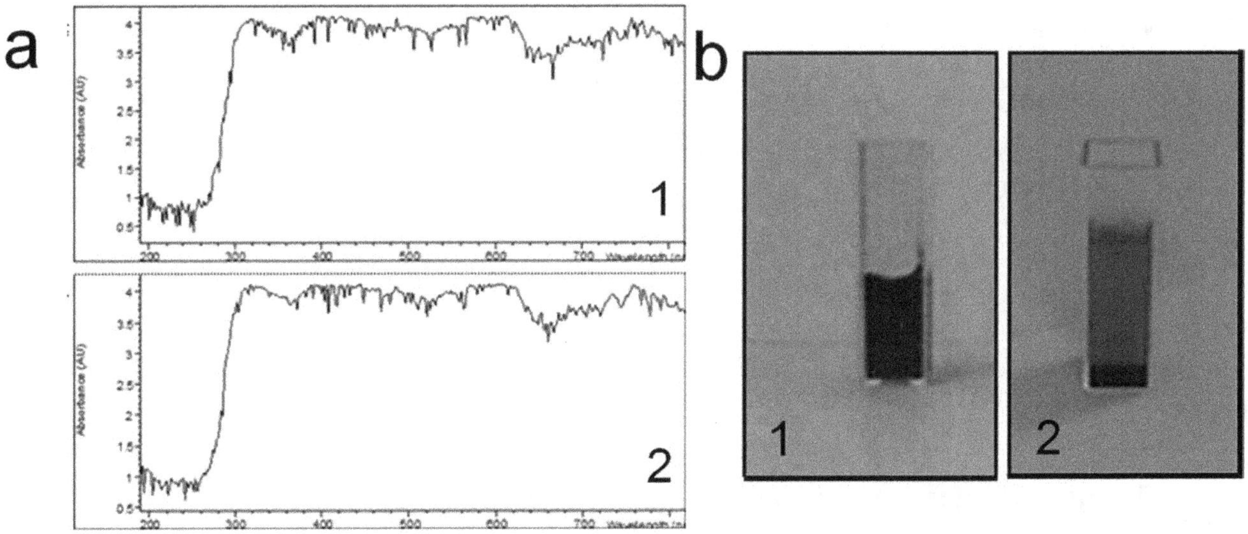

Figure 5. (a) UV-Vis spectra of water suspensions of CB-SP adduct: after sonication (1) and after centrifugation (2) and (b) water suspensions of CB-SP (1) and of CB (2), after centrifugation.

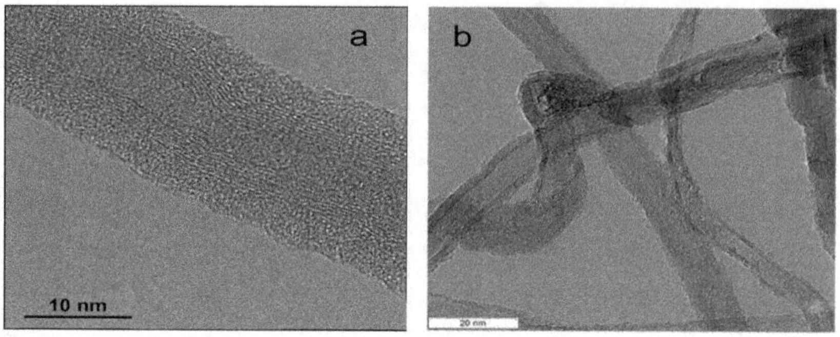

Figure 6. HRTEM micrographs of CNT-PU adduct (a), PE-CNT adduct (b). PU and PE contain serinol pyrrole as comonomer.

6. Applications of HSAG-SP adducts

As described in Section 4, HSAG and SP form stable adducts, since the aromatic moiety of the molecule is able to effectively interact with the condensed benzene rings of the graphene layers. Moreover, the hydroxy groups brought by SP promote the affinity of the adduct for polar environments. First objective was thus to prepare stable dispersions of HSAG-SP adducts in polar liquids, such as water.

6.1. Dispersions of HSAG-SP adducts in water

Water suspensions were prepared as described in Section 8, with HSAG and HSAG-SP-M adducts, with concentrations ranging from 0.1 to 30 g/L. In brief, the solid powder was suspended in water and ball milled (300 rpm, 2 hours). HSAG was observed to settle down, whereas stable suspensions were obtained with the adducts. Vials containing HSAG-SP water suspensions, with concentrations ranging from 0.10 to 10 g/L, are shown in **Figure 7**.

HSAG-SP suspensions were observed to follow the Lambert-Beer law, as it can be seen in the graph in **Figure 8a**. The UV-Vis spectra in **Figure 8b** confirm the stability of adducts' suspensions: high absorbance was recorded also upon centrifugation. It has been reported in Section 4.2 that fractionation of HSAG-SP adducts was obtained only after prolonged centrifugation, at least at 2000 rpm.

Suspensions with high HSAG-SP concentrations were prepared, up to 200 g/L. Suspensions with 10, 30 and 200 g/L concentrations are shown in **Figure 9**. Stability of the latter suspension was obtained by simply adding a small amount (1% w/w) of carboxymethylcellulose. Viscosity was observed to increase slightly, and the suspension was stable for more than a week.

6.2. Antistatic coating layers from HSAG-SP water suspensions

Water suspensions of HSAG-SP adducts are suitable for the preparation of coating layers with antistatic properties. The suspension at 200 g/L was indeed used to prepare coating layers on paper, with the help of a resin to ensure the obtainment of a continuous layer and good grip with the substrate. Ratio between the adduct and the resin ranged from 90/10 to 95/5. As it is shown in **Figure 9d**, the 90/10 layer was characterized by high homogeneity. Resistivity ranged from about 2 to about 4 kOhm/sq.

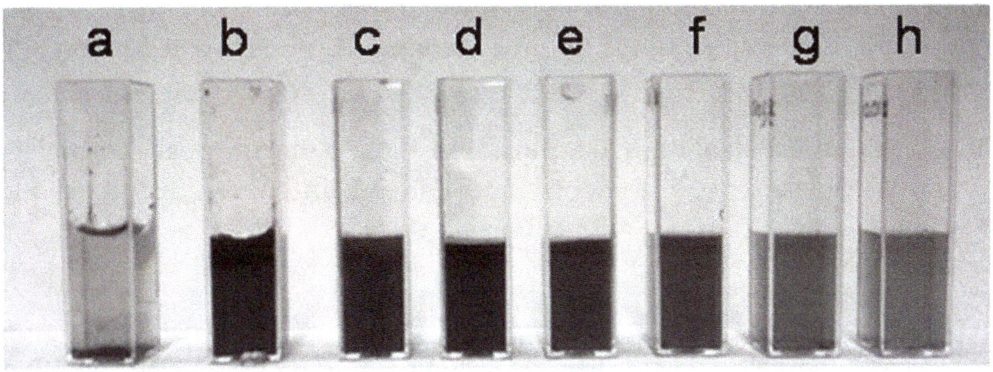

Figure 7. Water suspensions of HSAG (a) and HSAG-SP (b–h), with the following concentrations (g/L): 10 (b), 5.0 (c), 2.5 (d), 1.0 (e), 0.50 (f), 0.25 (g) and 0.10 (h).

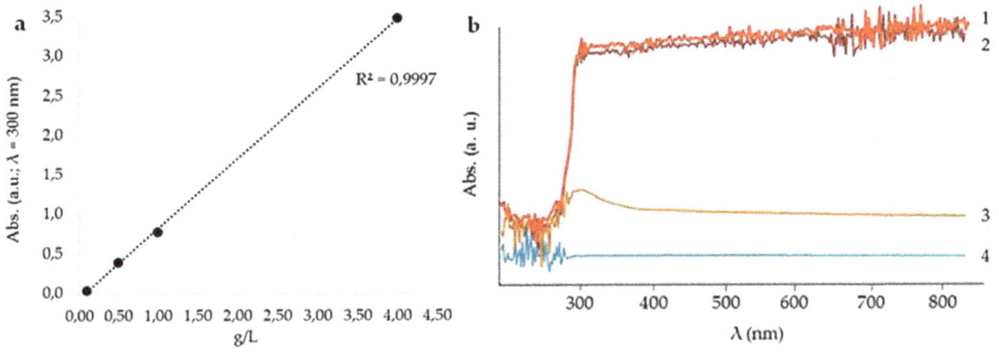

Figure 8. Dependence of UV-Vis absorbance of HSAG-SP-M (1 mg/mL) water suspensions on concentration (g/L) (a) and on storage, sonication and centrifugation (9000 rpm) (b): after sonication (1), after storage at rest for 1 week (2), after centrifugation at 9000 rpm for 10 min (3). UV-Vis spectrum of HSAG water suspension after centrifugation at 9000 rpm for 10 min as a reference (4).

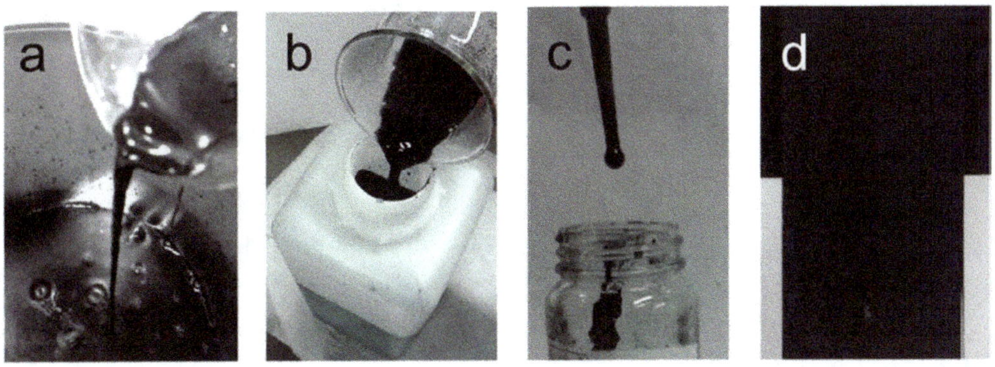

Figure 9. Aqueous suspensions of HSAG-SP with the following concentrations: (a) 10 g/L, (b) 30 g/L and (c) 200 g/L. Coating layer of HSAG-SP adduct on paper substrate (see text) (d).

6.3. Nanocomposites with few-layer graphene in natural rubber as the matrix

Poly(1,4-cis-isoprene) from *hevea brasiliensis* is known as natural rubber (NR) and is the most important rubber, accounting for about 60% of the global rubber market, with a worldwide production of more than 12 million tons per year [91, 92]. It is well known that NR comes from the tree in a latex. The chance of preparing stable water dispersions of HSAG was exploited to obtain homogeneous dispersions of HSAG in NR matrix. Details are in Section 8. Procedure is summarized in **Scheme 4**.

In **Figure 10**, TEM micrograph at high magnification of the nanocomposite formed by NR and HSAG-SP is shown. Stacks of only a few graphene layers and also isolated graphene layers are visible.

Scheme 4. Procedure for the preparation of NR/HSAG-SP nanocomposite from NR latex.

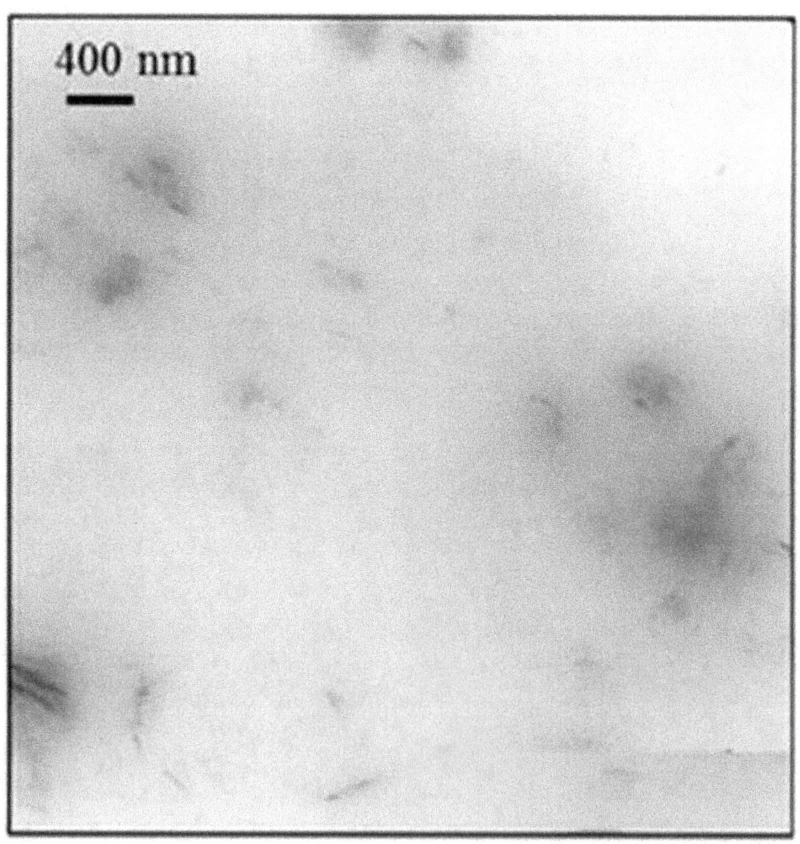

Figure 10. TEM micrograph of NR/HSAG-SP nanocomposite.

Latex pre-mixing of nanographite has been indicated [93–95] as the best procedure to obtain highly homogeneous distribution of nanoparticles. Graphene layers are placed around NR globules and form interconnected networks. Low electrical percolation threshold is achieved with low nanographite contents. It is well known that rubber composites are prepared through melt blending that should not only make the latex mixing step redundant, but should also destroy the interconnected network. However, even after melt blending, more homogeneous nanographite dispersion is documented in nanocomposites obtained first via latex blending. In the mentioned references, graphitic layers come from chemically reduced graphite oxide [93], thermally exfoliated graphite oxide [94] and expanded graphite [95] that means from graphites that require relevant treatments.

6.4. Dispersions of HSAG-SP adducts in polyols

Dispersions of HSAG-SP were as well prepared in polyols, as described in Section 8. The goal was to obtain highly homogeneous stable dispersion of graphene layers in a polyol, which means in a precursor of polyurethanes (PUs). PUs are important polymers that account for about 5% of worldwide polymer production and experience an increasing market penetration. Main drawback is their insufficient fire behaviour. Graphite layers are known to impart intumescent properties to polymer materials. However, due to their poor affinity for the polymeric matrices, they can be hardly dispersed in polyols and have been observed to settle down in short times. In the research reported here, three types of graphites (see **Table 4** in Section 8) were used and stable dispersions were obtained, with concentrations from 2 to 10, as mass%. For example, the suspension with 2% by mass as HSAG-SP concentration (SP content: 23.5% by mass) was analysed via UV-Vis spectrophotometry over a month period, observing stable absorption. It is worth commenting that unmodified graphites clearly settled down even after few days. Moreover, the same HSAG-SP suspension was centrifuged at 3000 rpm for 45 min. The supernatant suspension revealed the same UV absorption. In **Figure 11**, 2% by mass suspensions are shown. They were easily prepared (see in Section 8) and are suitable for large-scale productions.

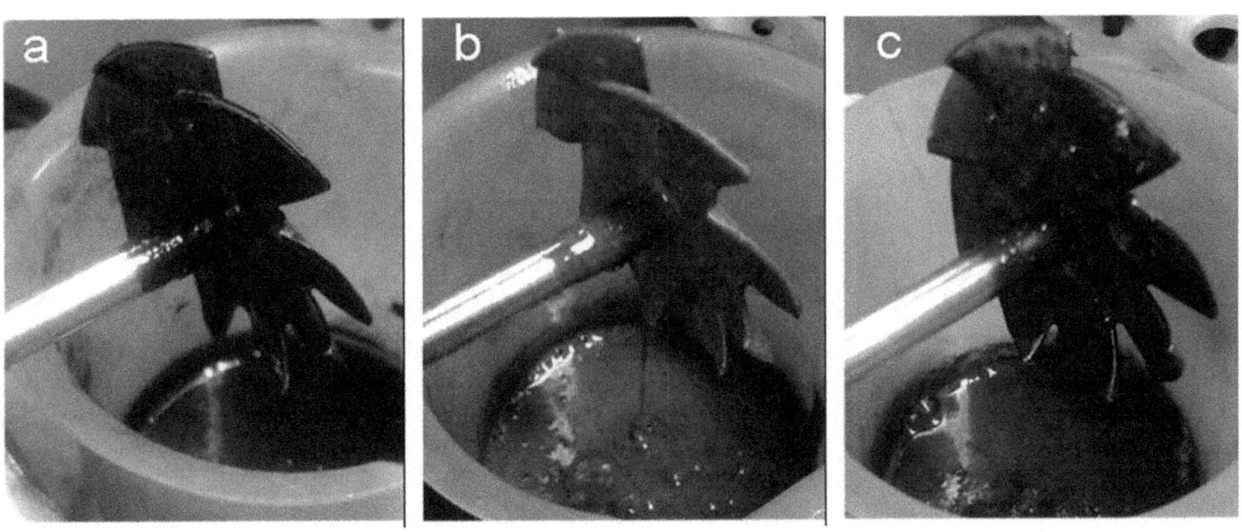

Figure 11. Suspensions in polyol of HSAG Nano 24 (a), flake graphite 3807 (b) and expanded graphite Timrex SFG6 (c).

7. Conclusions

Functionalization of graphene layers was performed with a serinol derivative containing a pyrrole ring: 2-(2,5-dimethyl-1H-pyrrol-1-yl)-1,3-propanediol, named as serinol pyrrole.

Synthesis of SP was characterized by very high atom economy and yield and thus by a high atom efficiency, close to 90%. A high surface area nanosized graphite was preferentially used as the graphitic substrate, but the reaction with SP was successful with other types of graphite and of sp^2 carbon allotropes, such as carbon black and carbon nanotubes.

Reaction was performed by simply mixing SP and the carbon allotrope and giving either mechanical or thermal energy.

SP and the graphitic substrate revealed strong interaction and formed very stable adducts. Such stability could be explained with the π-π interaction of the aromatic moieties, the pyrrole ring in SP and the C6 rings in the graphene layers. However, experimental indications seem to support the occurring of a chemical reaction between SP and the aromatic layers of the substrate, more exactly a cycloaddition reaction.

By centrifuging water suspensions of HSAG-SP adducts, stacks of few layers graphene were isolated. The procedure based on the preparation of water suspensions of HSAG-SP adducts and the subsequent centrifugation appears to be a promising, environmentally friendly method for obtaining layers of graphene.

HSAG-SP adducts allow to envisage a large variety of applications. They form very stable suspensions in water and polyols. Antistatic coating layers have been already prepared and work for the preparation of polyurethanes is in progress. Very homogeneous dispersion has been obtained for graphene layers in natural rubber matrix.

8. Experimental

8.1. Materials

Four grades of graphite have been modified and used in the course of our studies. Their features are summarized in **Table 4**.

Other reagents, described in detail in cited references, were purchased and used without further purifications.

8.2. Synthesis of 2-(2,5-dimethyl-1H-pyrrol-1-yl)-1,3-propanediol (serinol pyrrole, SP)

A mixture of hexan-2,5-dione (41.40 g; 0.36 mol) and serinol (30.00 g; 0.33 mol) was poured into a 100 mL round bottomed flask equipped with a magnetic stirrer. The mixture was then stirred, at room temperature, for 6 hours. The yield of the resulting intermediate compound 4a,6a-dimethyl-hexahydro-1,4-dioxa-6b-azacyclopenta[cd]pentalene was estimated to be 99%. Then the mixture was kept under vacuum for 2 hours and subsequently heated to 180°C for 50 min. After distillation under reduced pressure at 130°C and 0.1 mbar, serinol pyrrole (indicated as SP from now on) was isolated as yellow oil with 96% yield [68–70].

8.3. Synthesis of SP on high surface area graphite as the carbon allotrope

A total of 0.20 g of HSAG, 0.08 g of serinol (0.922 mmol) and 0.11 g of 2,5-hexanedione (0.922 mmol) were put in a 50 mL round bottom flask, which was also equipped with a magnetic stirrer. The reaction mixture was left to stir at 150°C for 120 min, then it was cooled to room temperature. A sample of black powder was put in a test tube and suspended with deuterated water (D_2O); the so obtained heterogeneous mixture was filtered through a PTFE 0.2 μm filter. The liquid passing through the filter was analysed by Proton Nuclear Magnetic Resonance spectroscopy (^{1}H-NMR). NMR spectrum revealed signals that could be attributed only to SP.

8.4. Preparation of HSAG-SP adducts

HSAG-SP adducts were prepared by reacting HSAG and SP with the help of either mechanical (HSAG-SP-M) or thermal (HSAG-SP-T) energy.

8.4.1. HSAG-SP-M adduct

HSAG (0.05 g, 66 mmol) and acetone (15 mL) were put in sequence in a 100 mL round bottom flask. The suspension was sonicated for 15 min, using a 2 L ultrasonic bath. After this time, a solution of SP (5.87 g, 66 mmol) in acetone (15 mL) was added and the mixture was sonicated for 15 min. The solvent was then removed under reduced pressure and the graphite/SP black powder was treated using a planetary ball mill whose grinding jar was loaded with six ceramic balls (diameter = 20 mm). After 6 hours of rotation at a rate of 300 rpm, the mixture was thoroughly washed with distilled water (6 × 100 mL) on a funnel with a sintered glass filter and dried in air overnight. A total of 9.80 g of black powder was recovered [70].

8.4.2. HSAG-SP-T adduct

HSAG (5.00 g, 66 mmol), acetone (15 mL) and a solution of SP (1.12 g, 6.6 mmol) in acetone (5 mL) were put in a 100 mL round bottom flask and sonicated as described in Section 8.4.1. Then, after removing the solvent under reduced pressure, the flask was equipped with a magnetic stirrer and the mixture was heated at 130°C for 6 hours under vigorous stirring. After cooling at room temperature, the mixture was repeatedly washed with distilled water (3 × 20 mL) on a funnel with a sintered glass filter and finally dried in air overnight. A total of 5.86 g of black powder was obtained.

8.5. Extraction of HSAG-SP adducts

HSAG-SP adducts were washed using a Soxhlet extractor. In a typical procedure, 3.00 g of black powder was put inside a thimble, which was loaded into the main chamber of the Soxhlet extractor. Acetone was used as the extraction solvent and was placed in a distillation flask. The flask was heated at 70°C, and the extraction was continually performed for 12 hours.

After this time, the thimble was removed from the main chamber of the Soxhlet extractor; the powder was recovered and then dried in an oven at 70°C for some hours [70].

8.6. Preparation of water suspensions of HSAG-SP adducts, evaluation of their stability and microscopic characterization

Water suspensions of HSAG-SP adducts at different concentrations (1 mg/mL; 0.5 mg/mL; 0.3 mg/mL; 0.1 mg/mL) were prepared. Each solution was sonicated for 10 min using an ultrasonic bath (260 W) and subsequently UV-Vis absorption was measured. The solution (10 mL) of each sample was put in a Falcon™ 15 mL conical centrifuge tubes and centrifuged at: 2000 rpm for 10 min and at 9000 rpm for 10, 30, 60 and 90 min. After each centrifugation step, supernatant fluids were picked up, transferred into a cuvette and analysed. UV-Vis absorptions and DLS analysis were measured immediately after each centrifugation and after 1 week storage.

HRTEM investigations on adducts' samples, isolated from the supernatant solutions after centrifugation, were carried out with a Philips CM 200 field emission gun microscope operating at an accelerating voltage of 200 kV. Few drops of the aqueous solutions were deposited on 200 mesh lacey carbon-coated copper grid and air-dried for several hours before analysis. During acquisition of HRTEM images, the samples did not undergo structural transformation. Low-beam current densities and short acquisition times were adopted. To estimate the number of stacked graphene layers and the dimensions of the stacks visible in HRTEM micrographs, the Gatan Digital Micrograph software was used.

8.7. Preparation of polyol suspensions of HSAG-SP adducts

The adducts of SP with Nano 24, 3807 and Timrex SFG6 (SP: graphite = 0.1 molar ratio) were prepared following the procedure described in Section 8.4.2; they were subsequently purified by suspending them in acetone (15 mL every gram of adduct), letting the suspension stir overnight and filtering it over a funnel with a sintered glass filter. The powder was subsequently washed several times with acetone until colourless washings and dried in air overnight.

Samples at 2, 5 and 10% (w/w) concentrations of the three SP-graphite adducts in polyol (Aropol 424™, supplier Pozzi Arosio SAS) were obtained by slowly pouring the chosen grey-black dry powder into a vessel containing the viscous liquid while vigorously stirring with a laboratory rod stirrer. After 30 min, the mixture was recovered and stored in PET bottles.

8.8. Preparation of CB-SP adducts

Carbon black (1.00 g) and acetone (15 mL) were transferred to a 100 mL round bottom flask and the obtained suspension was sonicated for 15 min, using a 2 L ultrasonic bath. After this time, a solution of SP (0.24 g) in acetone (15 mL) was added. The mixture was sonicated for 15 min. After removing the solvent under reduced pressure, a black powder of SP absorbed carbon black was obtained; 0.70 g of this powder were placed in a vial equipped with a magnetic stirrer. The reaction mixture was heated at 180°C for 2 hours and then let to cool at room temperature; the product was thoroughly washed with distilled water (3 × 100 mL) on a funnel with a sintered glass filter and finally dried in air overnight.

Author details

Maurizio Galimberti*, Vincenzina Barbera and Annalisa Sironi

*Address all correspondence to: maurizio.galimberti@polimi.it

Department of Chemistry, Materials and Chemical Engineering "G. Natta", Polytechnic University of Milan, Milan, Italy

References

[1] Novoselov K. S., Geim A. K., Morozov S. V., Jiang D., Zhang Y., Dubonos S. V., Grigorieva I. V., Firsov A. A. (2004) Electric field effect in atomically thin carbon films. Science 306: 666–669.

[2] Geim A. K., MacDonald A. H. (2007) Graphene: exploring carbon flatland. Phys. Today 60: 35–42.

[3] Geim A. K., Novoselov K. S. (2007) The rise of graphene. Nat. Mater. 6: 183–191.

[4] Zhang Y., Tan J. W., Stormer H. L., Kim P. (2005) Experimental observation of the quantum hall effect and Berry's phase in graphene. Nature 438: 201–204.

[5] Novoselov K. S., Geim A. K., Morozov S. V., Jiang D., Katsnelson M. I., Grigorieva I. V., Dubonos S. V., Firsov A. A. (2005) Two-dimensional gas of massless Dirac fermions in graphene. Nature 438 (7065): 197–200.

[6] Balandin A. A. (2011) Thermal properties of graphene and nanostructured carbon materials. Nat. Mater. 10: 569–581.

[7] Chen S., Wu Q., Mishra C., Kang J., Zhang H., Cho K., Cai W., Balandin A. A., Ruoff R. S. (2012) Thermal conductivity of isotopically modified graphene. Nat. Mater. 11: 203–207.

[8] Thostenson E. T., Li C. Y., Chou T. W. (2005) Nanocomposites in context. Comp. Sci. Technol. 65: 491–516.

[9] Stankovich S., Dikin D. A., Dommett G. H. B., Kohlhaas K. M., Zimney E. J., Stach E. A., Piner R. D., Nguyen S. T., Ruoff R. S. (2006) Graphene-based composite materials. Nature. 442: 282–286.

[10] Singh V., Joung D., Zhai L., Das S., Khondaker S. I., Seal S. (2011) Graphene based materials: past, present and future. Progr. Mater. Sci. 56: 1178–1271.

[11] Wang X., Zhi J., Mullen K. (2008) Transparent, conductive graphene electrodes for dye-sensitized solar cells. Nano Lett. 8: 323–327.

[12] Pushparaj V. L., Shaijumon M. M., Kumar A., Murugesan S., Ci L., Vajtai R., Linhardt R. J., Nalamasu O., Ajayan P. M. (2007) Flexible energy storage devices based on nanocomposite paper. Proc. Natl. Acad. Sci. USA. 104 (34): 13574–13577.

[13] Waje M. M., Wang X., Li W., Yan Y. (2005) Deposition of platinum nanoparticles on organic functionalized carbon nanotubes grown in situ on carbon paper for fuel cells. Nanotechnology 16 (7): 395–400.

[14] Yagoh H., Murayama H., Suzuki T., Tominaga Y., Shibuya N., Masuda Y. (2006) Simultaneous monitoring method of polycyclic aromatic hydrocarbons and persistent organic pollutants in the atmosphere using activated carbon fiber filter paper. Anal. Sci. 22 (4) 583–590.

[15] Zheng F., Baldwin D. L., Fifield L. S., Anheier Jr N. C., Aardahl C. L., Grate J.W. (2006) Single-walled carbon nanotube paper as a sorbent for organic vapor preconcentration. Anal. Chem. 78 (7): 2442–2446.

[16] Coleman J. N., Blau W. J., Dalton A. B., Muñoz E., Collins S., Kim B. G., Razal J., Selvidge M., Vieiro G., Baughman R. H. (2003) Improving the mechanical properties of single-walled carbon nanotube sheets by intercalation of polymeric adhesives. Appl. Phys. Lett. 82 (11): 1682–1684.

[17] Navalon S., Dhakshinamoorthy A., Alvaro M., Garcia H. (2014) Carbocatalysis by graphene-based materials. Chem. Rev. 114: 6179–6212.

[18] Chua C. K., Pumera M. (2015) Carbocatalysis: the state of "metal-free" catalysis. Chem. Eur. J. 21: 12550–12562.

[19] Dreyer D. R., Bielawski C. W. (2011) Carbocatalysis: heterogeneous carbons finding utility in synthetic chemistry. Chem. Sci. 2: 1233–1240.

[20] Hernandez Y., Pang S., Feng X., Müllen K. (2012) Graphene and Its Synthesis. Polymer Science: A Comprehensive Reference 8: 415-438, Eds. Matyjaszewski K., Möller M.; Elsevier B. V.

[21] Randviir E. P., Brownson D. A. C., Banks C. E. (2014) A decade of graphene research: production, applications and outlook. Mater. Today 17 (9): 426–432.

[22] Coleman J. N., Lotya M., O'Neill A., Bergin S. D., King P. J., Khan U., Young K., Gaucher A., De S., Smith R. J., Shvets I. V., Arora S. K., Stanton G., Kim H. Y., Lee K., Kim G. T., Duesberg G. S., Hallam T., Boland J. J., Wang J. J., Donegan J. F., Grunlan J. C., Moriarty G., Shmeliov A., Nicholls R. J., Perkins J. M., Grieveson E. M., Theuwissen K., McComb D. W., Nellist P. D., Nicolosi V. (2011) Two-dimensional nanosheets produced by liquid exfoliation of layered materials. Science 331: 568–571.

[23] Plummer J. (2015) Graphene synthesis: molten bed. Nat. Mater. 14: 1186–1186.

[24] Eigler S., Grimm S., Hof F., Hirsch A. (2013) Graphene oxide: a stable carbon framework for functionalization. J. Mater. Chem. A 1: 11559–11562.

[25] Dreyer D. R., Park S., Bielawski C. W., Ruoff R. S. (2010) The chemistry of graphene oxide. Chem. Soc. Rev. 39: 228–240.

[26] Collins W. R., Lewandowski W., Schmois E., Walish J., Swager T. M. (2011) Claisen rearrangement of graphite oxide: a route to covalently functionalized graphenes. Angew. Chem. Int. Ed. 50: 8848–8852.

[27] Sarkar S., Bekyarova E., Haddon R. C. (2012) Chemistry at the Dirac point: Diels-Alder reactivity of graphene. Acc. Chem. Res. 45 (4): 673–682.

[28] Swager T. M. (2012) Functional graphene: top-down chemistry of the π-surface. ACS Macro. Lett. 1: 3–5.

[29] Zaman I., Kuan H.-C., Meng Q., Michelmore A., Kawashima N., Pitt T., Zhang L., Gouda S., Luong L., Ma J. (2012) A facile approach to chemically modified graphene and its polymer nanocomposites. Adv. Funct. Mater. 22 (13): 2735–2743.

[30] Huang X., Qi X., Boeya F., Zhang H. (2012) Graphene-based composites. Chem. Soc. Rev. 41: 666–686.

[31] Chua C. K., Pumera M. (2013) Covalent chemistry on graphene. Chem. Soc. Rev. 42: 3222–3233.

[32] Englert J. M., Dotzer C., Yang G., Schmid M., Papp C., Gottfried J. M., Steinrück H.-P., Spiecker E., Hauke F., Hirsch A. (2011) Covalent bulk functionalization of graphene. Nat. Chem. 3: 279–286.

[33] Toth P. S., Ramasse Q. M., Velický M., Dryfe R. A. W. (2015) Functionalization of graphene at the organic/water interface. Chem. Sci. 6: 1316–1323.

[34] Brodie B. C. (1859) On the atomic weight of graphite. Philos. Trans. R. Soc. Lond. 14: 249–259.

[35] Lopez M. I., Croce A. E., Sicre J. E. (1994) Explosive decomposition of gaseous chlorine dioxide. J. Chem. Soc. Faraday Trans. 90 (22): 3391–3396.

[36] Poh H. L., Sanek F., Ambrosi A., Zhao G., Sofer Z., Pumera M. (2012) Graphenes prepared by Staudenmaier, Hofmann and Hummers methods with consequent thermal exfoliation exhibit very different electrochemical properties. Nanoscale. 4 (11): 3515–3522.

[37] Staudenmaier L. (1898) Process for the preparation of the graphitic acid. Ber. Dtsch. Chem. Ges. 31: 1481-1487.

[38] Hummers W. S., Offeman R. E. (1958) Preparation of graphitic oxide. J. Am. Chem. Soc. 80: 1339–1339.

[39] Ang P. K., Wang S., Bao Q., Thong J. T. L., Loh K. P. (2009) High-throughput synthesis of graphene by intercalation–exfoliation of graphite oxide and study of ionic screening in graphene transistor. ACS Nano 3 (11): 3587–3594.

[40] Marcano D. C., Kosynkin D. V., Berlin J. M., Sinitskii A., Sun Z., Slesarev A., Alemany L. B., Lu W., Tour J. M. (2010) Improved synthesis of graphene. ACS Nano 4 (8): 4806–4814.

[41] Chen J., Yao B., Li C., Shi G. (2013) An improved Hummers method for eco-friendly synthesis of graphene oxide. Carbon 64: 225–229.

[42] Chen J., Li Y., Huang L., Li C., Shi G. (2015) High-yield preparation of graphene oxide from small graphite flakes via an improved Hummers method with a simple purification process. Carbon 81: 826–834.

[43] Guerrero-Contreras J., Caballero-Briones F. (2015) Graphene oxide powders with different oxidation degree, prepared by synthesis variations of the Hummers method. Mater. Chem. Phys. 153: 209–220.

[44] Kovtyukhova N. I., Ollivier P. J., Martin B. R., Mallouk T. E., Chizhik S. A., Buzaneva E. V., Gorchinskiy A. D. (1999) Layer-by-layer assembly of ultrathin composite films from micron-sized graphite oxide sheets and polycations. Chem. Mater. 11 (3): 771–778.

[45] Posudievsky O. Yu., Khazieieva O. A., Koshechko V. G., Pokhodenko V. D. (2012) Preparation of graphene oxide by solvent-free mechanochemical oxidation of graphite. J. Mater. Chem. 22: 12465–12467.

[46] Galimberti M., Cipolletti V., Mauro M., Conzatti L. (2013) Nanocomposites of Poly(1,4-cis-Isoprene) with graphite oxide intercalation compounds. Macromol. Chem. Phys. 214 (17): 1931–1939.

[47] Boehm H. P., Clauss A., Fischer G. O., Hofmann U. (1962) Dünnste Kohlenstoff-Folien. Z. Naturforsch Pt. B. 17 (3): 150–153.

[48] Eda G., Fanchini G., Chhowalla M. (2008) Large-area ultrathin films of reduced graphene oxide as a transparent and flexible electronic material. Nat. Nanotechnol. 3 (5): 270–274.

[49] Gomez-Navarro C., Weitz R. T., Bittner A. M., Scolari M., Mews A., Burghard M., Kern K. (2007) Electronic transport properties of individual chemically reduced graphene oxide sheets. Nano Lett. 7 (11): 3499–3503.

[50] Stankovich S., Dikin D. A., Piner R. D., Kohlhaas K. A., Kleinhammes A., Jia Y., Wu Y., Nguyen S. T., Ruoff R. S. (2007) Synthesis of graphene-based nanosheets via chemical reduction of exfoliated graphite oxide. Carbon 45: 1558–1565.

[51] Li D., Muller M. B., Gilje S., Kaner R. B., Wallace G. G. (2008) Processable aqueous dispersions of graphene nanosheets. Nat. Nano. 3 (2):101–105.

[52] Fan X. B., Peng W., Li Y., Li X., Wang S., Zhang G., Zhang F. (2008) Deoxygenation of exfoliated graphite oxide under alkaline conditions: a green route to graphene preparation. Adv. Mater. 20 (23): 4490–4493.

[53] Abdolhosseinzadeh S., Asgharzadeh H., Kimb H. S. (2015) Fast and fully-scalable synthesis of reduced graphene oxide. Sci. Rep. 5: 10160–10166.

[54] You F., Wang D., Cao J., Li X., Dang Z.-M., Hu G.-H. (2014) In situ thermal reduction of graphene oxide in a styrene–ethylene/butylene–styrene triblock copolymer via melt blending. Polym. Int. 63: 93–99.

[55] Cote L. J., Cruz-Silva R., Huang J. (2009) Flash reduction and patterning of graphite oxide and its polymer composite. J. Am. Chem. Soc. 131: 11027–11032.

[56] Mauro M., Cipolletti V., Galimberti M., Longo P., Guerra G. (2012) Chemically reduced graphite oxide with improved shape anisotropy. J. Phys. Chem. C 116 (46): 24809–24813.

[57] Schniepp H. C., Li J.-L., McAllister M. J., Sai H., Herrera-Alonso M., Adamson D. H., Prud'homme R. K., Car R., A. Saville D. A., Aksay I. A. (2006) Functionalized single graphene sheets derived from splitting graphite oxide. J. Phys. Chem. B 110 (17): 8535–8539.

[58] He H., Klinowski J., Forster M., Lerf A. (1998) Structure of graphite oxide revisited. Chem. Phys. Lett. 287 (1–2): 53–56.

[59] Zhu Y., Murali S., Cai W., Li X., Suk J. W., Potts J. R., Ruoff R. S. (2010) Graphene and graphene oxide: synthesis, properties, and applications. Adv. Mater. 22: 3906–3924.

[60] Andreeßen B. A. S. (2011) Serinol: small molecule - big impact. AMB Express 1: 1–6.

[61] European Parliament (2013) Technology options for feeding 10 billion people. Science and Technology Options Assessment.

[62] Pagliaro M., Ciriminna R., Kimura H., Rossi M., Della Pina C. (2007) From glycerol to value added products. Angew. Chem. Int. Ed. 46: 4434–4440.

[63] Zhou C.-H., Beltramini J. N., Fana Y.-X., Lu G. Q. (2008) Chemoselective catalytic conversion of glycerol as a biorenewable source to valuable commodity chemicals. Chem. Soc. Rev. 37 (3): 527–549.

[64] Yang F., Hanna M. A., Sun R. (2012) Value-added uses for crude glycerol – a by-product of biodiesel production. Biotechnol. Biofuels 5: 13.

[65] Jérôme F., Pouilloux Y., Barrault J. (2008) Rational design of solid catalysts for the selective use of glycerol as a natural organic building block. Chem. Sus. Chem. 1 (7): 586–613.

[66] Knorr L. (1885) Effect of diethyl succinate on ammonia and primary amine bases. Chem. Ber. 18: 299.

[67] Paal C. (1885) Synthesis of thiophene and pyrrole derivatives. Chem. Ber. 18: 367.

[68] Barbera V., Citterio A., Galimberti M., Leonardi G., Sebastiano R., Shisodia S. U., Valerio A. M. (2015) Process for the synthesis of 2-(2,5-dimethyl-1h-pyrrol-1-yl)-1,3-propanediol and its substituted derivatives. WO2015189411 A1.

[69] Galimberti M., Barbera V., Citterio A., Sebastiano R., Truscello A., Valerio A. M., Conzatti L., Mendichi R. (2015) Supramolecular interactions of carbon nanotubes with biosourced polyurethanes from 2-(2,5-dimethyl-1H-pyrrol-1-yl)-1,3-propanediol. Polymer 63: 62–70.

[70] Galimberti M., Barbera V., Guerra S., Conzatti L., Castiglioni C., Brambilla L., Serafini A. (2015) Biobased janus molecule for the facile preparation of water solutions of few layer graphene sheets. RSC Adv. 5: 81142–81152.

[71] Barbera V., Musto S., Citterio A., Conzatti L., Galimberti M. (2016) Polyether from a biobased Janus molecule as surfactant for carbon nanotubes. Express Polym. Lett. 10 (7): 548–555.

[72] Broadbent H. S., Burnham W. S., Sheeley R. M., Olsen R. K. (1976) Novel heterotricyclic systems: 2,6-dioxa- and 2-oxa-6-thia-10-azatricyclo-[5.2.1.$0^{4,10}$] decanes; 2,6-dioxa-11-azatricyclo[5.3.1.$0^{4,11}$] undecane; and 9,13-dioxa-14-azatetracyclo[6.5.1.$0^{2,7}$.$0^{11,14}$]tetradeca-2,4,6-triene. J. Heterocycl. Chem. 13 (2): 337–348.

[73] Casagrande C., Fabre P., Veyssie M., Raphael E. (1989) "Janus Beads": realization and behaviour at water/oil interfaces. Europhys. Lett. 9 (3): 251–255.

[74] de Gennes P. G. (1992) Soft matter (Nobel lecture). Angew. Chem. Int. Ed. 31 (7): 842–845.

[75] Li F., Josephson D. P., Stein A. (2011) Colloidal assembly: the road from particles to colloidal molecules and crystals. Angew. Chem. Int. Ed. 50 (2): 360–388.

[76] Mori H., Müller A. H. (2003) New polymeric architectures with (meth)acrylic acid segments. Progr. Polym. Sci. 28 (10): 1403–1439.

[77] Agnelli S., Cipolletti V., Musto S., Coombs M., Conzatti L., Pandini S., Riccò T., Galimberti M. (2014) Interactive effects between carbon allotrope fillers on the mechanical reinforcement of polyisoprene based nanocomposites. Expr. Polym. Lett. 8 (6): 436–449.

[78] Reich S., Thomsen C. (2004) Raman spectroscopy of graphite. Phil. Trans. Roy. Soc. Lond. A 362 (1824): 2271–2288.

[79] Castiglioni C., Tommasini M., Zerbi G. (2004) Raman spectroscopy of polyconjugated molecules and materials: confinement effect in one and two dimensions. Phil. Trans. Roy. Soc. Lond. A 362 (1824): 2425–2459.

[80] Ferrari A. C. (2007) Raman spectroscopy of graphene and graphite: disorder, electron–phonon coupling, doping and nonadiabatic effects. Sol. State Commun. 143: 47–57.

[81] Pimenta M. A., Dresselhaus G., Dresselhaus M. S., Cancado L. G., Jorio A., Saito R. (2007) Studying disorder in graphite-based systems by Raman spectroscopy. Phys. Chem. Chem. Phys. 9 (11): 1276–1290.

[82] Graf D., Molitor F., Ensslin K., Stampfer C., Jungen A., Hierold C., Wirtz L. (2007) Spatially resolved Raman spectroscopy of single- and few-layer graphene. Nano Lett. 7 (2), 238–242.

[83] Casiraghi C., Hartschuh A., Qian H., Piscanec S., Georgi C., Fasoli A., Novoselov K. S., Basko D. M., Ferrari A. C. (2009) Raman spectroscopy of graphene edges. Nano Lett. 9 (4): 1433–1441.

[84] Casiraghi C., Pisana S., Novoselov K. S., Geim A. K., Ferrari A. C. (2007) Raman fingerprint of charged impurities in graphene. Appl. Phys. Lett. 91 (23): 233108–233110.

[85] Tommasini M., Castiglioni C., Zerbi G., Barbon A., Brustolon M. (2011) A joint Raman and EPR spectroscopic study on ball-milled nanographites. Chem. Phys. Lett. 516 (4): 220–224.

[86] Amarnath C. A., Hong C. E., Kim N. H., Ku B. C., Kuila T., Lee J. H. (2011) Efficient synthesis of graphene sheets using pyrrole as a reducing agent. Carbon 49 (11): 3497–3502.

[87] Galimberti M., Barbera V., Citterio A., Sebastiano R., Valerio A. M., Leonardi G. (2015) Adducts between carbon allotropes and serinol derivatives. WO2015EP72641.

[88] Galimberti M., Barbera V., Cipolletti V., Guerra S., Citterio A. (October 14 2014) Innovative biosourced compatibilizers for carbon and white fillers in rubber compounds. *Presented at the Fall 186th Technical Meeting of the Rubber Division of the American Chemical Society, Inc.* Nashville (TN). 16.

[89] Galimberti M., Barbera V., Truscello A., Sebastiano R., Valerio A. M. (2016) Polymer comprising repeating units consisting of a substituted pyrrole ring and products obtained by combining said polymers with carbon allotropes. WO2016023915 A1.

[90] Bahnmueller S., Meyer H., Dijkstra D., Henning W., Koning C. E., Van S. Y. H., Mulders S. F. M. (2010) Polyurethane compounds having carbon nanotubes. WO2010102763 A1.

[91] Barlow F.W. (2001) Basic elastomer technology, chapter 9. 235-258. Eds. Rubber Division American Chemical Society, The University of Akron

[92] Natural Rubber Statistics 2016 [Internet]. Available from http://www.lgm.gov.my/nrstat/nrstats.pdf

[93] Potts J. R., Shankar O., Du L., Ruoff, R. S. (2012). Processing–morphology–property relationships and composite theory analysis of reduced graphene oxide/natural rubber nanocomposites. Macromolecules 45 (15): 6045–6055.

[94] Potts J. R., Shankar O., Murali S., Du L., Ruoff, R. S. (2013) Latex and two-roll mill processing of thermally-exfoliated graphite oxide/natural rubber nanocomposites. Compos. Sci. Technol. 74: 166–172.

[95] Luo Y., Zhao P., Yang Q., He D., Kong L., Peng Z. (2014) Fabrication of conductive elastic nanocomposites via framing intact interconnected graphene networks. Compos. Sci. Technol. 100: 143–151.

6

Tribological Aspects of Graphene-Aluminum Nanocomposites

Prashantha Kumar H.G. and Anthony Xavior M.

Abstract

Graphene is a new class of material in carbon group with strong sp^2—hybridized 2D-sheet like nanomaterial. In order to make an effective utilization of their astounding properties, they are currently used in the form of reinforcements in various proportions in metals and its alloys to fabricate the nanocomposites. Graphene is incorporated in oil and grease at nano range that results in higher load-carrying capacity compared with that of raw grease and oils without additives, which shows that graphene possesses self-lubricating capacity. Graphene is a planar sheet-like structure (2D), with more contact surface area in the developed composites that can make them suitable for industrial applications with well-established tribological performance. The novelty of this work focuses on the role of graphene addition in enhancing the wear performance aluminum composites to replace the conventional materials by graphene composite combinations. The current chapter explains the processing and tribological performance of graphene-aluminum composites and its effect with various hybrid combinations of MWCNT/SiC/Al_2O_3. Dispersion of graphene is carried out through ultrasonic liquid processor followed by ball-milling aluminum powder. Thus prepared precursors are vacuum-pressed and microwave-sintered. Graphene in the nanocomposites has resulted in significantly improving the tribological properties, where it gives the wear resistance by creating a solid, lubricant layer between the sliding surfaces.

Keywords: graphene, aluminum, nanocomposites, tribology

1. Introduction

Recent research advancements in the field of materials, development of strengthened materials without compromising weight factor, are very much fascinating. Due to evolution in the materials and processing methods, it is required to study the new class of materials and its unusual blends to improve the properties [1–3]. The significance of composites materials technology is the conglomeration of important features of base metals/matrix and the reinforcements. The technology enables achieving high-strength materials without losing the ductility and density, besides, overcoming the drawbacks of base materials by providing viability to add one or more high-strengthened particles and platelets. The necessity for mending the various properties and improved tribological performance will be governed by its applications in addition to various operating conditions [4, 5]. Most of the metal matrix composites are fabricated through ductile materials such as titanium, copper, aluminum, nickel, and its alloy. Such composite materials are extensively taking a role in miniature components of automobile, defense, and huge aerospace structural applications [6, 7]. Ceramic particulates such as silicon carbide, alumina, and boron nitride are the commonly employed reinforcements for the metal matrix composites (MMCs) [8, 9]. The particulate reinforcement trend gave rise to enhanced material properties under appropriate contents that are processed through advanced fabrication techniques compared with its monolithic [10]. Further, MMCs can be sorted out according to the type of reinforcements, namely whiskers (fibers), platelets, or particulates and its physical form with continuous and orientation configurations. In general, whiskers-reinforced and continuous fiber MMCs with advanced fabrication techniques become a promising cost-effective and light-weight engineering material from the past decades [11, 12]. In the recent years, the trend of research is on combinations of advanced materials and new forming methods to synthesis metal matrix composites (MMCs). However, substantial improvement of the aluminum and its alloy-based composites has conquered various manufacturing and industrial sectors. Further, these composites have been widely adapted in various aerospace and automobile industries because of light weight and high strength to weight ratio with improved thermal and electrical properties that are processed through the classical ingot (casting) route [13, 14].

Currently, demands for alternate materials are massive, mainly in various engineering sectors and military applications to facilitate higher payloads and more economical [15, 16]. In order to attain successful fabrication of this material with enriched properties, various permutations of new and advanced materials with cutting-edge processing method are to be adopted [17, 18]. Particulate technology (powder metallurgy) is one such feasible technology that provides the integration of various novel and new combination of reinforcements that can fuse through innovative processing method. The principal element in any composite material assimilating with nanoparticles comprises homogeneous distribution, type of boundary bond, and the interaction of nano reinforcement [19–21].

The materials' research field had been abstracted to find various carbon and its allotropes as reinforcements. Graphene being 2D-sheet like nanomaterial is endowed with extreme strength (tensile strength 130 GPa) in a carbon group that was coined by Andre Geim and Konstantin Novoselov experimentally during 2004 [22, 23]. Further, graphene is an elementary building block for several graphitic materials such as three-dimensional material graphite, single and multiwalled carbon nanotubes (CNTs, one-dimensional), fullerenes (C60, zero-dimensional) [24]. It is strong sp^2-hybridized 2D-nano scale material (sheet like morphology) that is well recognized for its extreme strength. Moreover, it got inspiring attention because of its exceptional physical and mechanical properties, which includes elastic modulus (up to 0.5–1 TPa); thermal conductivity (up to 5.3 × 103 $Wm^{-1} K^{-1}$); tensile strength (up to 130 GPa) in the field of research [25]. **Figure 1** shows a distinctive molecular model of single-atomic-layer graphene flake with many other graphitic forms. Carbon nanotubes (CNTs) also have emerged as materials with unique properties greater than those of any available conventional material [26]. Carbon nanotubes (defect-free), both multiwalled nanotubes (MWNTs) and single-wall nanotubes (SWNTs) have superior tensile strengths (~90 GPa) and elastic modulii (~0.6 TPa). Carbon nanotube (CNT), graphite, and fullerenes were used in base metals such as Al, Cu, Mg, and their alloys [27, 28]. Some of the results are found to have significant improvement in the mechanical properties on CNT-reinforced MMCs. But these composites subjecting to various challenges, including uniform distribution of CNTs in the matrix materials, lower surface contact area, and wettability, have been among the major concerns [29].

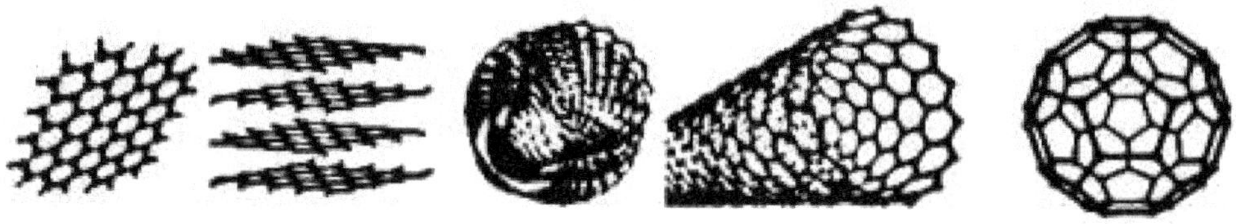

Figure 1. Structure of allotropes of carbon: (a) Graphene; (b) Graphite; (c) MWCNT; (d) SWCNT; (e) Fullerence.

Initially, graphene preparations are carried out by breaking the graphite into graphene through a liquid phase exfoliation or mechanical cleavage method. An other equally common method is chemical vapor deposition (CVD) [30]. Compared to other physical forms of carbon materials such as CNTs and graphite, graphene has been anticipated to outperform its competitors due to its distinctive properties and has great latent in developing the nanocomposites. So the requirement was limited to microscale aggregate without concerning about mechanical properties and strengths. But the excellent mechanical properties of carbon-based graphene are yet to be taken advantage of in metal and its alloy matrix composites [31, 32]. The challenges in synthesis and production of its alloy—graphene composites remain similar to that of ceramic-graphene system or polymer-graphene systems, which have poor dispersion, agglomeration, and individually exfoliated graphene in the matrix. An additional challenge for metal and its alloy—graphene composites could be the unusual reactions at the interfaces [33]. Aluminum-graphene and various aluminum/CNT/SiC/Al_2O_3 combination composites are fabricated through advanced processing techniques such as spark plasma sintering (SPS), mechanical alloying, powder metallurgy (PM), hot press and extrusion [34, 35], melt-blending [36], and hot isostatic pressing (HIP) [12]. But the important factor is the mechanism of various planar/tubular/spherical-shaped structured reinforcement bonding, specific surface area (SSA), and its proposed structural and tribological performances [37]. Further, hybrid combinations of graphene are combined with other reinforcements, and its effect on various mechanical and tribological performances of aluminum matrix is not addressed so far. The current chapter deals with the composite made out of aluminum reinforced with the graphene and its hybrid combinations with CNT/SiC/Al_2O_3 that are processed through ultrasonic liquid processor, ball milling, and consolidated through vacuum hot press followed by microwave sintering. Thus developed composites are tested for its tribological potential. Comparison and analysis are carried out to explore the effect of friction, wear, and structural damage on the added carbon allotropes during various testing conditions.

2. Processing and fabrication technique of graphene-aluminum composites

Figure 2 shows the morphology of aluminum, SiC, Al_2O_3, MWCNT, and graphene, which is characterized through scanning electron microscope. **Figure 2(a)** and **(b)** shows the TEM micrographs of graphene and MWCNT, respectively. Graphene is in flake form, single-layered morphology in which the layers are stacked upon each other. **Figure 2**(c–e) shows the aluminum, SiC, Al_2O_3, respectively. Aluminum powders that are produced through gas atomization method were procured commercially with a density of 2.7 g/cm^{-3}. Also, SiC particles with a density of 3.20 g/cm^{-3}, Al_2O_3 particles with a density of 3.8 g/cm^{-3}, CNT with a density of 1.6 g/cm^{-3} and graphene with average platelets (flake) size >10 μm, and a density of 1.5–2.0 g/cm^{-3} are considered for the fabrication as reinforcements. Aluminum-graphene nanocomposites with hybrid combinations are produced through a number of ways. At the first step, dispersion of graphene is carried out through an ultrasonic dispersion method with various weight percentages. SiC/CNT/Al_2O_3 was added separately to a solvent containing graphene that is dispersed in acetone. Dispersion (ultrasonication) is carried out until it turns to complete block solution indicating that no graphene sediments were left in the beaker also; all flakes are individually exfoliated. Further, graphene mixtures are dried in hot air oven and subjected to ball milling. Ball milling is carried out for 90 min with the ball mill ratio 16:1 at 200 rpm rotating speed for hybrid combination graphene weight fraction precursor. Encapsulation and homogeneous dispersion of graphene on SiC/Al_2O_3 and homogeneity of mixture are achieved at optimized processing parameters. **Figure 2**(f–h) shows the portion of the ball-milled graphene-SiC/Al_2O_3 powder mixtures that confirm the encapsulation of SiC and Al_2O_3, respectively. Apparent density of aluminum, SiC, Al_2O_3, MWCNT, and graphene is listed in **Table 1**, which are measured according to ASTM B703-10 using Arnold density meter, and tap density is carried out according to ASTM B527-14 by tapping apparatus giving 250 taps under dry conditions. Aluminum metal powder is added parallel to ball-milled hybrid-graphene precursors, and ball-milling process is continued with the same milling ratio for another 30 min. After this, the precursor (mixture of graphene and its hybrid combinations) is dried in hot air oven for 24 hours. Further, precursors are consolidated in vacuum hot press at 240 N/mm^2 in chromium hot-work tool steel (40–48 HRC die case, at 480°C). Boron nitride spray (Momentive Performance Materials Inc. USA) type lubrication is used as a lubricating agent between the precursor and the die walls. Thus prepared compacts are (each comprising two sets) subjected to microwave sintering at 510°C in a nitrogen inert gas (flushing) atmosphere for 30 min followed by furnace cooling to attain room temperature. The process followed for the synthesis of aluminum-graphene and its hybrid combinations is illustrated in **Figure 3**.

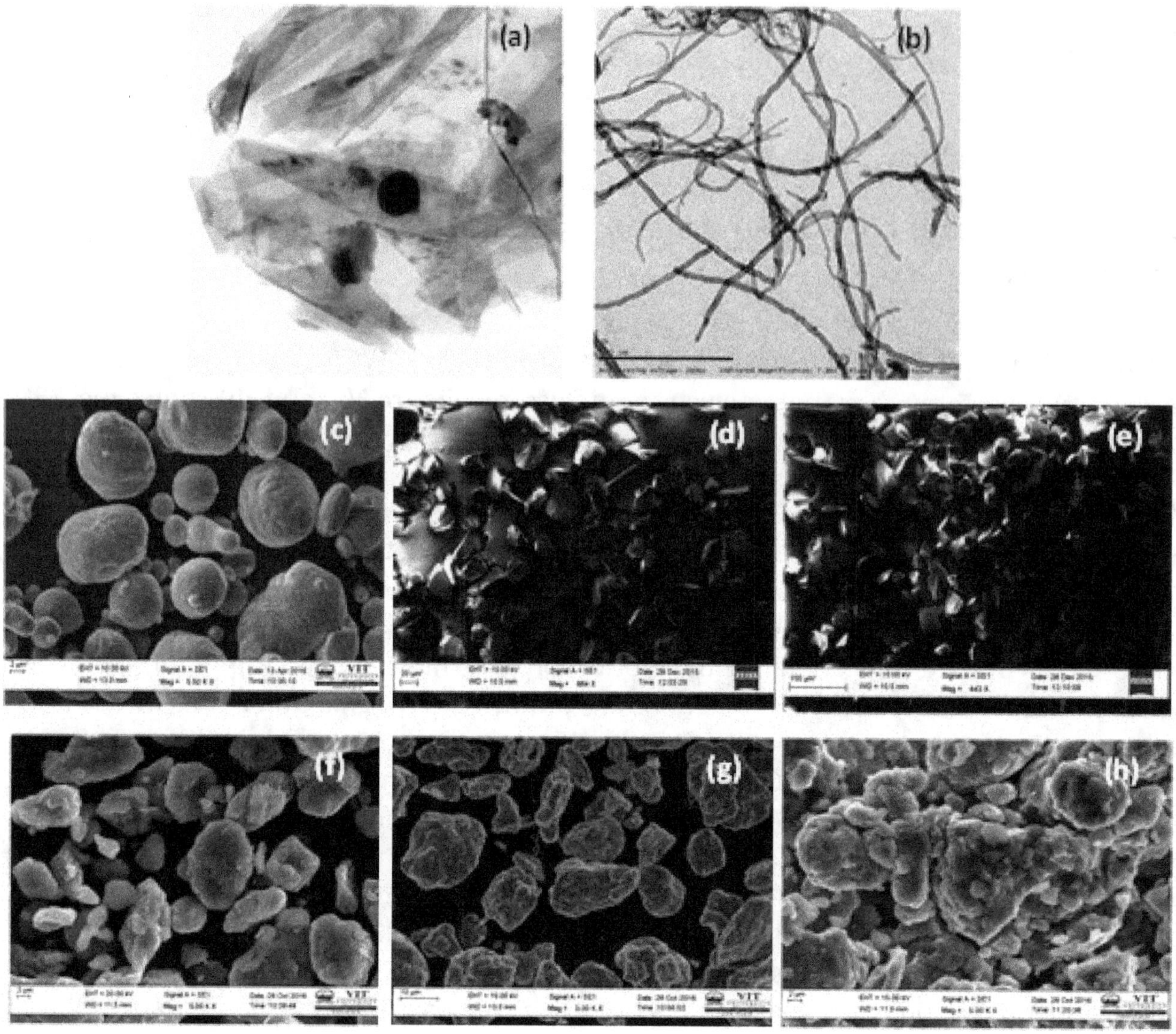

Figure 2. TEM image of (a) Graphene (b) CNT; SEM image (c) Aluminum powder morphology (d) SiC and (e) Al_2O_3; Ball-milled aluminum with (f) MWCNT (g) SiC and (h) Al_2O_3; precursors.

(g/cm^3)	**Apparent density (ρ_a)**		**Tap density (ρ_t)**
	Max	**Min**	
Graphene	0.088 ± 0.002	0.071 ± 0.002	0.1288 ± 0.02
AA 6061	1.118 ± 0.05	0.95 ± 0.05	3.68 ± 0.05
SiC	2.600 ± 0.06	1.9 ± 0.04	3.83 ± 0.03
CNT	0.076 ± 0.002	0.069 ± 0.002	0.1288 ± 0.02
Al_2O_3	2.580 ± 0.06	2.0 ± 0.04	3.98 ± 0.03

Table 1. Measured apparent and tap density values.

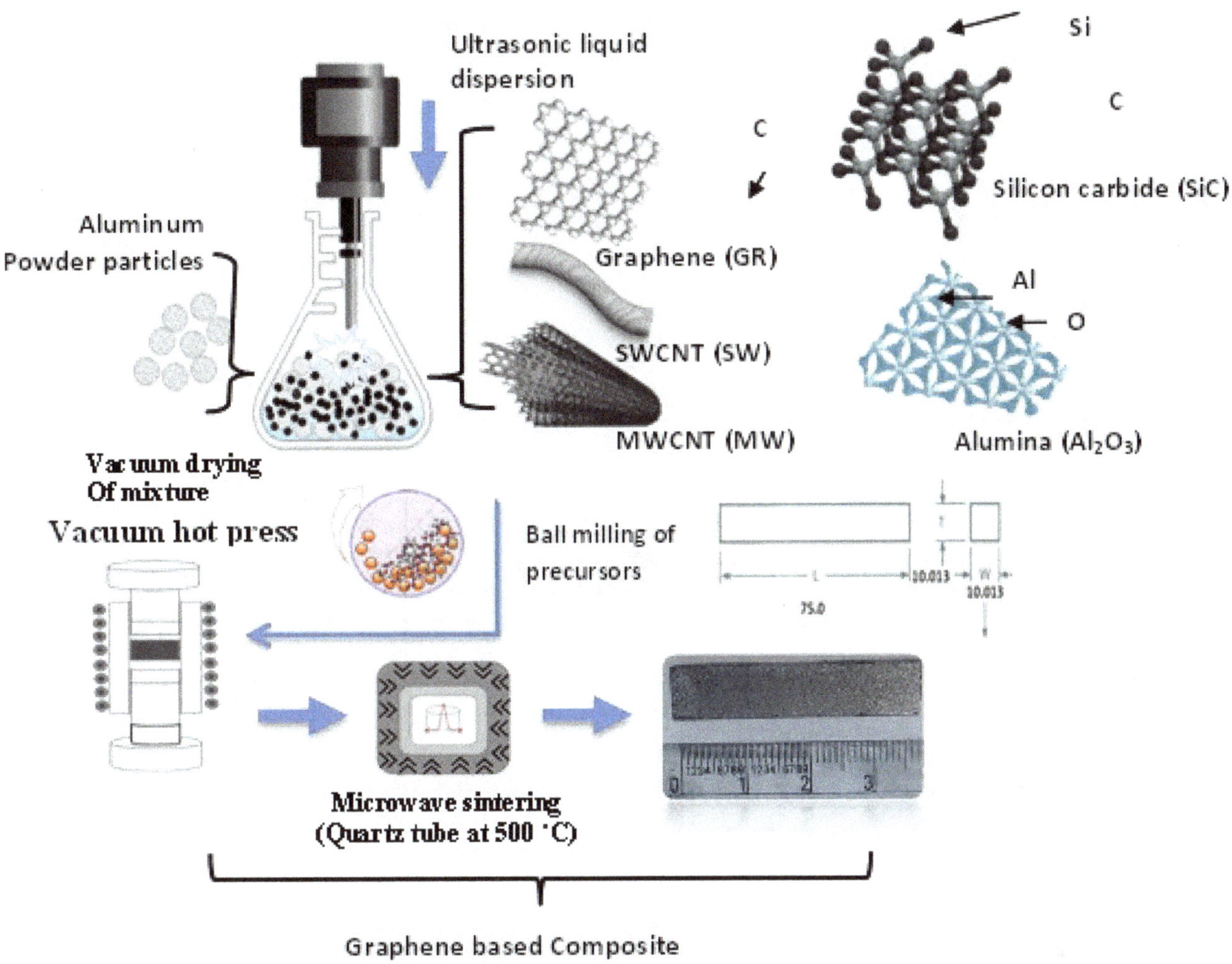

Figure 3. Powder processing through ultrasonic liquid processor and ball milling followed by vacuum hot press and microwave sintering.

3. Testing methods

Vickers hardness values are measured for all aluminum-graphene and other hybrid composite samples using 200 gf after fine polishing of the samples that are sintered at microwave furnace. The mean values of at least five measurements from different areas on the composites sample were taken to analyze. Theoretical densities (g/cm^{-3}) of nanocomposites measured according to the rule of hybrid mixtures (RoHM) relation shown in Eq. (1) and the values of both theoretical and sintered density (g/cm^{-3}) are summarized.

$$D_{Theorerical} = D_{matrix} * V_{Matrix} + D_{Particulate} * V_{Particulate} + D_{Graphene} * V_{Graphene} \quad (1)$$

where D is the density and V is the volume fraction.

The tribological experiment was carried out according to ASTM G 99-95 standards on pin-on-disc friction and wear testing machine, with disc material: EN-31 steel (hardness: 60HRC). It consists of a pin holder that is connected to the loading lever and measures the wear loss by LVDT. The fresh specimen is used for each test with a size of 8 mm diameter and 25 mm length prepared by machining. Wear tracks of the disc are polished before and after every wear test using emery paper till it gives a roughness value of 1.6 Ra. The polished pin is cleaned in an ultrasonic bath containing methanol. Mass loss or wear loss values were determined by weighing the samples before and after the test. The surface roughness (Ra), max peak (Rp), and max valley (Rv) values are measured through a surface roughness tester.

4. Microwave sintering mechanism of graphene-aluminum nanocomposites

In aluminum nanocomposites, increase in strength during solid-state sintering (microwave sintering) is attributed to diffusion type of material transport mechanism that takes place due to a vacancy concentration gradient. When the electromagnetic wave is incident on the nanocomposites, most of it will be absorbed, converted into heat, and partly will be reflected as depicted in **Figure 4**. This phenomenon also depends on temperature and chemical potential gradient. For the effective grain, a large number of high energized movements of atoms should be there at the vacancies, but this movement is inhibited by the graphene creating a barrier against grain growth by wrapping. So densification increases during microwave sintering due to increase in the diffusion rate that is directly proportional to the sintering temperature, and it can be related by Arrhenius equation as shown in Eq. (2).

$$N = N_0 \, e^{\frac{-q}{RT}} \tag{2}$$

where N is the number of vacancy sites, N_0 is the total number of lattice sites, q is the activation energy, RT is the average kinetic energy.

Also, under compressive loading conditions, graphene is subjected to deformation and buckling when it is aligned parallel to compressive load. Further, there is no influence on compressive strength if flakes are aligned perpendicular to compressive load applied. In both the cases, nanocomposites are subjected to more strain rate during deformation compared with monolithic aluminum, and this can be arrested by adding SiC/Al_2O_3 particles (harder particle) in both compressive and diametrical loading conditions. It is also possible to achieve the improved strength when the flake is oriented approximately at 45° to the loading direction. The chances of achieving the strength are possible only in particular directions, but achieving

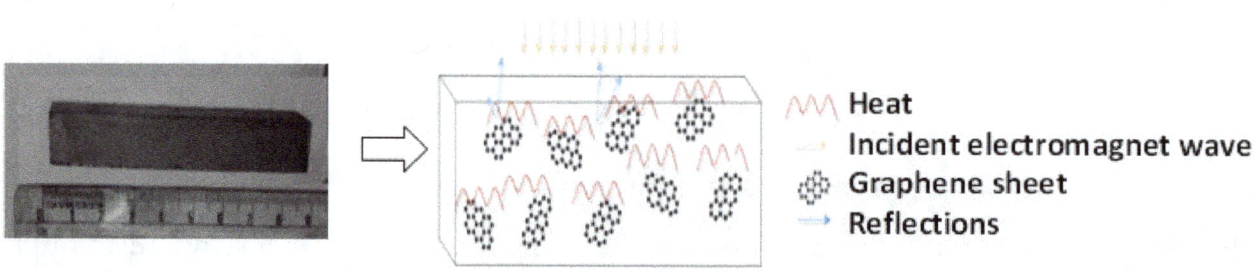

Figure 4. Possible dissipation route of incident electromagnetic wave in the aluminum/graphene composite.

in all directions is quite incredible when graphene alone is considered in the nanocomposites. But combinations of SiC/Al_2O_3 and graphene have advantages in enhancing the strength of the developed nanocomposites. Experiments are repeated on five samples of each compositions and maximum compressive strength achieved is considered for the analysis. It is expected that addition of graphene and SiC/Al_2O_3 leads to dislocation of the pile, resulting in increasing its strengths. Moreover, flake morphology of graphene is contributed in obstructing the particle movement that leads to narrowing the distance between them, and this is found to be the reason for increase in hardness and density.

5. Tribological performances of aluminum-graphene composites

5.1. Effect of graphene

The hardness values of various weight fractions of graphene-based aluminum composites synthesized by uniaxial hot compaction (in mushy state) and microwave sintered are found to have drastic improvements (**Figure 5(a)**) due to addition of graphene and found highest for 0.3 (98 HV), 0.6 (92 HV), and 0.9 (83 HV) weight fractions of graphene compared to aluminum (65 HV). The increase in the hardness confirms the formation of more refined compacted structure [38], and further increase in the hardness can be estimated by rule of mixture, i.e., increase in the graphene content increases the hardness due to reinforcement effect by filling the micro voids [39]. Further, relative density ($\rho_{exp/}\rho_{th}$) of aluminum and all aluminum-graphene composites was above 98%. **Figure 5(b)** relates the wear loss (gms) of pure aluminum and aluminum-graphene nanocomposites with various weight fractions (in weight percentage) of graphene as a function of various load conditions. It is inferred that upon addition of graphene the wear loss decreased self-effacingly. Further, while increasing the graphene content, significant decrease in the wear loss was observed, and minimum wear loss is observed for the 0.3 and 0.6 weight fraction composites. This extensive reduction in the wear loss is attributed to the strengthening of composites by an addition of graphene [40, 41], and its tribological properties form the thin dry lubricant layer between mating metal. This behavior is inconsistent with the work done by the other researchers [42, 43]. While considering the effect of load on the composites, it was inferred that there is an increasing trend of wear loss upon increasing the load. This is due to increase in the corresponding friction values between the mating parts. On the other hand, the presence of more content (≥1.2 wt.%) of graphene may lead to agglomeration, thereby resulting in porosity and cracks that lead to a weakening of structural properties [44]. From **Figure 5(c)**, it is inferred that there is a decrease in the coefficient friction for an increase in graphene content in the composite compared to that of aluminum-base material. The friction value is found to be 0.25–0.35, which is very less compared to aluminum composite friction values viz 0.32–0.39 irrespective of applied normal loads. In general, the friction value depends on the mating materials and relative hardness. Further, the decrease in the coefficient of friction values is attributed to the formation of graphene solid lubricant layer between the wear surfaces, which gives rise to the smooth surface, thereby reducing the contact area [45]. When there is an increase in the normal load, the effective contact area increases between the pin (aluminum-graphene) and the disc (EN-31Steel), which leads to an increase in the frictional forces. **Figure 5(d)** is the comparative plots of variation of surface roughness (Ra) values with respect to aluminum base and aluminum-graphene composites for various graphene contents. The plots are obtained as a function of load, and it can be inferred that the surface roughness, Ra value is (2.54–1.9 for aluminum, 1.7–1.33 for the aluminum-graphene composites) found to be lesser than that of the base material for all the normal load applications. The addition of graphene has a positive effect, where graphene becomes a sacrificial layer that when smeared (dry lubricant) on the wearing surface leads to a decrease in the max peak, Rp (3.6–2.8 for aluminum, 3.3–2.8 for aluminum-graphene) and max valley Rv (4.1–3.1 for aluminum, 3.5–2.7 for aluminum-graphene composites) values shown in the **Figure 5(e)** and **(f)**, respectively. The decrease in Rp and Rv values leads to fewer asperities contact and in turn produces a relatively smooth surface.

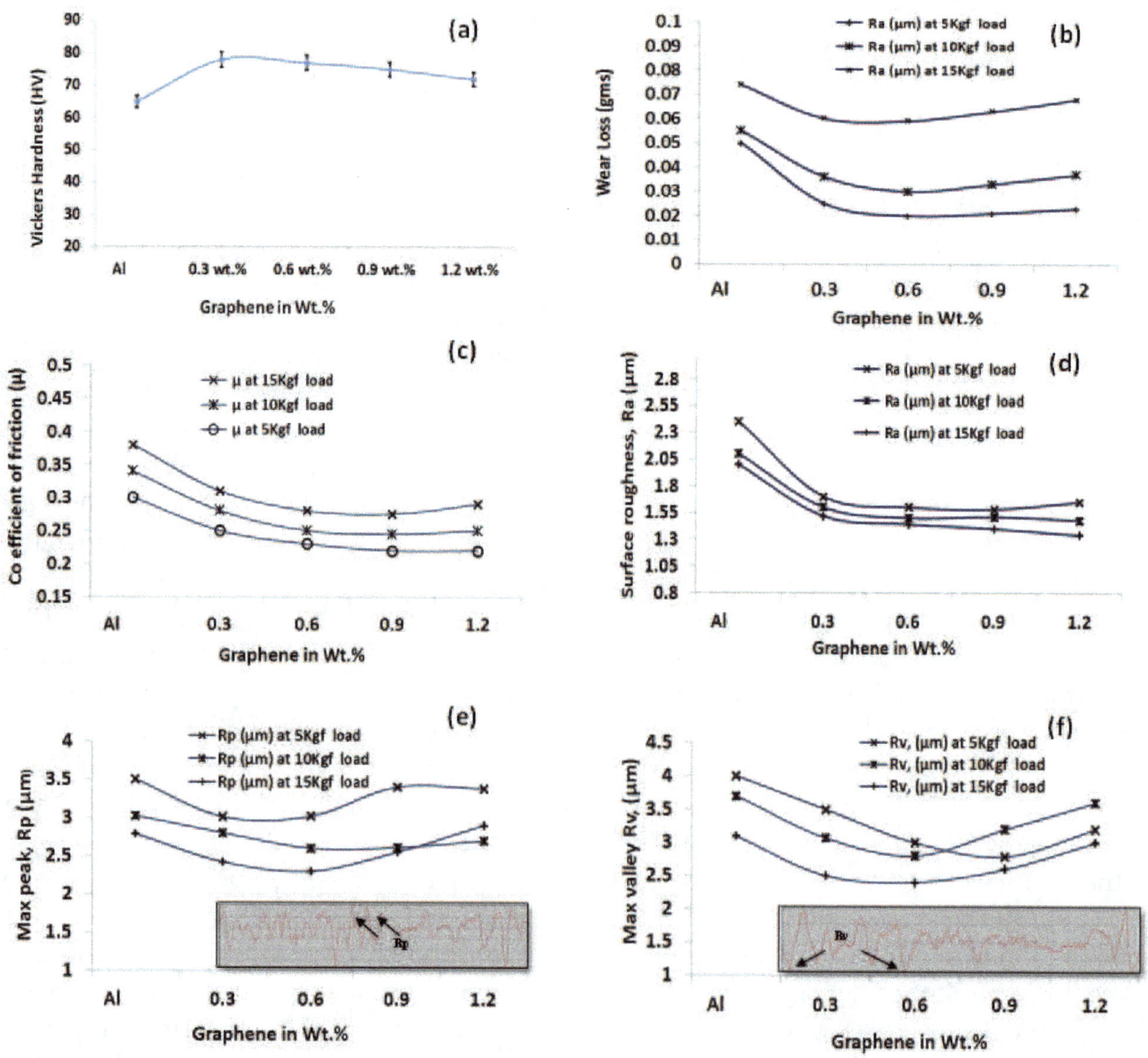

Figure 5. (a) Variation of hardness values; (b) Wear loss; (c) Co-efficient of friction; (d) Surface roughness; (e) Max peak; (f) Max valley with varying the graphene content as a function of load.

Figure 6(a_1–a_5) shows the SEM morphology of worn surfaces immediately after higher normal load conditions (15 kgf) for various aluminum-graphene volume fractions and base under dry sliding conditions. Several macro and deep grooves are noticed in the unreinforced (**Figure** 6(a_1)) compared to graphene-reinforced composites. Further, microgrooves are seen in the aluminum −0.3 wt.% graphene (**Figure 6**(a_2)) and aluminum –0.6 wt.% graphene (**Figure 6**(a_3)) composites. Very few or no grooves are seen on the 0.9 wt.% (**Figure 6**(a_4)) and 1.2 wt.% (**Figure** 6(a_5)) graphene composites. The smeared graphene layer is (puckering effect shown in **Figure** 7) a beneficial (as a dry lubricant) layer that prevents the abrasive wear by protecting the surface same like as seen in carbon nanotubes [46]. It is also expected that at higher volume fractions, graphene may pull out due to more loosely packing/weak bonding with the matrix, which creates thin graphene film that is unnoticeable.

The SEM analysis of wear debris is projected in **Figure** 6(b_1–b_5). Most of the debris are in spherical shape (looks like the metallic particle) (**Figure** 6(c_1)), which are fragments of aluminum under dry wear conditions. **Figure** 6(b_2–b_4) predicts few metallic and more sheets rolled up morphology due to the presence of graphene and tangential/shear force on embedded metal-graphene layers that are formed on the tribo surface. Further, it confirms the formation of solid lubricating layer and failure, subsequently. **Figure 6**(b_5) shows the clear sheet morphology made of fragments of graphene that are detached, crushed/ embedded and that which contributed to the formation of the solid film between the mating parts [47].

Figure 6($c_{1\_}c_5$) shows the effect of graphene on the surface morphology under dry wear conditions that are analyzed through AFM imaging. The samples without the graphene content (**Figure 6**(c_1)) show more asperities penetration depth compared to aluminum-graphene composites (**Figure 6** (c_2–c_5)) that are tested under 15 kgf load condition. The wear rate of the sample is directly proportional to asperities depths, which are seen more in the unreinforced aluminum compared to aluminum-graphene composites (small rough spots), and it is in good agreement with the current research work. Further, the formation of graphene film prevents the welding of asperities by local heating, which may lead to the formation of scoring or scuffing by providing smooth and low surface roughness.

Marginal change of wear rate trends is observed (**Figure 8(a)**) as a function of sliding speed, and it is not in good agreement with the other researchers work [48, 49] for the aluminum-graphene (various weight fractions) composites that are carried out under different (0.4, 0.8, and 1.2 m/s) sliding speed. But the addition of graphene always brings down the wear rates by lowering the frictional heat, and it is observed to be optimum at 0.6 weight fraction for all the sliding speed conditions. Also, it can be noticed that the wear debris on wearing surface is more on increasing the sliding speed. Most debris are seen as flakes and that may give rise to rubbing abrasion media, which lead to insignificant increasing on the wear rate upon increasing the sliding speeds. Further accumulation of debris leading to agglomeration that was found at some points tested under high sliding speed. The comparison of surface roughness suggests that addition of graphene decreases the roughness values, and it has less significance to the increase in the sliding speed. The occurrence of plastic deformation together with debris morphology indicates the ploughing wear. These are seen at higher sliding speeds (1.2 m/s) and higher graphene concentration (1.2 wt.%)

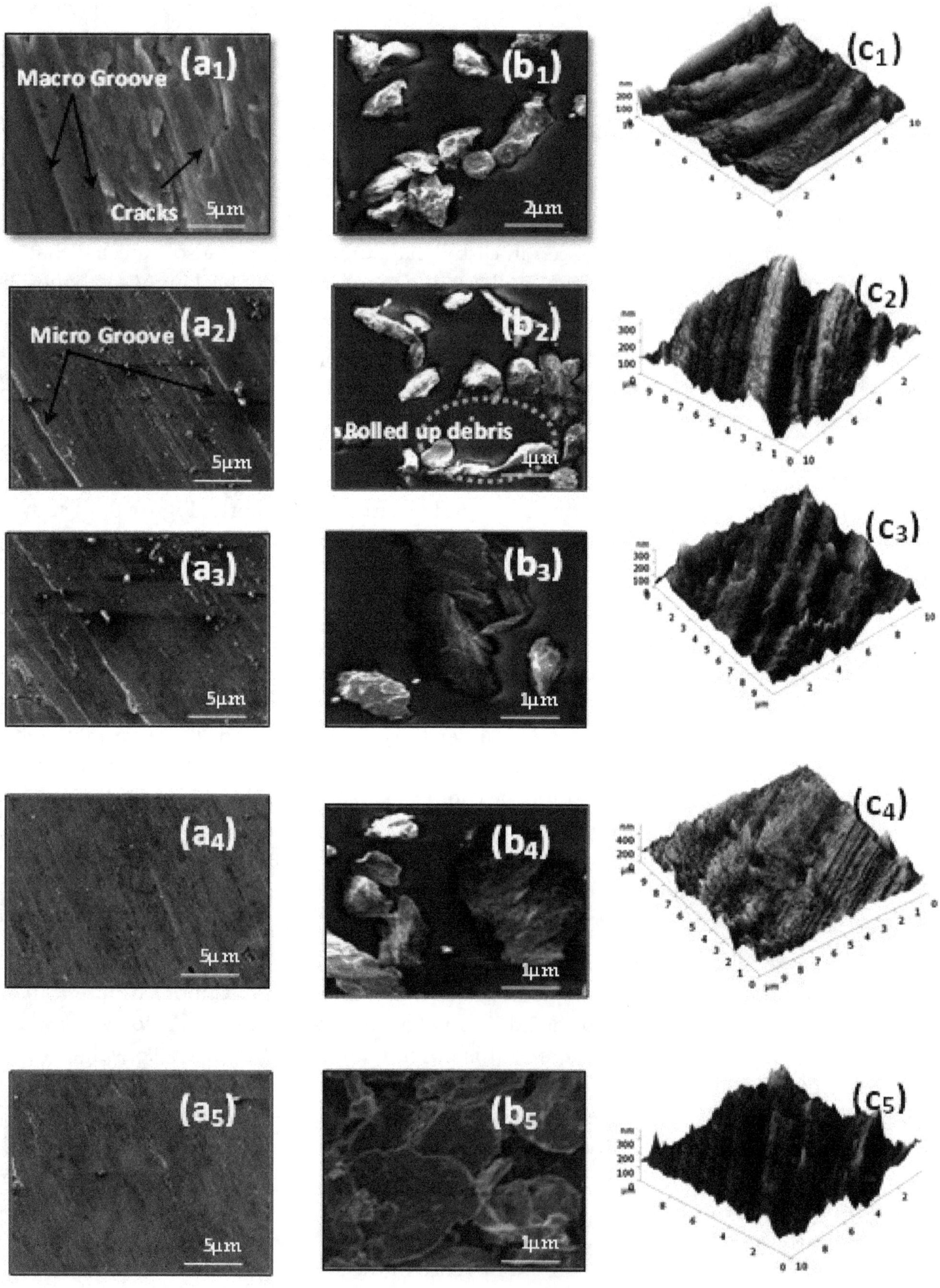

Figure 6. (a) SEM of worn out surface; (b) SEM of wear debris; (c) AFM image of worn out surface.

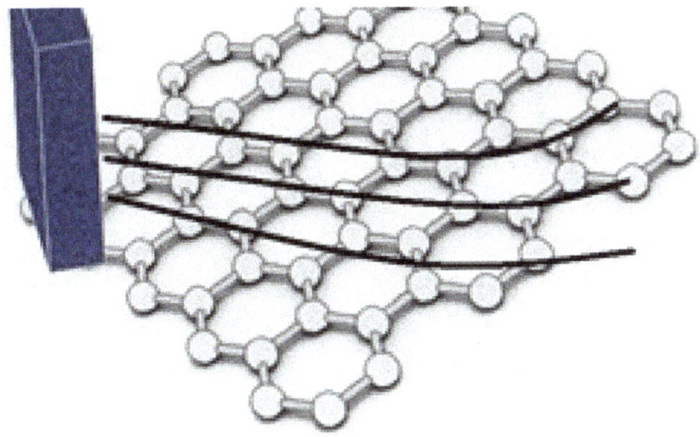

Figure 7. Puckering influence on hexagonal carbon sheet.

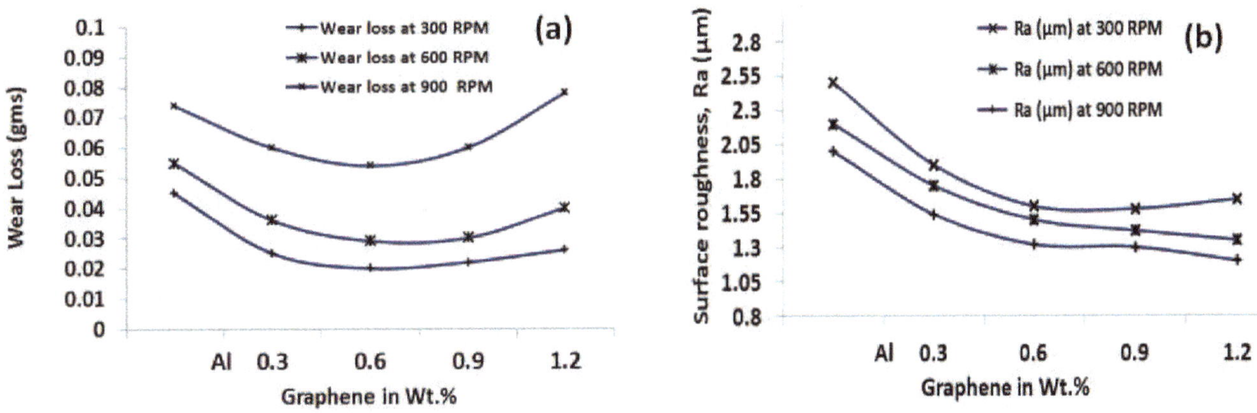

Figure 8. (a) Wear lass; (b) Surface roughness with varying graphene content as a function of speed.

composites (**Figure 9(a)**). Although more black shades on the rubbed surface confirm the presence of more graphene content that peels off and accumulates during rubbing at high frictional force and also same sheet-like debris are seen in **Figure 9(b)**.

5.2. Effect of graphene hybrid combinations with MWCNT

The improvement in the hardness confirms the formation of a more refined compacted structure for the graphene and graphene/MWCNT ball-milled precursors. Addition of graphene acts more as a grain growth inhibitor compared to MWCNT due to difference in the specific surface areas (SSAs) leading (pinning at boundaries) to formation of fine grains that increase the hardness. Further, the combination of graphene/MWCNT leads to slippage and clustering that leads to least effect on grain growth formation within the composites. But the combination of graphene/MWCNT processed through ball milling leads to breaking of MWCNT clustering, flattening, and embedding on the aluminum particulate through plastic deformation. In addition, during ball milling, graphene flakes get flattened and become a monolayer, and interatomic distance between carbon atoms will increase and get packed by matrix atoms [50, 51].

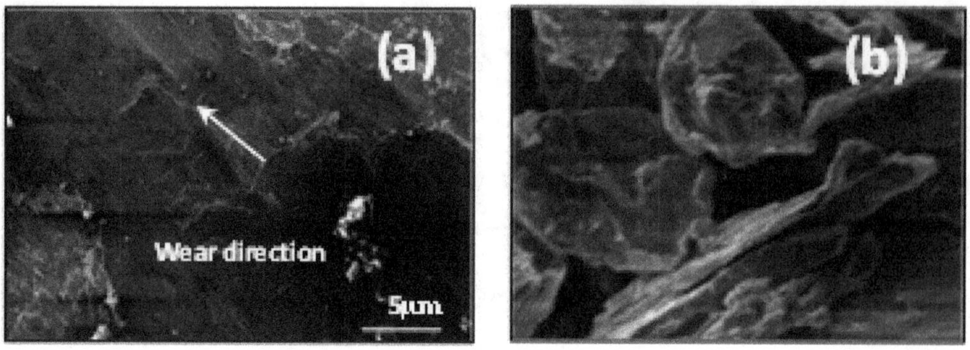

Figure 9. SEM of (a) Worn out surface; (b) Wear debris.

It is inferred that upon addition of graphene on aluminum and aluminum-graphene/MWCNT (flattened) combinations will give the least wear loss compared to monolithic, MWCNT, and aluminum-graphene/MWCNT (tubular) composites. This extensive reduction in the wear loss is attributed to the strengthening of composites by a combination of graphene/MWCNT, processing method, and it is in good agreement with the improved hardness values. Combinations of MWCNT/graphene and MWCNT alone may lead to agglomeration, thereby resulting in porosity and cracks that lead to weakening of structural properties. The friction value is found to be very less for aluminum-graphene/MWCNT composites compared to monolithic aluminum irrespective of applied normal loads. Also, aluminum-graphene/MWCNT (flattened) is found to have least friction values compared to other combinations. This difference in the coefficient of friction values is attributed to the formation of graphene and flattened MWCNT during ball milling as solid lubricant layer between the wear surfaces. In general, the friction value depends on the mating materials and relative hardness. The effective contact area increases between the pin of various aluminum-graphene/MWCNT composites and the disc (EN-31Steel) due to increase in the normal load that leads to increase in the frictional forces. **Figure 10**(a–e) shows the SEM morphology of worn surfaces immediately after higher normal load conditions (15 kgf) for various aluminum-graphene/MWCNT and base aluminum under dry sliding conditions. Several macro and deep grooves are noticed for the monolithic aluminum (**Figure 11(a)**) compared to various alus. **Figure 11**(b–e) represents the worn-out surface of various aluminum-graphene/MWCNT composites in which the smeared graphene layer and MWCNT (flattened) are a beneficial (as a dry lubricant) layer in the composites, which prevent the abrasive wear by protecting the surface.

From the wear rate and friction coefficient, values of aluminum and aluminum-graphene/MWCNT composites are carried out under different (0.5, 1.0, and 1.5 m/s) sliding speed at a sliding distance of 1200 m and constant applied load (45 N). But the addition of graphene and combination with flattened MWCNT during ball milling has a significant advantage in bringing down the wear rates by lowering the frictional heat. The decreasing trend of wear rate is observed as a function of sliding speed. Also, the embedded graphene and flattened MWCNT give planar edges compared to spherical edge of the CNT in the composites, which are prone to high stress interacting sites. Further, accumulation of debris leading to agglomeration was found at some points tested under high sliding speed. Most debris are seen as flakes and that may give rise to rubbing abrasion media, which lead to insignificant increasing

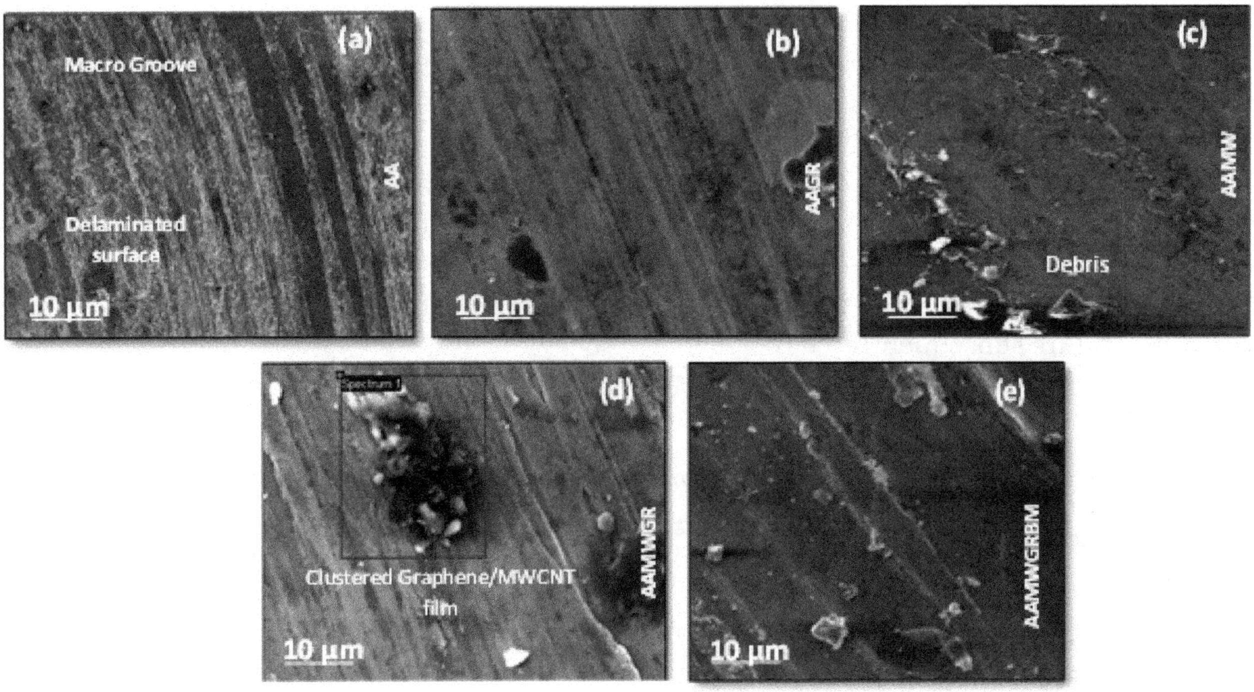

Figure 10. SEM of aluminum-graphene/MWCNT composites worn-out surface varying with the load (N) of (a) AA (b) AAGR (c) AAMW (d) AAMWGR (e) AAMWGRBM.

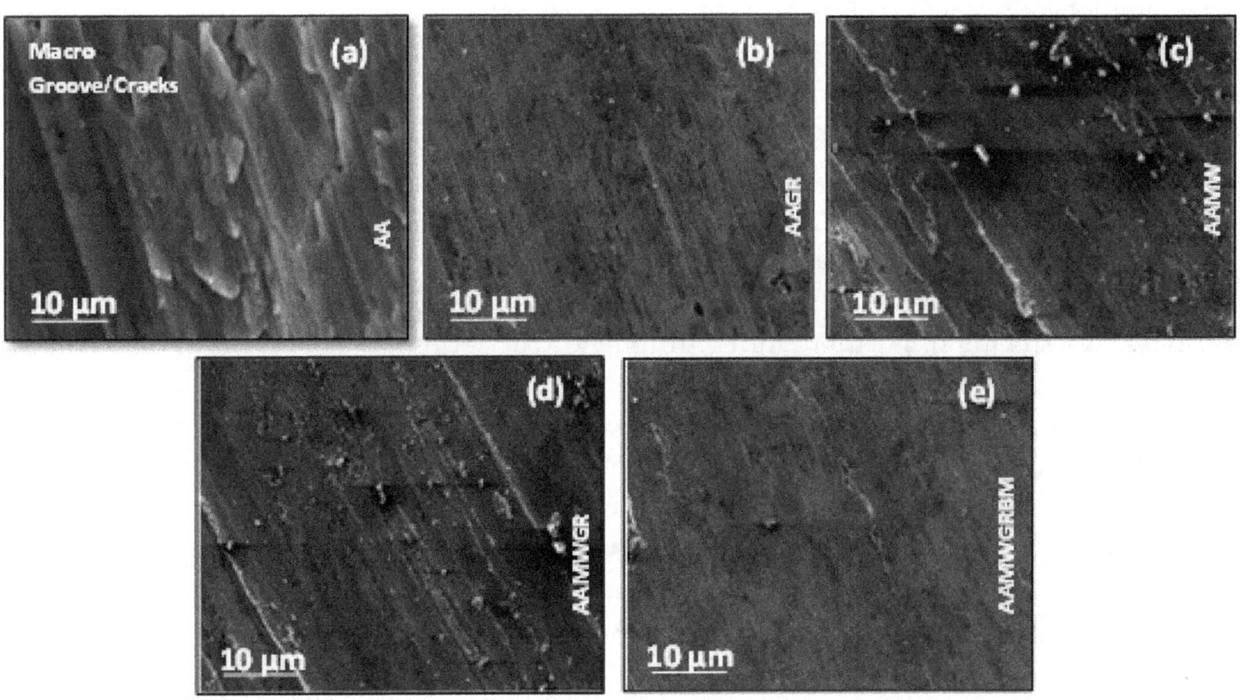

Figure 11. SEM of aluminum-graphene/MWCNT composites worn-out surface varying with the sliding speed (m/s) of (a) AA (b) AAGR (c) AAMW (d) AAMWGR (e) AAMWGRBM

on the wear rate upon increasing the sliding speeds. Most of the wear debris are in spherical shape (delaminated from the wearing surface), which are fragments of aluminum under dry wear conditions. This infers that the deformation is pure by ductile on the surface. Clear sheet morphology made of fragments of sheet graphene and flattened MWCNT combination that are crushed/embedded on tribo surface contributed to the formation of the solid film between the mating parts. Also, few metallic and more sheets rolled up morphology due to the presence of MWCNT, graphene/MWCNT (tubular) combinations, and tangential/shear force on embedded metal on the tribo surface. Further, it confirms the formation of solid lubricating layer and failure subsequently.

5.3. Effect of graphene hybrid combinations with SiC and Al_2O_3

The Vickers hardness values for various concentrations of graphene and SiC/A12O3 in Al aluminum nanocomposites are found to be improved. This shows that the nanocomposite processed through ultrasonic dispersion and microwave sintering technique exhibits a substantial improvement in micro hardness values compared to base aluminum. It is also being found that the values are superior to the results obtained from other processing methods on the same material [51, 52]. The grouping of SiC/Al_2O_3 and graphene is capable of excellent microwave absorption capacity [53]. Also, when graphene is coated/encapsulated with SiC/Al_2O_3 at lower concentrations (up to 0.5 wt.%) that lead to improved microwave absorption of materials, which ascribed to a change in dielectric parameters. Importantly, graphene and SiC/Al_2O_3 in microscale-ranged particles posses higher surface area that is crucial to make interfacial bonding effective and to enhance the overall densification by minimizing the porosity level.

Addition of graphene and SiC/A12O3 has a positive outcome in the synthesized composites in which wear loss is found to be reduced self-effacingly. Extensive wear resistance is due to strengthening of aluminum-SiC/Al_2O_3-graphene composites by a combination and addition of graphene and encapsulation with SiC, which is in good agreement with the improved hardness and other properties. Further, coefficient of frictional values is found to be less for the composite compared to aluminum base friction values and is independent of applied normal loading. In general, surface roughness values are constituted by mating materials and hardness (relative). Further, added graphene will act as a solid lubricant between the wearing surfaces, and also the presence of added SiC/Al_2O_3 particles creates the point contact by reducing the third body abrasion. The addition of graphene was found to result in a positive outcome in which coated/encapsulated SiC/Al_2O_3 will become an efficient lubricating layer by smearing on the mating and wearing surface that will create the fine and smooth surface that reduces the roughness values. **Figure 12** shows the optical micrographs of aluminum-graphene-SiC/Al_2O_3 nanocomposites wear out surface, which are rubbed in unlubricated environments. Many macrocracks or grooves are observed (**Figure 12(a)**) for the samples without and very less microsurface (**Figure 12(b)**) on graphene and SiC/Al_2O_3 (**Figure 12**(c–e)) reinforced composites. So it can be concluded that the combination of graphene and SiC/Al_2O_3 with coating/encapsulation leads to enhancing the tribological properties at optimized values.

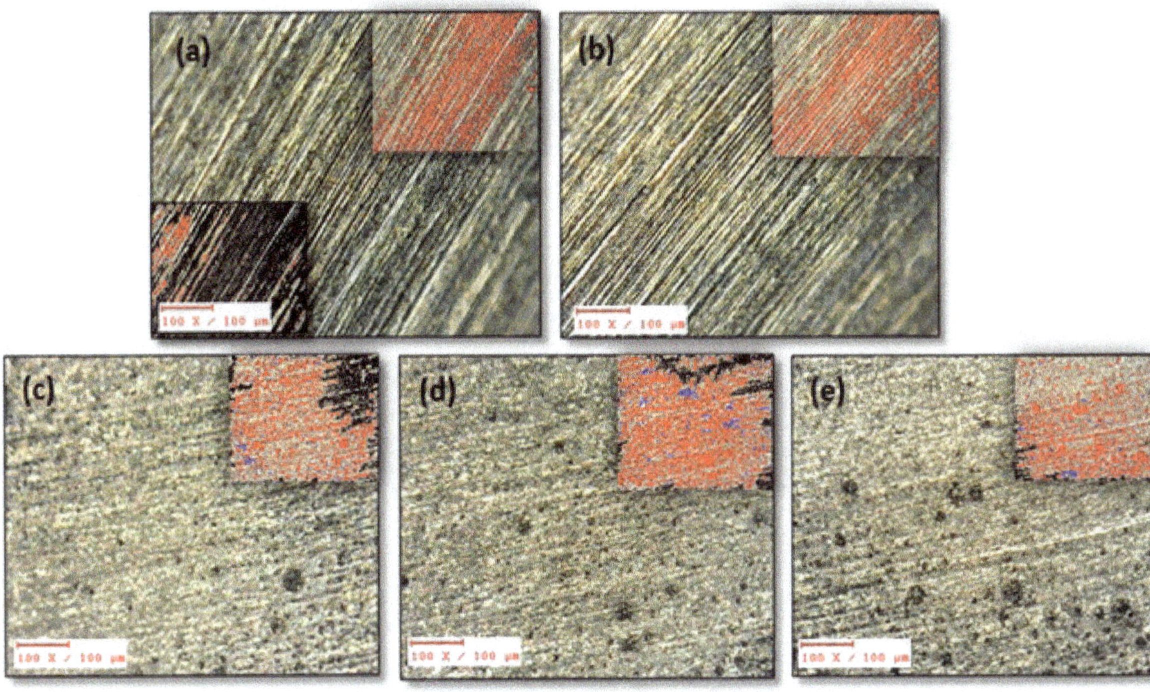

Figure 12. Optical micrographs of worn-out surface of (a) Aluminium (b) Aluminum + Graphene, (c) Aluminum + Graphene + SiC (d) Aluminum + Graphene + Al_2O_3 (e) Aluminum + Graphene + Al_2O_3 + SiC.

6. Conclusion

One of the common engineering research challenges in materials is to determine ways and means to improve the wear resistance between any coupling surfaces, thereby reducing the energy and wear losses. In the current research work, aluminum with graphene is processed separately through vacuum hot press (gives high improved density compacts >98%) followed by microwave sintering successfully. Addition of graphene will provide more number of nucleation's regions. Aluminum-graphene composites are observed to be superior in hardness compared to monolithic material owing to uniform dispersion and higher dispersion strengthening mechanism. Graphene exhibits more grain refinement, which improves the fracture toughness of the composite. This feature has reduced wear losses, and therefore, graphene-reinforced composites are suitable for various tribological applications. The significant differences in the friction coefficients due to high protective nature of graphene will lower the shear force and reduce the material losses. Further, graphene and MWCNT combinations in composite become a sacrificial layer which smeared (dry lubricant) on the wearing surface and enable the self-lubricating properties of aluminum-graphene/MWCNT composites. Graphene/MWCNT (flattened–ball-milled) combination is a favorable tribological application, coupled with improved hardness, strength, and surface roughness values compared to individual MWCNT or graphene/MWCNT (tubular)-reinforced aluminum composites applications, where the component is exposed to wear and friction. Addition of graphene in SiC/Al_2O_3 and encapsulation enables significant improvement in hardness compared to base aluminum. Microwave sintering method can efficiently increase the diffusion of ions in the composite and thus gear up the sintering process, leading to finer grain growth and the densification. Also, combination of SiC/Al_2O_3 and graphene in the nanocomposites has resulted in significant improvement on tribological properties, where it gives the wear resistance by creating a solid lubricant layer between the sliding surfaces.

Author details

Prashantha Kumar H.G.* and Anthony Xavior M.

*Address all correspondence to: prashanthakumar.hg@gmail.com

School of Mechanical Engineering, VIT University, Vellore, Tamil Nadu, India

References

[1] Surappa MK. Aluminium matrix composites: challenges and opportunities. Sadhana. 2003;**28**(1–2):319-334. DOI: 10.1007/BF02717141.

[2] Hunt W, Herling DR. Aluminum metal matrix composites. Advanced Materials & Processes. 2004;**162**(2):39-44. DOI: AC05-76RL01830.

[3] Prasad SV, Asthana R. Aluminum metal-matrix composites for automotive applications: tribological considerations. Tribology Letters. 2004;**17**(3):445-453. DOI: 10.1023/B:TRIL.0000044492.91991.f3.

[4] Logsdon WA, Liaw PK. Tensile, fracture toughness and fatigue crack growth rate properties of silicon carbide whisker and particulate reinforced aluminum metal matrix composites. Engineering Fracture Mechanics. 1986;**24**(5):737-751. DOI: 10.1016/0013-7944(86)90246-8.

[5] Wang W, Shi QY, Liu P, Li HK, Li T. A novel way to produce bulk SiCp reinforced aluminum metal matrix composites by friction stir processing. Journal of Materials Processing Technology. 2009;**209**(4):2099-2103. DOI: 10.1016/j.jmatprotec.2008.05.001.

[6] Miracle DB. Metal matrix composites – from science to technological significance. Composites Science and Technology. 2005;**65**(15):2526-2540. DOI: 10.1016/j.compscitech.2005.05.027.

[7] Rawal SP. Metal-matrix composites for space applications. JOM. 2001;**53**(4):14-17. DOI: 10.1007/s11837-001-0139-z.

[8] Ibrahim IA, Mohamed FA, Lavernia EJ. Particulate reinforced metal matrix composites—a review. Journal of Materials Science. 1991;**26**(5):1137-1156. DOI: 10.1007/BF00544448.

[9] Shorowordi KM, Laoui T, Haseeb AS, Celis JP, Froyen L. Microstructure and interface characteristics of B_4C, SiC and Al_2O_3 reinforced Al matrix composites: a comparative

study. Journal of Materials Processing Technology. 2003;**142**(3):738-743. DOI: 10.1016/S0924-0136(03)00815-X.

[10] Nair SV, Tien JK, Bates RC. SiC-reinforced aluminium metal matrix composites. International Metals Reviews. 1985;**30**(1):275-290. DOI: 10.1016/j.jmatprotec.2008.05.001.

[11] Tzamtzis S, Barekar NS, Babu NH, Patel J, Dhindaw BK, Fan Z. Processing of advanced Al/SiC particulate metal matrix composites under intensive shearing—a novel Rheo-process. Composites Part A: Applied Science and Manufacturing. 2009;**40**(2):144-151. DOI: 10.1016/j.compositesa.2008.10.017.

[12] Lloyd DJ. Particle reinforced aluminium and magnesium matrix composites. International Materials Reviews. 1994;**39**(1):1-23. DOI: 10.1179/imr.1994.39.1.1.

[13] Hasselman DP, Donaldson KY, Geiger AL. Effect of reinforcement particle size on the thermal conductivity of a particulate-silicon carbide-reinforced aluminum matrix composite. Journal of the American Ceramic Society. 1992;**75**(11):3137-3140. DOI: 10.1111/j.1151-2916.1992.tb04400.x/full.

[14] Zweben C. Advances in composite materials for thermal management in electronic packaging. Journal of the Minerals Metals and Materials Society. 1998;**50**(6):47-51. DOI: 10.1007/s11837-998-0128-6.

[15] Sannino AP, Rack HJ. Dry sliding wear of discontinuously reinforced aluminum composites: review and discussion. Wear. 1995;**189**(1):1-9. DOI: 10.1016/0043-1648(95)06657-8.

[16] Nardone VC, Prewo KM. On the strength of discontinuous silicon carbide reinforced aluminum composites. Scripta Metallurgica. 1986;**20**(1):43-48. DOI: 10.1016/0036-9748(86)90210-3.

[17] Ciftci I, Turker M, Seker U. Evaluation of tool wear when machining SiC p-reinforced Al-2014 alloy matrix composites. Materials & Design. 2004;**25**(3):251-255. DOI: 10.1016/j.matdes.2003.09.019.

[18] Mishra RS, Ma ZY, Charit I. Friction stir processing: a novel technique for fabrication of surface composite. Materials Science and Engineering A. 2003;**341**(1):307-310. DOI: 10.1016/S0921-5093(02)00199-5.

[19] Srivatsan TS, Ibrahim IA, Mohamed FA, Lavernia EJ. Processing techniques for particulate-reinforced metal aluminium matrix composites. Journal of Materials Science. 1991;**26**(22):5965-5978. DOI: 10.1007/BF01113872.

[20] Urquhart AW. Novel reinforced ceramics and metals: a review of Lanxide's composite technologies. Materials Science and Engineering A. 1991;**144**(1–2):75-82. DOI: 10.1016/0921-5093(91)90211-5.

[21] Tjong SC. Novel nanoparticle-reinforced metal matrix composites with enhanced mechanical properties. Advanced Engineering Materials. 2007;**9**(8):639-652. DOI: 10.1002/adem.200700106.

[22] Kumar HP, Xavior MA. Graphene reinforced metal matrix composite (GRMMC): a review. Procedia Engineering. 2014;**31**(97):1033-1040. DOI: 10.1016/j.proeng.2014.12.381.

[23] Jang BZ, Zhamu A. Processing of nanographene platelets (NGPs) and NGP nanocomposites: a review. Journal of Materials Science. 2008;**43**(15):5092-5101. DOI: 10.1007/s10853-008-2755-2.

[24] Yu MF, Lourie O, Dyer MJ, Moloni K, Kelly TF, Ruoff RS. Strength and breaking mechanism of multiwalled carbon nanotubes under tensile load. Science. 2000;**287**(5453):637-640. DOI: 10.1126/science.287.5453.637.

[25] Lee C, Wei X, Kysar JW, Hone J. Measurement of the elastic properties and intrinsic strength of monolayer graphene. Science. 2008;**321**(5887):385-388. DOI: 10.1126/science.1157996.

[26] Young RJ, Kinloch IA, Gong L, Novoselov KS. The mechanics of graphene nanocomposites: a review. Composites Science and Technology. 2012;**72**(12):1459-1476. DOI: 10.1016/j.compscitech.2012.05.005.

[27] Baradeswaran A, Perumal AE. Wear and mechanical characteristics of Al 7075/graphite composites. Composites Part B: Engineering. 2014;**56**:472-476. DOI: 10.1016/j.compositesb.2013.08.073.

[28] Ruoff RS, Lorents DC. Mechanical and thermal properties of carbon nanotubes. Carbon. 1995;**33**(7):925-930. DOI: 10.1016/0008-6223(95)00021-5.

[29] Odom TW, Huang JL, Kim P, Lieber CM. Atomic structure and electronic properties of single-walled carbon nanotubes. Nature. 1998;**391**(6662):62-64. DOI: 10.1038/34145.

[30] Suk JW, Kitt A, Magnuson CW, Hao Y, Ahmed S, An J, Swan AK, Goldberg BB, Ruoff RS. Transfer of CVD-grown monolayer graphene onto arbitrary substrates. ACS Nano. 2011;**5**(9):6916-6924. DOI: 10.1021/nn201207c.

[31] Wang J, Li Z, Fan G, Pan H, Chen Z, Zhang D. Reinforcement with graphene nanosheets in aluminum matrix composites. Scripta Materialia. 2012;**66**(8):594-597. DOI: 10.1016/j.scriptamat.2012.01.012.

[32] Liu J, Yan H, Jiang K. Mechanical properties of graphene platelet-reinforced alumina ceramic composites. Ceramics International. 2013;**39**(6):6215-6221. DOI: 10.1016/j.ceramint.2013.01.041.

[33] Mokdad F, Chen DL, Liu ZY, Xiao BL, Ni DR, Ma ZY. Deformation and strengthening mechanisms of a carbon nanotube reinforced aluminum composite. Carbon. 2016;**104**:64-77. DOI: 10.1016/j.carbon.2016.03.038.

[34] Xu CL, Wei BQ, Ma RZ, Liang J, Ma XK, Wu DH. Fabrication of aluminum-carbon nanotube composites and their electrical properties. Carbon. 1999;**37**(5):855-858. DOI: 10.1016/S0008-6223(98)00285-1.

[35] Park JG, Keum DH, Lee Y. Strengthening mechanisms in carbon nanotube-reinforced aluminum composites. Carbon. 2015;**95**:690-698. DOI: 10.1016/j.carbon.2015.08.112.

[36] Stein J, Lenczowski B, Fréty N, Anglaret E. Mechanical reinforcement of a high-performance aluminium alloy AA5083 with homogeneously dispersed multi-walled carbon nanotubes. Carbon. 2012;**50**(6):2264-2272. DOI: 10.1016/j.carbon.2012.01.044.

[37] Bastwros MM, Esawi AM, Wifi A. Friction and wear behavior of Al-CNT composites. Wear. 2013;**307**(1):164-173. DOI: 10.1016/j.wear.2013.08.021.

[38] Abadi SB, Khavandi A, Kharazi Y. Effects of mixing the steel and carbon fibers on the friction and wear properties of a PMC friction material. Applied Composite Materials. 2010;**17**(2):151-158. DOI: 10.1007/s10443-009-9115-5.

[39] Berman D, Erdemir A, Sumant AV. Graphene: a new emerging lubricant. Materials Today. 2014;**17**(1):31-42. DOI: 10.1016/j.mattod.2013.12.003.

[40] Lee CS, Kim YH, Han KS, Lim T. Wear behaviour of aluminium matrix composite materials. Journal of Materials Science. 1992;**27**(3):793-800. DOI: 10.1007/BF00554055.

[41] Lin J, Wang L, Chen G. Modification of graphene platelets and their tribological properties as a lubricant additive. Tribology Letters. 2011;**41**(1):209-215. DOI: 10.1007/s11249-010-9702-5.

[42] Chattopadhyay PK, Chattopadhyay S, Das NC, Bandyopadhyay PP. Impact of carbon black substitution with nanoclay on microstructure and tribological properties of ternary elastomeric composites. Materials & Design. 2011;**32**(10):4696-4704. DOI: 10.1016/j.matdes.2011.06.050.

[43] Krishnan BP, Rohatgi PK. Modification of Al-Si alloy melts containing graphite particle dispersions. Metals Technology. 2013; 11(1) 41-44. DOI: 10.1179/030716984803274297.

[44] Hocheng H, Yen SB, Ishihara T, Yen BK. Fundamental turning characteristics of a tribology-favored graphite/aluminum alloy composite material. Composites Part A: Applied Science and Manufacturing. 1997;**28**(9):883-890. DOI: 10.1016/S1359-835X(97)00055-9.

[45] Chu HS, Liu KS, Yeh JW. An in situ composite of Al (graphite, Al 4 C 3) produced by reciprocating extrusion. Materials Science and Engineering A. 2000;**277**(1):25-32. DOI: 10.1016/S0921-5093(99)00562-6.

[46] Kim IY, Lee JH, Lee GS, Baik SH, Kim YJ, Lee YZ. Friction and wear characteristics of the carbon nanotube-aluminum composites with different manufacturing conditions. Wear. 2009;**267**(1):593-598. DOI: 10.1016/j.wear.2008.12.096.

[47] Bakshi SR, Keshri AK, Agarwal A. A comparison of mechanical and wear properties of plasma sprayed carbon nanotube reinforced aluminum composites at nano and macro scale. Materials Science and Engineering A. 2011;**528**(9):3375-3384. DOI: 10.1016/j.msea.2011.01.061.

[48] Yousif BF. Design of newly fabricated tribological machine for wear and frictional experiments under dry/wet condition. Materials & Design. 2013;**48**:2-13. DOI: 10.1016/j.matdes.2012.06.046.

[49] Findik F. Latest progress on tribological properties of industrial materials. Materials & Design. 2014;**57**:218-244. DOI: 10.1016/j.matdes.2013.12.028.

[50] Choi HJ, Shin JH, Bae DH. The effect of milling conditions on microstructures and mechanical properties of Al/MWCNT composites. Composites Part A: Applied Science and Manufacturing. 2012;**43**(7):1061-1072. DOI: 10.1016/j.compositesa.2012.02.008.

[51] Bastwros M, Kim GY, Zhu C, Zhang K, Wang S, Tang X, Wang X. Effect of ball milling on graphene reinforced Al6061 composite fabricated by semi-solid sintering. Composites Part B: Engineering. 2014;**60**:111-118. DOI: 10.1016/j.compositesb.2013.12.043.

[52] Leonelli C, Veronesi P, Denti L, Gatto A, Iuliano L. Microwave assisted sintering of green metal parts. Journal of Materials Processing Technology. 2008;**205**(1):489-496. DOI: 10.1016/j.jmatprotec.2007.11.263.

[53] Oghbaei M, Mirzaee O. Microwave versus conventional sintering: a review of fundamentals, advantages and applications. Journal of Alloys and Compounds. 2010;**494**(1):175-189. DOI: 10.1016/j.jallcom.2010.01.068.

7

Synthesis of Graphene on Metal/SiC Structure

Petr Machac

Abstract

The chapter deals with the synthesis of graphene on metal/SiC substrates. The graphene synthesis is pursued at a relatively low temperature. The method can be used for the graphene transfer from SiC to dielectric materials. Annealing of the structure results in a chemical reaction of a metal with SiC forming silicides and carbon-rich products at the boundary between metal and silicon carbide. Carbon atoms segregate at the top of metal/metal silicide layer during the cooling period of the process. The chapter is divided into the following sections: Introduction, Structure preparation, Graphene preparation from the structure Ni/SiC, Graphene preparation from the structure Co/SiC, Application of other metals, Influence of additive materials, and Conclusion.

Keywords: graphene, nickel, cobalt, Raman spectroscopy, graphene transfer

1. Introduction

Graphene is a 2D form of graphite which can be used in electronics and other branches of technique due to its interesting parameters [1–3]: high electron mobility—in literature, the published value was at 1.5×10^8 m^2/Vs, very low resistivity (1×10^{-4} Ωm), metallic and semiconductive character, low absorption of white light (2.3%), low noise, quantum Hall effect, and so on. These important parameters predestine graphene for substitution of silicon in future microelectronics. Application of graphene is expected for construction of field effect transistor (FET) transistors, transparent conductive electrodes, gas and biosensors, lithium batteries, memory structures, super capacitors, and other structures [2]. The so-called single-layer graphene (SLG) shows the widest application; so the preparation of few layer graphenes (FLG) can be usually simpler.

Graphene is grown by many technological processes. The method of mechanical exfoliation or cleavage was one of the first methods, used by Geim and Novoselov, for the preparation of the graphene films [4–6]. This method comes from mechanical splitting of thin pieces of graphite from the oriented pyrolytic graphite. The main disadvantage of the method is that it produces only small pieces of graphene—graphene flakes.

Today, many methods are used for graphene growth. Graphene is very frequently prepared by a chemical vapor deposition (CVD) method [7–9]. Thin foils of Cu, Ni, or other metals are used as substrates. The method is scalable (graphene films with large dimension can be prepared), but the graphene film must be transferred onto a dielectric substrate for the next application (e.g., SiO_2/Si). Temperature or plasma CVD processes are used in practice.

The next method is a high-temperature decomposition of SiC, which is sometimes called as an epitaxial growth of graphene (EG) [10–12]. In the method silicone atoms sublimate from the surface of SiC substrate at high temperature—1100–1600°C in high vacuum and remaining carbon creates graphene. The method can be used for industrial growth of graphene due to full-wafer technology. Careful control of the sublimation process has recently led to the growth of very thin graphene film over the SiC surface, with only single graphene layer.

The vacuum sublimation of SiC usually produces graphene films with small crystallinity (30–200 nm) [13, 14] due to surface SiC roughening and creation of deep pits. The preparation of graphene by decomposition of SiC in an argon atmosphere of about 100 kPa gives better layers. This method gives SLG films with greater domain sizes. Graphene parameters can be also improved by increasing the growth temperature (up to 2000°C) since SiC decomposition occurs at 1500°C under argon atmosphere rather than at 1150°C in vacuum.

A special sort of method of graphene preparation is the so-called transfer-free method [15]. The method comes from a metal/C/SiO_2/Si structure. The synthesis of graphene is based on a metal-catalyzed crystallization of amorphous carbon (a-C) by thermal annealing. Polymer layer [16] and thin SiC layer [17] are used very frequently as the carbon source instead of a-C. Carbon atoms diffuse into a metal layer at elevated temperatures followed by their precipitation as graphene during the cool-down step.

The graphene growth on SiC substrates at relatively low temperature is very perspective [18]. The technique applies the Ni/SiC system as a basic structure. The method is very promising for the transfer of graphene layers from the SiC substrate to other substrates (mainly on dielectric one). By the annealing of the Ni/SiC system, carbon-rich products can be obtained at the Ni-SiC interface, and the graphene film is segregate on the top of the Ni layer.

This method has been developed by various groups of authors in many ways. The Ni(200 nm)/SiC structure was applied in the work [18]. Graphene was grown on the structure surface by annealing at 750°C (the speed of 25°C per second), with follow-up cooling with no specified velocity. The prepared graphene showed the FLG character. The results of the mentioned work are in discrepancy with the results of works [19, 20], where only thin Ni films (from single to tens nm) were applied. Heating speed was slightly lower (4°C per second), and cooling velocity was approximately 20°C per second. Prepared graphene had a character of FLG and SLG. Authors of work [21] elaborated with the Ni/SiC structure, where the thickness of nickel layer did not exceed the value of 100 nm. The annealing of the structure produced at 1100°C for 300 s (heating and cooling velocities were not specified; a rapid thermal process was used). After the annealing process, the created silicide layer was etched off, and a thin graphene film of FLG type remained on the silicon carbide substrate. Lastly, the modificated method described in the study [22] is also valuable. A

50 nm-thick amorphous silicon carbide layer was used here. The layer was deposited onto the SiO_2/Si substrate, and after that, the SiC layer was covered with 500 nm of Ni. Annealing was done by an Rapid Thermal Annealing (RTA) process at 1100°C for 30 s. The formed silicide was etched off, and the graphene film was detected on the SiO_2 surface. Co as a metal can be used too instead of Ni [23]. The segregation method can be applied in case of metal combination for reduction of carbon solubility and thus modification of the segregation process itself. Ni/Cu metallization was used in Ref. [24] for the preparation of graphene films in silicon carbide-based Micro Electro Mechanical Systems (MEMS) and Nano Electro Mechanical Systems (NEMS) devices.

2. Structure preparation

N-type 4H-SiC substrate wafers, Si-face polished, 4° off-axis, and doping level 4×10^{18} cm^{-3} (supplied by SiCrystal A.G.) were used in our experiments. Majority of metallization were deposited using e-beam evaporator at 135°C in the vacuum of 2×10^{-4} Pa; alternatively, a DC magnetron sputtering apparatus with Ar plasma was used. Prior to the deposition of metals, the wafers were cleaned by the following wet chemical process: 5 min in acetone (ultrasound bath), 5 min in NH_4OH:H_2O:H_2O_2 (5:1:1) (ultrasound bath), 5 min in 5% HF (ultrasound bath), 10 min in boiling water, and finally drying by nitrogen.

Graphene films were produced by thermal treatment of metal/SiC structures in a small vacuum chamber equipped with a resistively heated table (Boralectric heating element). First off, samples were degassed at 300°C for 5 min and then annealed at 600–1100°C in a pressure bellow 3×10^{-4} Pa. Temperature was measured with an optical pyrometer. Heating rate was approximately 17.5°C/s and cooling rate was 15°C/s.

Samples were analyzed by means of the Raman spectroscopy using a DXR Raman microscope spectrometer of the company Thermo Fisher Scientific equipped with confocal Olympus microscope. Solid-state Nd:YAG laser (wavelength 532 nm, maximum power 10 mW) was used as excitation source. Measurement conditions were 7 mW power, ten accumulations of 10 s. scans, grating with 900 lines/mm, and aperture 50 μm slit. A multichannel thermoelectrically cooled Charge Coupled Device (CCD) camera was used as detector. Magnification 50x provided measurement spot size ~1 μm^2. For X-ray photoelectron spectroscopy (XPS), measurements were applied in two apparatus. Majority of the measurement were done in an ESCAProbe P apparatus from Omicron NanoTechnology Ltd. (vacuum 10^{-8} Pa, Al anode, energy of monochromated X-ray source 1486.7 eV, analyzed area with size of 1 mm^2, depth profiling by Ar ions sputtering). Alternatively, for XPS measurement was apply a NanoESCA apparatus from Omicron NanoTechnology Ltd. equipped with a photoemission electron microscopy navigation technique. The apparatus obtained these radiation sources: Hg-lamp (5.2 eV), HeI (21.2 eV), and a monochromatic X-ray source Al K-alpha (1486.7 eV). The spectra processing and evaluation were produced by CasXPS software. Peak component fitting was done by symmetric Gaussian (70%)-Lorentzian (30%) peaks and Shirley background. Atomic force microscopy (AFM) analysis was conducted in a Veeco CP II apparatus in the tapping mode. Photos on structures were made on an optical microscope Jenavert G0685, magnification 2000. Electrical measurements were done on a computer-controlled workplace by a standard van der Pauw method.

3. Graphene preparation from the structure Ni/SiC

Growth of graphene films was done on a set of Ni/SiC structures with the different thickness of nickel layers (1, 5, 10, 50, 200, and 300 nm) and in the different type of thermal forming [25, 26]. Raman spectra were measured for all samples after annealing. The structure with the thinnest nickel layer showed the Raman spectrum with a very small G band, while the D and 2D bands were absent. Apparently, due to a shortage of Ni in this sample, the necessary quantity of carbon atoms for growth of graphene films by SiC decomposition was not released. So, this structure was omitted in the other text.

3.1. Raman spectroscopy

In **Figure 1**, an example of Raman spectra of structures differing in the thickness of the deposited nickel layer is shown. Annealing was carried out in the same way for all samples (heating temperature 1080°C, annealing period 10 s). All Raman spectra in **Figure 1** contain the main characteristic bands of carbon materials designated as D (1350 cm^{-1}), G (1580 cm^{-1}), and 2D (2700 cm^{-1}). The integral intensity ratio of the D and G bands (I_D/I_G) can be applied for the quantification of defects in graphene. An equation for the calculation of the crystallite size L_a in the graphene layer using any laser radiation can be written as [27]

$$L_a(nm) = \frac{560}{E^4(eV)}\left(\frac{I_D}{I_G}\right)^{-1} \tag{1}$$

where E is the energy of laser which was applied in the Raman analysis. The integral intensity ratio of the 2D and G bands (I_{2D}/I_G) can be applied for the calculation of carbon layer number in the graphene layer. Alternatively, the value of the 2D band full width at half maximum (FWHM) can be used [28]. SLG shows the value of I_{2D}/I_G in the range 3–3.5 (ideally 4), and FWHM of the 2D band is then 25 cm^{-1}.

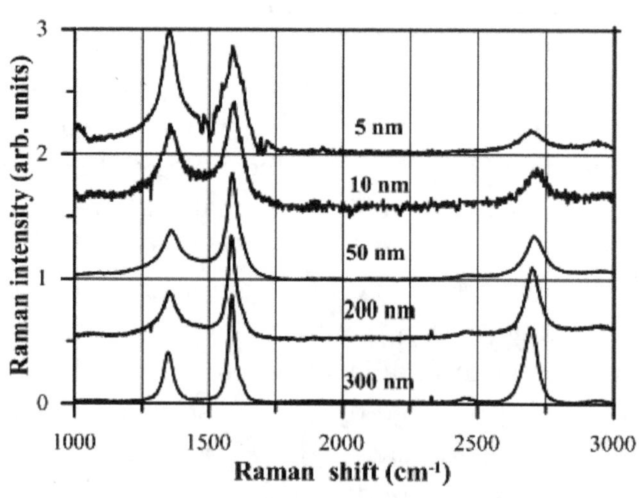

Figure 1. Raman spectra of Ni/SiC samples with Ni thickness in the range of 5, 10, 50, 200, and 300 nm. (Reprinted with permission from Thin Solid Films No. 3995281012757.).

It is obvious that the ratio I_{2D}/I_G from the Raman spectra of all samples on **Figure 1** increases, and opposite to the fact, the ratio I_D/I_G decreases, while the thickness of Ni layer is growing. The resulting graphene film then has less defects or bigger crystals, respectively, and contains smaller number of carbon monolayers for samples with a thicker Ni layer. The graphene film on the Ni(200)/SiC structure shows $I_D/I_G = 0.41$, $I_{2D}/I_G = 0.66$ and FWHM = 59 cm^{-1}. Using the data from Ref. [28] and the formula (1), the number of carbon monolayers was determined to be 3.8 and crystallite size $L_a = 45$ nm. The graphene film prepared on the Ni(300)/SiC structure is similar—the number of carbon monolayers is slightly lower (3.5), and crystallite size is lower too (40 nm).

Further, an influence of temperature of the annealing process on parameters of the graphene films was tested. **Figure 2** illustrates dependence of the I_{2D}/I_G ratio on the annealing temperature, which was gained from Raman spectra of the graphene films on the Ni(200 nm)/SiC structures. The films were prepared by annealing of the structure in the temperature changing from 710 to 1080°C. The annealing time was in all cases 10 s. From the obtained dependence, it is evident that the graphene film with minimal number of carbon monolayers was obtained in the case of the highest annealing temperature (1080°C).

We studied also the quality of carbon films that were formed at the boundary between the metallization and silicon carbide substrate (lower graphene). In these experiments the annealed metallization was etched off firstly. The etching was produced in the mixture of acids HNO_3:HF = 3:1 for 10 min. Analysis of the structures was carried out by the Raman measurement again. **Table 1** obtains the results—the values of ratios I_{2D}/I_G and I_D/I_G of the structures before (Ni surface) and after (SiC surface) the etching. It is evident from the results that the lower graphene film analyzed on the SiC surface showed for the most experiments higher I_{2D}/I_G ratio than in the graphene film on the metallization surface. Evidently, the graphene films on the SiC surface show a lower number of carbon monolayers. The great disadvantage of lower graphene is the much higher I_D/I_G ratio and so higher concentration of defects (smaller crystals). The crystallite size for the graphene film from the last line of **Table 1** is 67 nm before the etching and 16 nm after the etching of the nickel metallization.

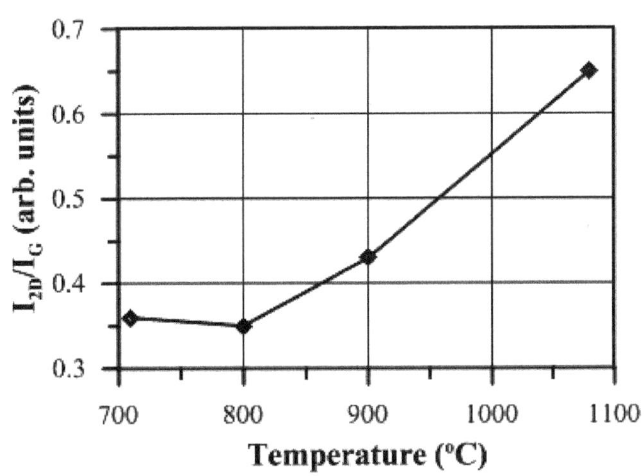

Figure 2. Ratio of 2D and G band intensities in Raman spectra of graphene on the Ni(200)/SiC structures annealed at various temperatures. (Reprinted with permission from Thin Solid Films No. 3995281012757.).

Parameters of graphene growth			Ni surface		SiC surface	
Thickness of Ni layer [nm]	Annealing temperature [°C]	Annealing time [s]	I_D/I_G	I_{2D}/I_G	I_D/I_G	I_{2D}/I_G
200	1080	10	0.40	0.65	1.20	0.46
200	1080	60	0.43	0.55	1.24	0.65
200	1080	180	0.27	0.50	1.10	0.67

Reprinted with permission from Thin Solid Films No. 3995281012757.

Table 1. Results of Raman analysis of the graphene structures (the ratios of band intensities) on the nickel surface and after the metallization was etched off (graphene on the SiC surface).

3.2. Thickness of graphene layers

The thickness of graphene films was calculated mainly from a dependency between the number of carbon monolayers and the I_{2D}/I_G ratio along with the FWHM obtained from Raman spectra [28]. An example is shown in **Table 2**. It shows values of the number of carbon monolayers (N) that were determined concurrently from the I_{2D}/I_G ratio and from the FWHM. There is a quite good conformity among the values. The values of N determined from the I_{2D}/I_G ratio are a bit higher than the values determined from the FWHM.

The XPS analysis was used for verification of the results in the case of the Ni(200 nm)/SiC structure (annealing temperature 1080°C, annealing time 10 s). The process is based on the attenuation of photoelectrons that are excited from the material under the graphene film [3]. Thickness of the graphene film can be determined from the formula

$$d = -\lambda \ln(I/I_o), \tag{2}$$

where I and I_o are intensities of XPS signal of a selected element that was measured on the surface of the graphene film and from the material under the graphene film, respectively. λ represents inelastic mean free path of electrons in the graphene film. Signal of Ni_2 p3/2 with the electron energy of 853 eV was selected in our experiments. The value of $\lambda = 2.4$ nm was gained from literature [29]. The surface of the graphene film was analyzed at first (I = 400 electron/s), and then the film was progressively sputtered off with Ar ions. Intensity of the Ni_2 p3/2 signal $I_o = 4500$ electron/s was reached when the graphene film was completely removed. Thickness

Parameters of graphene growth			I_{2D}/I_G	N	FWHM	N
Thickness of Ni [nm]	Annealing temperature [°C]	Annealing time [s]				
200	1080	0	0.53	5	66	4.7
200	1080	10	0.65	3.8	59	3.4

Reprinted with permission from Thin Solid Films No. 3995281012757.

Table 2. Number of carbon monolayers N for the chosen graphene structures.

of the graphene film 2.5 nm was calculated from the formula (2). The number of carbon monolayers in the prepared graphene film can be gained by dividing the obtained thickness of graphene film by the thickness of the monolayer of carbon (0.335 nm). In graphene films, however, the distance of carbon monolayers can change from 0.55 to 0.70 nm [30]. So we can conclude that the number of carbon monolayers in the graphene film on the Ni(200 nm)/SiC structure shows value from 3.6 to 4.5 and it corresponds to the data in **Table 2**.

3.3. Graphene morphology

The morphology of structure surfaces was studied by the AFM measurement. **Figure 3** shows the surface morphology of the Ni(300 nm)/SiC structure annealed at 1000°C for 120 s. Massive reaction of the nickel film with SiC substrate occurred in the annealing process, and the reaction was not homogeneous which is confirmed by large roughness R_a = 31 nm. The prepared graphene film lays on the metallization surface, and consequently it shows large number of defects and thus has low crystallinity.

3.4. Graphene transfer

Figure 4 shows an example of Raman spectra of a structure after annealing at 950°C, annealing period 30 s. The solid line represents spectrum after the annealing; the dashed line represents spectrum of the graphene film transferred onto the SiO_2/Si substrate. The transfer was done by the etching of the silicide layer by the mixture of HF and HNO_3 acids (ratio 3:1) and with help of polymethyl methacrylate (PMMA). It is possible to estimate that graphene from **Figure 4** contains four carbon layers, its crystallite size is 43 nm, and the transfer increases its value on 82 nm. The difference is probably done by smoother surface of SiO_2.

3.5. Electrical parameters

The basic electronic parameters of prepared graphene were measured. Experiments were done with the metallization prepared by the evaporation of nickel (thickness of 300 nm). For the measurements it is necessary to have a dielectric substrate; therefore, a semi-insulating SiC plate

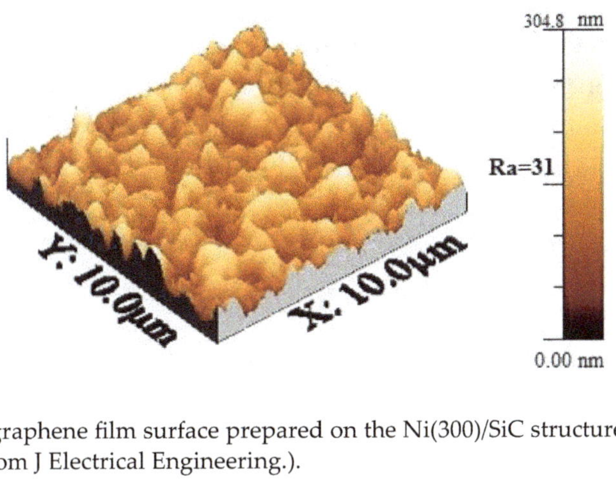

Figure 3. AFM picture of the graphene film surface prepared on the Ni(300)/SiC structure annealed at 1000°C for 120 s. (Reprinted with permission from J Electrical Engineering.).

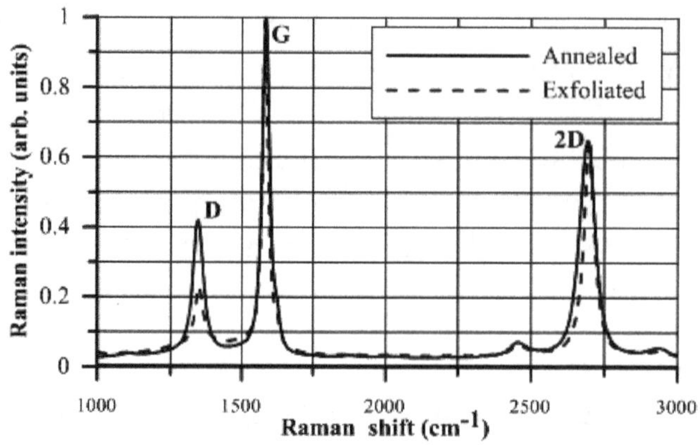

Figure 4. Raman spectra of Ni/SiC sample after the annealing and after the exfoliation. (Reprinted with permission from J Electrical Engineering.).

(SI-SiC) was used, and the graphene film on the interface between SiC and the metallization was tested (after the annealing, the silicide layer was etched off by HNO_3 acid, and the graphene/SI-SiC structure was obtained). For the measuring of electrical parameters by van der Pauw method, the Au(30)/Cr(10) contacts prepared by evaporation were applied. Obtained results are shown in **Table 3** (ρ_s is surface resistivity, μ_H is Hall mobility, and c_s is the concentration of charge carriers). Hall mobility of the prepared graphene films is very low probably due to large concentration of defects in graphene layers.

3.6. Conclusion

The synthesis of graphene on silicon carbide substrates by the Ni-silicidation reaction shows two significant advantages:

- It needs a relatively low temperature for graphene growth (up to 1100°C) compared to the preparation of epitaxial graphene on SiC (minimally 1300°C).
- It needs short process time (from several seconds to several minutes) compared to the CVD method (from several minutes to several hours).

The method is scalable—so it can be applied for the growth of large-area graphene films. However, the technique has a disadvantage—the graphene films are prepared on the conductive plate. For application in microelectronics (unipolar transistors and other devices), the transfer on the dielectric substrate (SiO_2/Si) is necessary. Alternatively, it is possible to apply

Annealing conditions		ρ_s (Ω)	μ_H (cm^2/Vs)	c_s (cm^{-2})
T (°C)	t (s)			
1050	60	527 ± 2.5	300 ± 7.3	$3.85 \times 10^{13} \pm 7.3 \times 10^{11}$

Reprinted with permission from J Electrical Engineering.

Table 3. Electrical parameters of graphene on SI SiC.

semi-insulating SiC substrate together with the lower graphene. The formed silicide layer must be etched off. The lower graphene film can be applied directly for the preparation of electronic devices.

The following conclusion can be done from the results. In our experiments, the Ni films of assorted thicknesses ranging from 1 to 300 nm were applied. Graphene was grown on all samples except one with Ni 1 nm thick. The Ni(200 nm)/SiC structure produced the best results. The quality of lower graphene films growth on the SiC surface was studied after removing of the annealed metallization. In majority of cases, lower graphene films were thinner, but they showed lower crystallinity, which was caused by rough SiC surface after the reaction of Ni with silicon carbide. Thermal forming of the Ni/SiC structures was carried out at the temperature range from 710 to 1080°C and for diverse times. The temperature of 1080°C and time of 10 s showed the best process conditions. The graphene film with character of FLG (the number of carbon monolayers in the range from 3 to 4) was produced on the base of the Ni(200 nm)/SiC structure. Longer annealing period had no noticeable influence to the graphene film parameters. Thickness of the graphene films was checked up by the XPS analysis; the value of 2.5 nm was obtained for the mentioned sample, which is similar to the value gained from the results of Raman spectroscopy. Very good agreement in results was given by both independent methods.

4. Graphene preparation from the structure Co/SiC

Graphene films were prepared by thermal processing of the Co/SiC structures, which differed in thickness of the cobalt layer (10, 50, 100, 200, and 300 nm) [31–33]. Co deposition was performed using a DC magnetron sputtering apparatus at room temperature. Alternatively, the Co layer with thickness 300 nm was deposited by the e-gun evaporation. For studying basic parameters of the prepared graphene films, Raman spectroscopy was chosen.

4.1. Raman spectroscopy

Raman spectra of several Co/SiC structures, which differ in thickness of the Co layer [31], are shown on **Figure 5**. The spectra were normalized to the same size of the G band, in order to compare D and 2D bands. All structures were formed at temperature of 1080°C, and the annealing period was 10 s. The graphene film was not prepared for the structure with 10 nm-thick cobalt layer. The resulting layer shows character of amorphous carbon with very low thickness (the signal is very weak—especially, the 2D band and the spectrum are noisy). With increasing thickness of the Co layer, the 2D band intensity grows up. This gives evidence of decreasing number of carbon monolayers in the formed graphene film. There is about five carbon monolayers for the structure Co(50 nm)/SiC. In the case of the structure with the thickest cobalt layer, a three-layered graphene film has been created. Similarly, crystallinity of the formed graphene films improves when thickness of the Co layer is growing. In the case of the structure Co(300 nm)/SiC, the crystal size in the graphene film is approximately 110 nm.

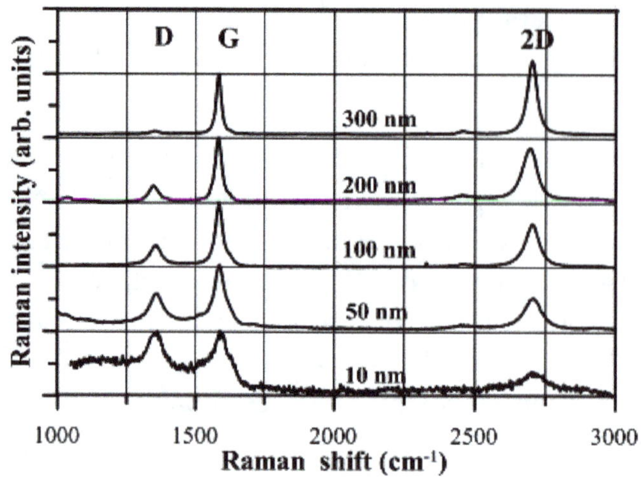

Figure 5. Raman spectra of several Co/SiC structures differing in Co layer thickness. (Reprinted with permission from J Mater Sci: Mater Electron, No. 3995250684677.).

An interesting effect was noticed at the structure Co(300 nm)/SiC. In the case of shorter annealing periods and in dependence on the graphene growth temperature, two different phases have emerged on the surface of the structures—a dark and a light. With the increasing of annealing period, the area of the dark-phase domain was increasing, and, finally, it was spreading over the whole area of the structure. Quality of the structure surface was nevertheless worse. **Figure 6** shows a photograph of the structure Co(300 nm)/SiC annealed at 990°C for 60 s. From the analysis in **Figure 7**, it is obvious that the dark phase contains the graphene film and cobalt silicides, while the light one does not include graphene and silicides. The XPS analysis has been used for verification of the phenomena [31]. The reaction of Co with SiC starts probably at defects and then gradually propagates over the surface of the structure. The effect does not occur at structures with a thinner cobalt layer, and their surface is smoother and homogenous [31].

Figure 6. Example of a structure surface with Co layer 300 nm thick annealed at 990°C for 60 s. (Reprinted with permission from J Mater Sci: Mater Electron, No. 3995250684677.).

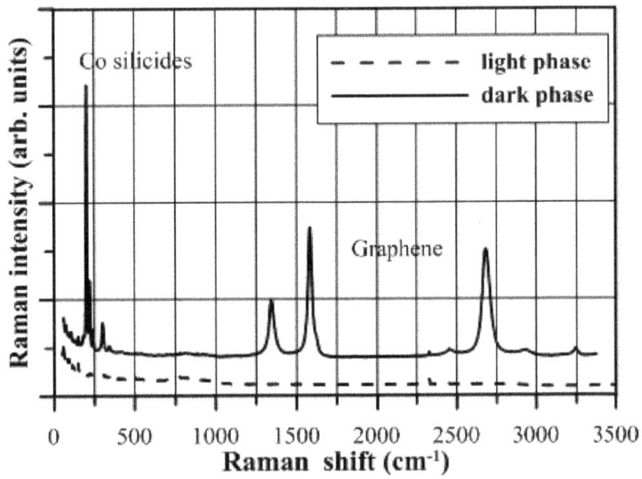

Figure 7. Raman spectra of the Co(300 nm)/SiC structure annealed at 990°C for 60 s. (Reprinted with permission from J Mater Sci: Mater Electron, No. 3995250684677.).

Slightly, other results were obtained in the case of cobalt layers prepared by evaporation [32]. The metallization was deposited onto SiC substrates at elevated temperature (135°C). The substrates were treated in situ by DC Ar plasma before the metal deposition. The layer thickness was 300 nm (the best thickness from previous research) [31]. **Figure 8** shows dependence of the I_{2D}/I_G ratio on annealing time of the structure annealed at 850°C. The results confirm that the sample annealed for 10 s is the best and shows character of nearly BLG. Contrary to the earlier work [31], there has not been observed formation of two phases on the sample surface during graphene growth. Samples' surface was homogenous after annealing, and the graphene phase spread all over the surface.

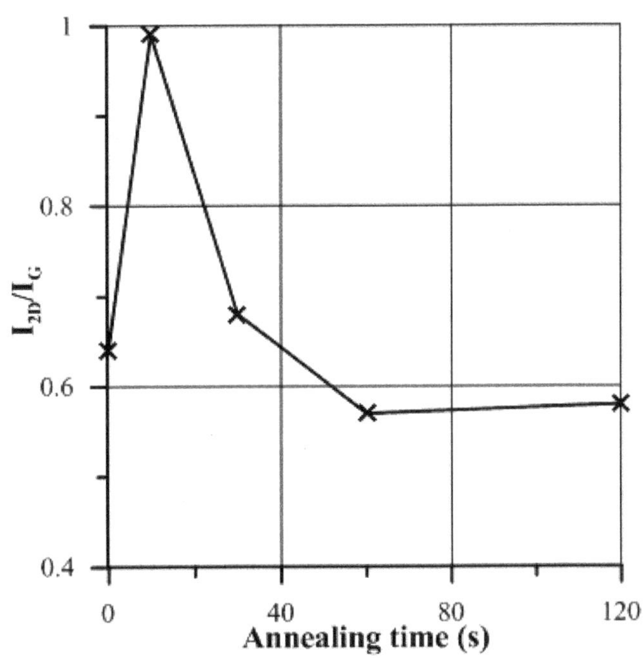

Figure 8. Ratio of 2D and G band intensities in Raman spectra of graphene prepared by annealing at 850°C and at various times. (Reprinted with permission from Applied Surface Science No. 3995260811830.).

4.2. Graphene exfoliation

Graphene films grown on the structure metal/SiC are not suitable for direct application in microelectronics. This is because of graphene having been produced by the reaction of a metal with silicon carbide leading to formation of silicides and carbon. The graphene film is created on a layer of silicides with large electrical conductivity, and thus it does not allow constructing electronic structures (such as unipolar transistor). A certain opportunity is to apply the lower graphene film which is created at the SiC-metal boundary. This graphene film shows however worse parameters than the surface graphene film (the graphene film on the silicide layer) [25]. Preparation of the lower graphene was tested with the Co/SiC structure as well [31]. **Table 4** shows results, which, apart from basic parameters of the technological process, contain also the values of I_{2D}/I_G and I_D/I_G before etching (the surface graphene) and after etching (the lower graphene on the surface of SiC substrate). It is evident from the table that the lower graphene shows in majority cases worse values of the I_{2D}/I_G and I_D/I_G ratios.

Another possibility to form a graphene film on a nonconductive substrate is to transfer the graphene film from the Co/SiC structure onto a dielectric substrate (SiO_2/Si). **Table 4** shows an example of obtained results—the last two columns in the table. It is evident from the table that the transfer of graphene has smaller influence on its parameters than the preparation of the lower graphene. In addition, the silicon carbide surface after the etching-off of the metallic layer evinces greater surface roughness than with the transferred graphene.

Graphene films prepared on the metal/SiC structure are not appropriate for direct measurements of electrical parameters since they are situated on a conductive silicide layer. So it is necessary to transfer the graphene film on a dielectric substrate. **Table 5** gives the example of an obtained result [32]. The table includes parameters (sheet resistance, mobility, and hole concentration) of the graphene film prepared by the annealing of the Co(300 nm)/SiC structure at 1050°C for 120 s. The graphene film has been transferred onto the SiO_2/Si substrate.

4.3. Influence of cooling rate

Authors of a number of studies, which deal with growing of graphene by synthesis on metal/SiC structures [18, 20, 23, 34], postulated that parameters of graphene films are dependent on the cooling rate (CR) of the structures after the annealing is stopped. We have tried to verify the statement experimentally. The structures annealed at 1050°C for 120 s were chosen for

Technological parameters			Surface graphene		Lower graphene		Exfoliated graphene	
Co layer thickness [nm]	Annealing temperature [°C]	Annealing time [s]	I_{2D}/I_G	I_D/I_G	I_{2D}/I_G	I_D/I_G	I_{2D}/I_G	I_D/I_G
300	1080	120	0.76	0.34	0.75	0.66	0.63	0.06
100	1080	120	0.76	0.10	0.88	0.37	0.73	0.20

Table 4. Basic parameters of graphene films—surface graphene, lower graphene, and exfoliated graphene.

Sheet resistance [Ω]	Mobility [cm^2/Vs]	Hole concentration [cm^{-2}]
529	433	2.73×10^{13}

Table 5. Electrical parameters of the graphene layer after the exfoliation process.

the experiment. The tested structures were mounted on a special molybdenum tray prior to the annealing process, and the tray was resistively heated by large current. The tray enables us to control cooling temperature within the CR from 2 to 70°C/s. The highest CR is given by switching off a power supply, and the structure cools down independently. The CR was set by controlling the passing current. **Figure 9** shows the results. The graph represents dependence of the resulting I_{2D}/I_G and I_D/I_G ratios of prepared structures on the CR. It is obvious from the graph that the CR has nearly no influence onto the I_{2D}/I_G ratio (the number of carbon monolayers in the graphene film). However, the CR has a significant influence on the I_D/I_G ratio (the number of defects in the prepared graphene film). The faster cooling is the more defects in the graphene films are created.

4.4. XPS analysis

The analysis was performed on the structure Co(300 nm)/SiC which was prepared by the evaporation process [32]. The structure for the XPS analysis is special, unlike regular structures. The cleaning, annealing, and analyzing of the sample were done in the NanoESCA apparatus. The sample was inserted into the apparatus, and the process of the experiment was done as follows:

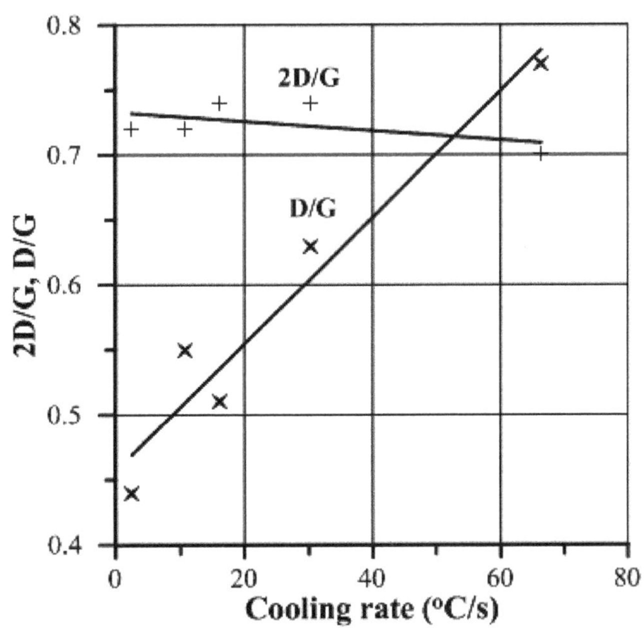

Figure 9. I_{2D}/I_G and I_D/I_G ratios for different cooling rates. (Reprinted with permission from Thin Solid Films No. 3995281012757.).

- Degassing at 250°C for 12 h.
- Cleaning by a 1 keV Ar sputtering at 2×10^{-4} Pa for 20 min.
- Annealing at 850°C for 10 s.

A vacuum of 5×10^{-9} Pa was kept in the apparatus chamber during the process.

XPS analysis was performed in the sample's as-received state, after the sputtering and finally after the annealing. **Figure 10** shows all three survey spectra. The spectrum of as-received sample surface includes oxygen, carbon, and cobalt. Co and O probably create a native cobalt surface oxide, and carbon probably comes from an atmospheric contamination [35–37]. The sample surface after the sputtering is evidently free from C, and presumably the oxygen concentration likewise decreased. Silicon peaks appeared, and O is desirably absent at the sample after the annealing. The appeared Si is a product of the reaction between Co and SiC which yields Co silicides and free C [38, 39]. So Co and C are detected simultaneously. **Table 6** summarizes the atomic concentrations of all elements.

Detail spectra, which contain only narrow scans, are presented in the next paragraphs and figures with aim to give us a deeper chemical investigation. Firstly, it should be pointed out

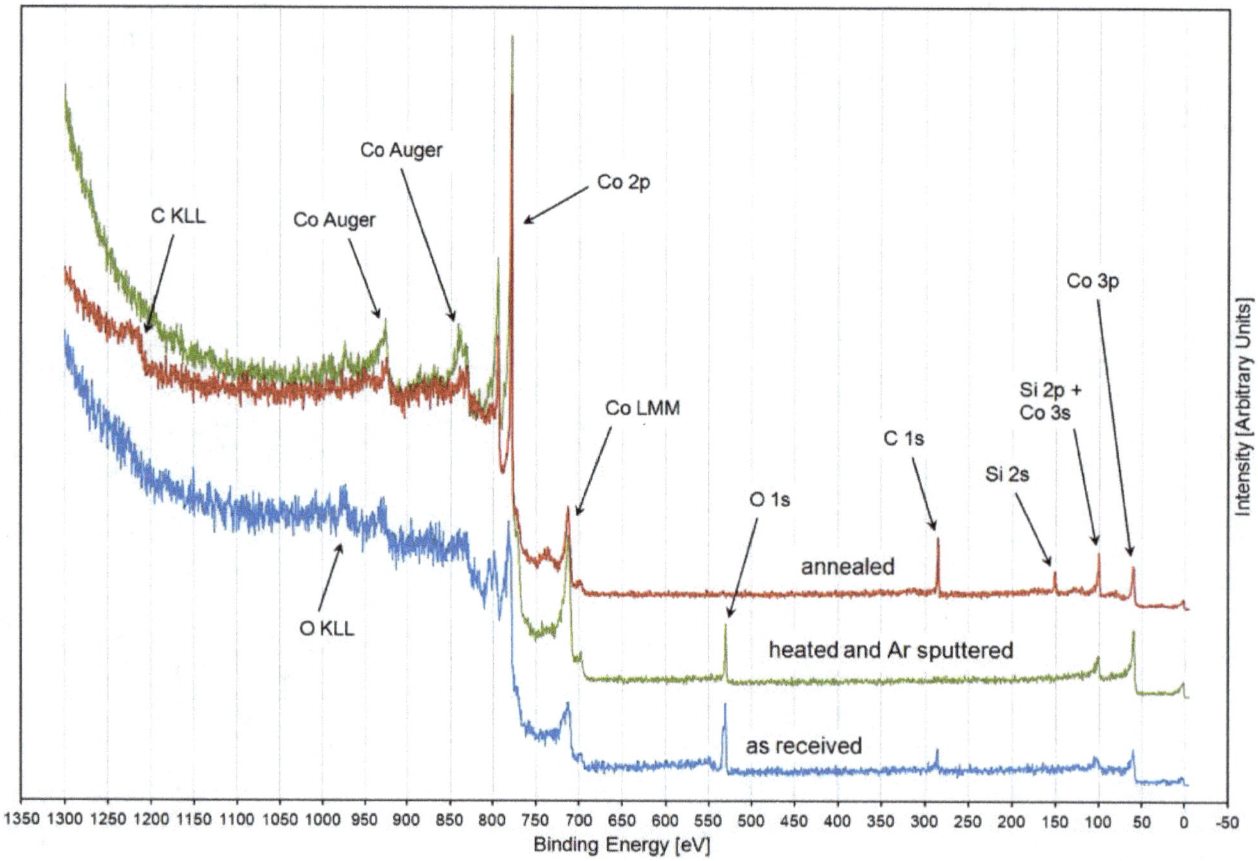

Figure 10. Survey spectra in comparison. Main found peaks are denoted. (Reprinted with permission from Thin Solid Films No. 3995281012757.).

Atomic concentration [%]	Si	C	O	Co
As received	0	24	40	36
Heated and Ar sputtered	0	0	23	77
Annealed	21	43	0	36

Reprinted with permission from Thin Solid Films No. 3995281012757.

Table 6. Quantification results of the atomic concentration of elements from XPS analysis.

that all spectra probably shifted to higher binding energies roughly by 0.5 eV. No software spectrum correcting was produced.

Figure 11 shows Co 2p peaks fitted by components on the basis of Ref. [37]. In the as-received state, a metallic Co component at around 779 eV, a Co^{II} and Co^{III} components at around 781 eV, and satellites at slightly higher binding energies can be identified, all confirming the native Co oxide [35–37]. As it can be anticipated, the quantity of oxide-based components was decreased by the Ar sputtering. There seems to be only metallic Co detected at the annealed sample. Cobalt is represented by an asymmetric main peak and weak plasmons [37]. Identification of

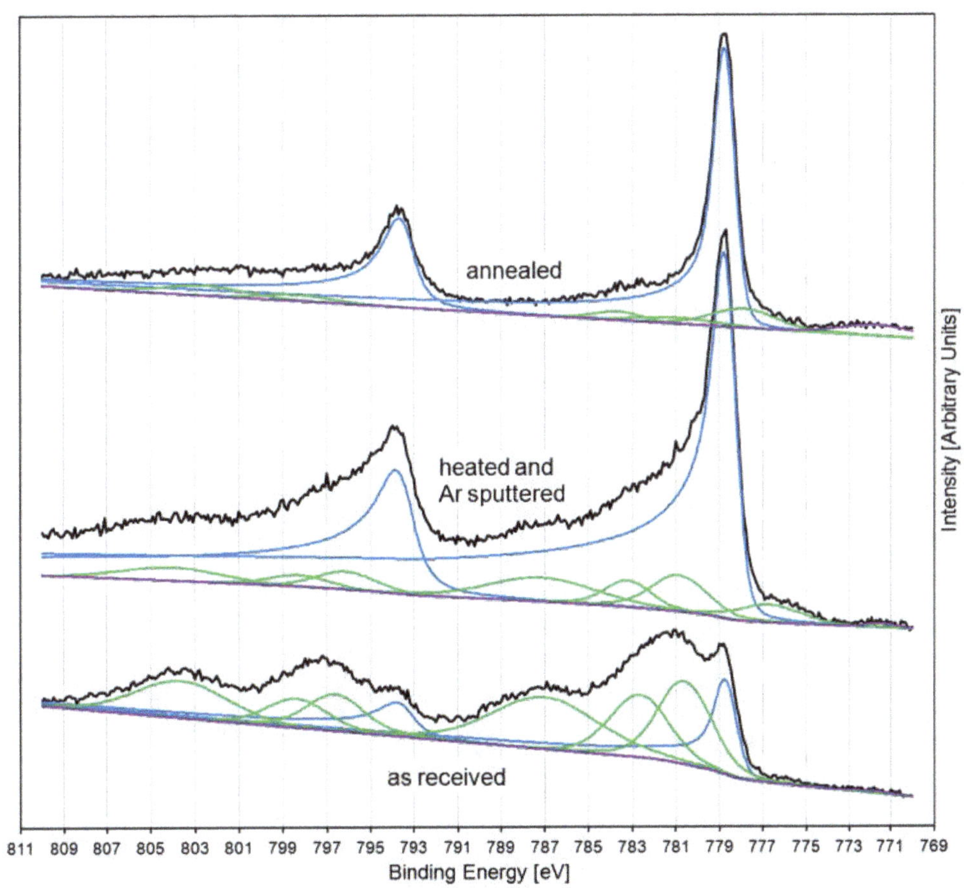

Figure 11. Co 2p in comparison. (Reprinted with permission from Thin Solid Films No. 3995281012757.).

silicide bonds by XPS is complicated [40]. But, we can suppose that Co and Si in the annealed sample create silicides. The silicides are standard reaction products in the Co/SiC system under similar reactions, and they are detected by other analytic techniques, for example, XRD [38, 39].

Information obtained from detail O 1s spectra support so far drawn suppositions. **Figure 12** shows that there are at least two chemical states of O in the as-received sample. The component at around 530.5 eV comprises O bonds in the Co oxides [35, 37]. The second component at around 532.5 eV probably belongs to disrupt oxide structure, Co hydroxides, water, and possibly impurities [35, 37]. The sputtering of the sample decreases the quantity of impurities and the oxides/hydroxides. O disappeared completely at the annealed sample.

Detail C 1s spectra, first off in comparison, are shown on **Figure 13** and with fitted components on **Figures 14** and **15**. The C 1s peak at the as-received sample can be fitted by three components; the first one positioned at 285.5 eV belongs to adventitious carbon, the second one at 287 eV, and the third one at 289.5 eV both belong to bonds between carbon and oxygen [37]. The components are standard for contamination carbon. The components can be easily removed by the argon sputtering.

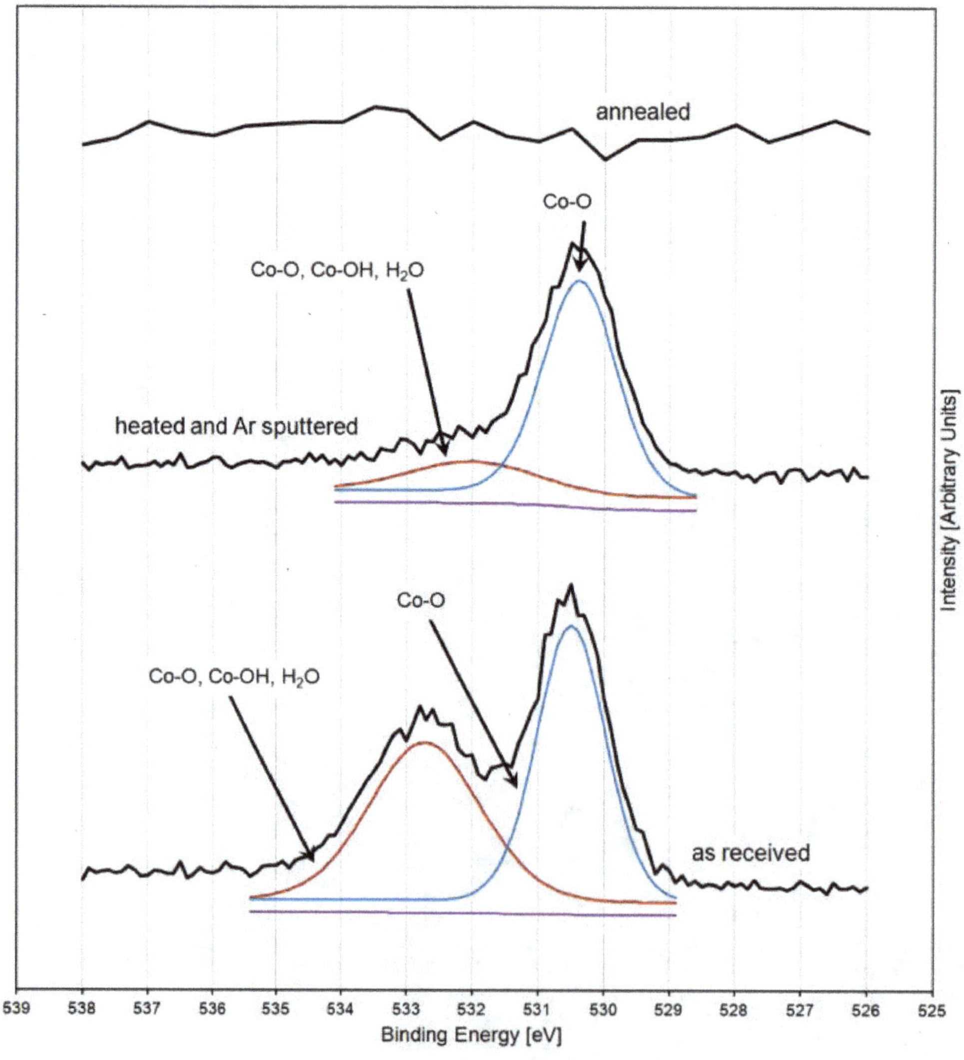

Figure 12. Fitted O 1s peaks in comparison. (Reprinted with permission from Thin Solid Films No. 3995281012757.).

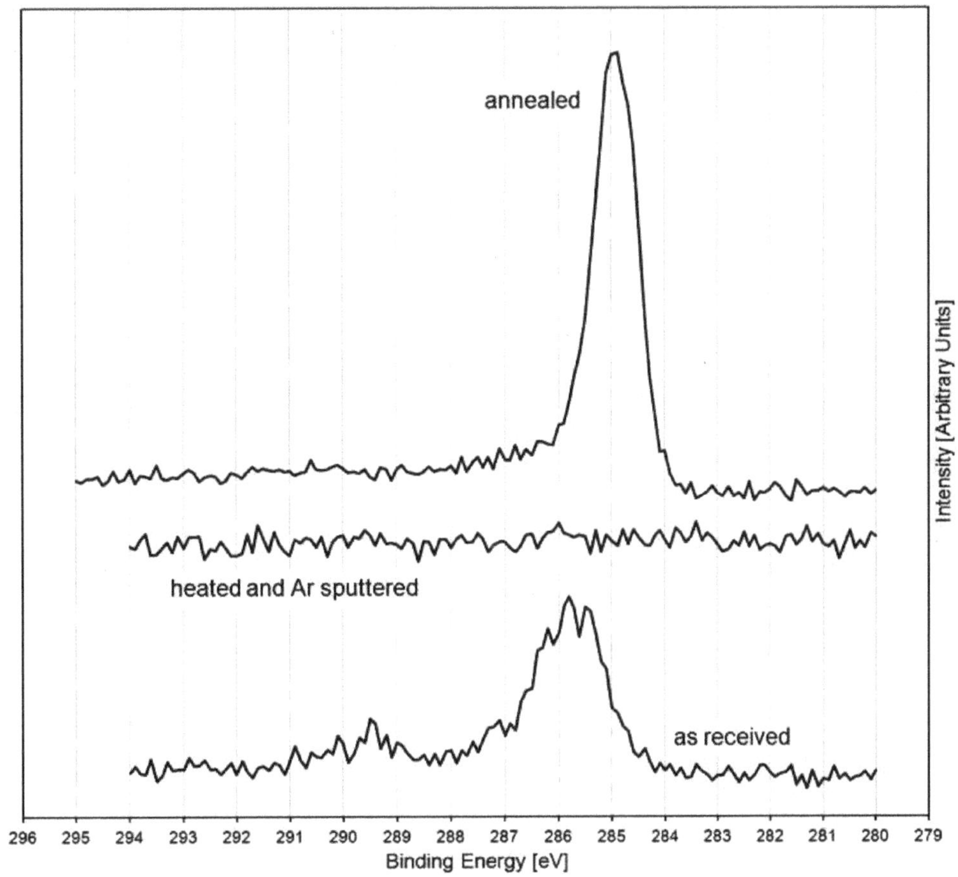

Figure 13. C 1s in comparison. (Reprinted with permission from Thin Solid Films No. 3995281012757.).

Figure 15 shows the most important carbon peak for us which was produced by the annealing. The peak is fitted with a single distinctive component positioned at 285 eV which has a relatively low FWHM (below 1 eV). C–C or C=C bonds are characteristic for the component and represent the formation of graphitic carbon phase during the annealing process [41, 42]. The peak was fitted by one symmetric component. Nevertheless, the peak shows slight asymmetry which is a mark of the metallic character of graphene [43]. The observed asymmetry of the C 1s peak in addition supports the presence of graphitic carbon.

4.5. Conclusion

The process of graphene growth was ground on an optimization of the thermal forming of the Co/SiC structure. The process was produced within a temperature range from 750 to 1080°C, and annealing time was changed in a range from 0 to 120 s. We were testing qualities of the prepared graphene films on samples' surface by the Raman spectroscopy. Nevertheless, the graphene film is also created at the metallization-SiC boundary; parameters of this so-called lower graphene are always worse than in the case of the surface graphene. This is produced by the reaction of SiC with Co leading to a large roughness of the silicon carbide surface at the interface. From the results, it is evident that the applied method produces graphene film with parameters of a bilayer one.

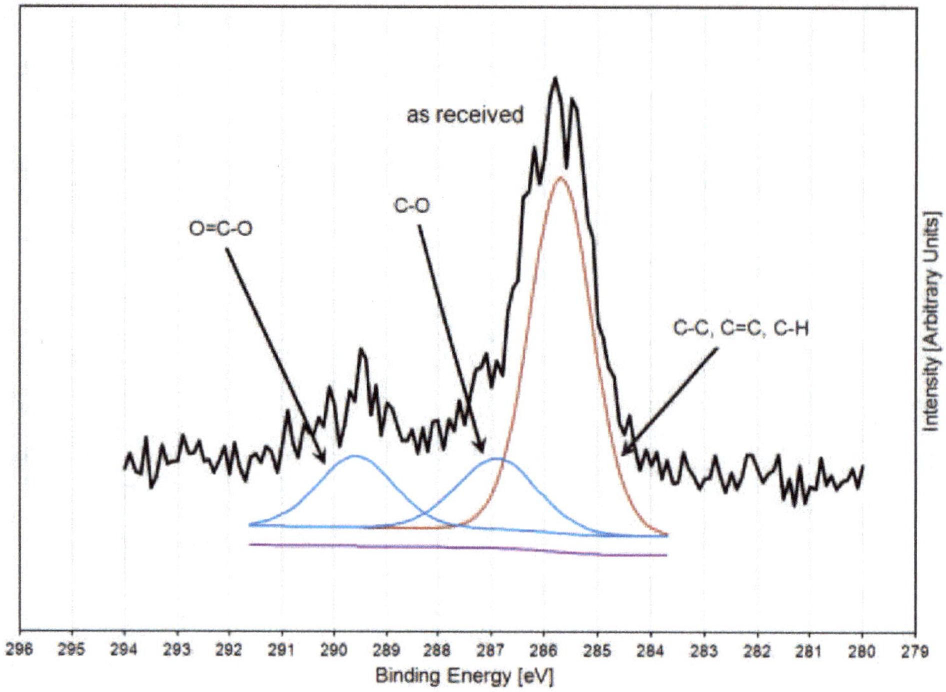

Figure 14. Fitted C 1s peak of the as-received sample. For clarity, peak components and background are vertically slightly offset. (Reprinted with permission from Thin Solid Films No. 3995281012757.).

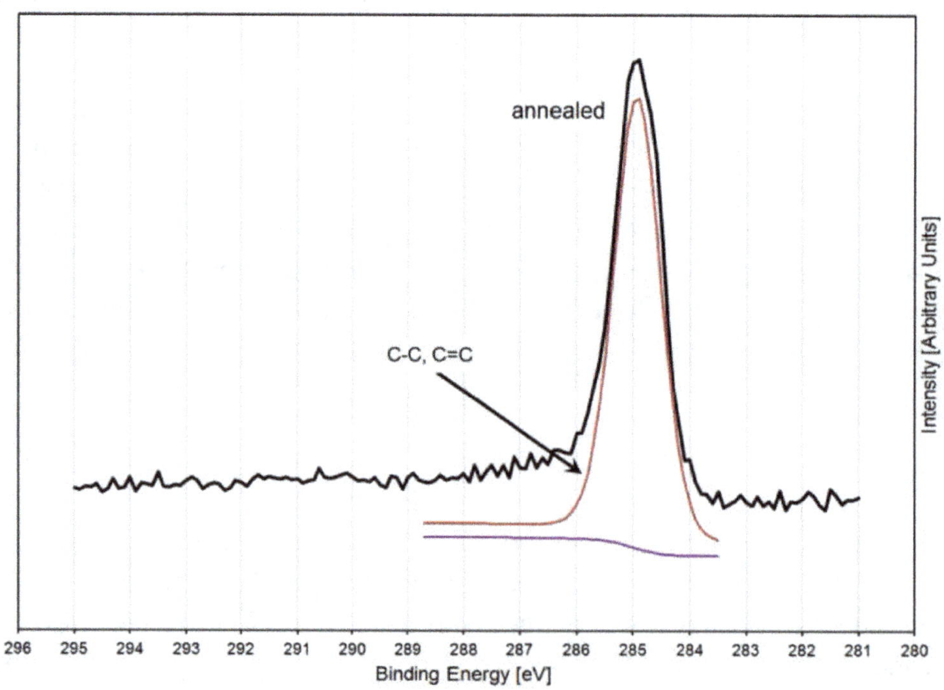

Figure 15. Fitted C 1s peak of the annealed sample. For clarity, peak components and background are vertically slightly offset. (Reprinted with permission from Thin Solid Films No. 3995281012757.).

At the Co(300 nm)/SiC structures prepared by the sputtering deposition process, the creation of two surface phases has been observed at shorter annealing period. In the case of the dark phase, the reaction between SiC and Co creates cobalt silicides and simultaneously the graphene film. Silicides and graphene are not created in the case of the light phase. By increasing the annealing time, the light-phase area shrinks, and finally it disappears. The XPS depth profiling detected a layer with increased amount of carbon atoms in the dark phase. The layer presumably represents carbon atoms that precipitate toward the surface of metallization. The precipitation is completed, and graphene is formed on the structure surface after the long experiment.

By applying the Co evaporation together with the plasma modification instead of the sputtering, the surface of the structure was homogenous and covered by the graphene film after annealing. We presuppose that the growing of the graphene film on the whole area can be attributed to a positive influence of the plasma treatment of the structure surface that was carried out just before the cobalt evaporation.

As mentioned above, it is generally postulated that parameters of the graphene films prepared by segregation depend very strictly on the CR of the structures after finishing the annealing process. Our results are different from this statement. The number of carbon monolayer in graphene has been practically independent on the CR; however, the graphene defectivity (the I_D/I_G ratio) is rapidly increasing. The possible explanation of the phenomena can be seen in the fact that the surface carbon layer (the graphene film) is formed not only by the segregation process after finishing the annealing as originally asserted but also during the annealing period. We assume that the CR modifies morphology and crystallic structure of the silicide layer and thus it increases the concentration of defects in the graphene film [44].

For the measurement of the electrical parameters, we apply the graphene film transferred onto the SiO_2/Si structure. The transfer of graphene slightly increased the graphene defectivity. The phenomenon is caused by interruption of covalent bonds between Co and carbon atoms in graphene. Unsaturated bonds are formed within the graphene due to etching off of Co, and so the number of defects in graphene film is increased [45]. Obtained value of sheet resistance is in agreement with data published in literature [45–47]. The charge carrier mobility is very small taking into consideration the maximal published values, but the graphene films with the same number of carbon monolayers produced by the CVD method [46, 48] show the charge mobility with similar or only slightly higher value. It is probably caused by the presence of significant number of defects which are absent at the best graphene structures prepared by mechanical exfoliation [4]. Grain size and cracks in the metal play a great role at the graphene transfer from polycrystalline metal substrates [46, 49]. Due to positive sign of the Hall constant, our graphene films show hole conductivity which is attributed to residues of PMMA used for the graphene transport [50].

XPS analysis of the Co/SiC structure together with its annealing and cleaning was performed in situ in the Ultra High Vacuum (UHV) apparatus. Contrary to our standard preparation process [31], special cleaning of the tested structure was performed by degassing and by Ar ions bombardment. The process significantly reduced the surface contamination. The UHV in situ structure cleaning and subsequent analysis provided clean and very well-defined experimental conditions. In **Figure 15**, it is obvious that the Co/SiC reaction formed the graphene film on the surface of the structure as intended.

5. Application of other metals

The mentioned method is suitable for many other materials, not only for Ni and Co. Selected metals have to react with silicon carbide by the way silicides and carbon rich products are formed at the metal-SiC interface. The accumulation of graphite at the top of the metal is produced during the cooling period of the growth process. The stability or reactivity of SiC versus the number of metals was studied on the basis of phase diagrams [51]. Pd, Ge [33, 52], and Fe [53] were tested for graphene growth. In majority cases FLG was prepared.

6. Influence of additive materials

The above-described process of graphene growth is influenced by solubility of carbon within the used metal. In practice, nickel and cobalt are most frequently used; solubility of carbon in these metals is relatively high, and consequently carbon diffusion is easy. By modifying the carbon solubility in metal, it is possible to influence the quantity of carbon, which gets into the volume of metallization during the reaction of the metal with SiC, and in this way, the parameters of formed graphene are possible to influence.

Preparation of graphene was carried out on the Ni/SiC and Co/SiC structures, where the following materials were added into the metallization: germanium, copper, and gold [54]. Solubility of carbon in these materials is very low compared to Ni and Co. The main goal of the work was then to study the influence of admixtures onto parameters of prepared graphene films. Adding the admixture materials to nickel did not improve the prepared graphene films very much. Certain improvement was observed for the structures with copper, where the presence of copper in the metallization manifested itself by lowering the defect quantity of graphene. Another situation is for the metallization with cobalt. Germanium and gold did not improve the graphene parameters; however, copper proved its significant meaning. BLG (bilayer graphene) with very low defect quantity was detected on the Co structure with molar content of Cu 20% (annealing at 1000°C for 10 s).

7. Conclusion

The text of this chapter was taken from the publication of the author and his other colleagues [25, 26, 31–33, 54]. The chapter is focused on graphene preparation by silicidation of the metal/SiC structure, where nickel and cobalt were studied in the form of the metal. Graphene is formed due to SiC decomposition by annealing of the said structure at the temperature of 800–1000°C. SiC reacts with the metal to form silicides and free carbon, which diffuses into the layer of the metal and its silicides. While cooling down the structure, carbon gets segregated onto the surface of the metallization in the form of graphene. Graphene gets formed also at the metal/SiC boundary; this lower graphene has poorer features. When using SI-SiC, it can however be used, for example, for construction of graphene transistors directly, without the need to transfer it onto a dielectric substrate.

Generally, we can say that graphene prepared from the structure containing cobalt reaches better parameters than for the structure with nickel. For the best result, preparation of bilayer graphene with low failure rate can be considered. Raman spectroscopy was chosen for the basic diagnostic method, for it is an easily accessible, nondestructive, and fast analysis.

Within the text, optimizing of graphene preparation by annealing is gradually discussed, including the questions related to the influence of cooling rate onto graphene parameters. Contrary to the results published in literature, it was found that the influence of cooling rate is negative. Further, we followed the question of the thickness of the prepared graphene using the XPS analysis. This analysis was also used for verification of graphene present on the used substrates. Surface of graphene films was studied using AFM. Transfer of graphene was carried out by means of PMMA onto the SiO_2/Si substrates. Basic electrical parameters were measured at graphene on dielectric substrates. In the closing part of the chapter, the influence of additional materials, which decrease carbon solubility in the metallization and which thus influence features of the prepared graphene film, was discussed. It was proven that the most suitable material is copper.

Graphene preparation by synthesis from the metal/SiC structure is a promising and simple method, which does not require high temperatures, complicated technological apparatuses, or handling with dangerous gasses. Therefore, the concerned method is of good prospects at graphene preparation for a number of applications. The mentioned method is very perspective for future microelectronics, because it can be used for direct growth of graphene films on semiconductor substrates (especially SiC).

Acknowledgements

The work was pursued under a financial support from the Czech Science Foundation, Project No. 17-00607S.

Author details

Petr Machac

Address all correspondence to: petr.machac@vscht.cz

Department of Solid State Engineering, University of Chemistry and Technology, Prague, Czech Republic

References

[1] W. Choi, I. Lahiri, R. Seerlabozina, Y.S. Kang, 2010 Synthesis of graphene and its applications: a review. Critical Reviews in Solid State and Materials Sciences 35 (1-2010) 52–71.

[2] M.J. Allen, V.C. Tung, R.B. Kaner, 2010 Honeycomb carbon: a review of graphene. Chemical Reviews 110 (1-2010) 132–145.

[3] T. Seyller, K.V. Emtsev, K. Gao, F. Speck, L. Ley, A. Tadich, L. Broekman, J.D. Riley, R.C.G. Leckey, O. Rader, A. Varykhalov, A.M. Shikin, 2006 Structural and electronic properties of graphite layers grown on SiC(0001). Surface Science 600 (18-2006) 3906–3911.

[4] K.S. Novoselov, A.K. Geim, S.V. Morozov, D. Jiang, Y. Zhang, S.V. Dubonos, I.V. Grigorieva, A.A. Firsov, 2004 Electric field effect in atomically thin carbon films. Science 306 (5696-2004) 666–669.

[5] K.S. Novoselov, D. Jiang, F. Schedin, T.J. Booth, V.V. Khotkevich, S.V. Morozov, A.K. Geim, 2005 Two-dimensional atomic crystals. Proceedings of the National Academy of Sciences of the United States of America (PNAS) 102 (30-2005) 10451–10453.

[6] H. Shioyama, 2001 Cleavage of graphite to graphene. Journal of Materials Science Letters 20 (6-2001) 499–500.

[7] A.N. Obraztsov, E.A. Obraztsova, A.V. Tyurnina, A.A. Zolotukhin, 2007 Chemical vapor deposition of thin graphite films of nanometer thickness. Carbon 45 (10-2007) 2017–2021.

[8] L.G. De Arco, Y. Zhang, A. Kumar, and C. Zhou, 2009 Synthesis, transfer, and devices of single- and few-layer graphene by chemical vapor deposition. IEEE Transactions on Nanotechnology 8 (2-2009) 135–138.

[9] Z.Y. Juang, C.Y. Wu, A.Y. Lu, C.Y. Su, K.C. Leou, F.R. Chen, C.H. Tsai, 2010 Graphene synthesis by chemical vapor deposition and transfer by a roll-to-roll process. Carbon 48 (11-2010) 3169–3174.

[10] W.A. de Heer, C. Berger, X. Wu, P.N. First, E.H. Conrad, X. Li, T. Li, M. Sprinkle, J. Hass, M.L. Sadowski, M. Potemski, G. Martinez, 2007 Epitaxial graphene. Solid State Communications 143 (1-2-2007) 92.

[11] S. Tanaka, K. Morita, H. Hibino, 2010 Anisotropic layer-by-layer growth of graphene on vicinal SiC(0001) surfaces. Physical Review B 81 (4-2010) 041406(R).

[12] M.L. Bolen, S.E. Harrison, L.B. Biedermann, M.A. Capano, 2009 Graphene formation mechanisms on 4*H*-SiC(0001). Physical Review B 80 (11-2009) 115433.

[13] K.V. Emtsev, A. Bostwick, K. Horn, J. Jobst, G.L. Kellogg, L. Ley, J.L. McChesney, T. Ohta, S.A. Reshanov, J. Röhrl, E. Rotenberg, A.K. Schmid, D. Waldmann, H.B. Weber, T. Seyller, 2009 Towards wafer-size graphene layers by atmospheric pressure graphitization of silicon carbide. Nature Materials 9 (3-2009) 203–207.

[14] T. Ohta, N.C. Bartelt, K. Thürmer, G.L. Kellogg, 2010 Role of carbon surface diffusion on the growth of epitaxial graphene on SiC. Physical Review B 81 (12-2010) 121411(R).

[15] M.E. Ayhan, G. Kalita, R. Papon, R. Hirano and M. Tanemura, 2014 Synthesis of transfer-free graphene by solid phase reaction process in presence of a carbon diffusion barrier. Materials Letters 129, 76–79.

[16] Q. Zhuo, Q. Wang, Y. Zhang, D. Zhang, Q. Li, C. Gao, Y. Sun, L. Ding, Q. Sun, S. Wang, J. Zhong, X. Sun, S. Lee, 2015 Transfer-free synthesis of doped and patterned graphene films. ACS Nano 9 (1-2015) 594–601.

[17] A. Delamoreanu, C. Rabot, C. Vallee and A. Zenasni, 2014 Wafer scale catalytic growth of graphene on nickel by solid carbon source. Carbon 66, 48–56.

[18] Z.Y. Juang, C.Y. Wu, C.W. Lo, W.Y. Chen, C.F. Juany, J.C. Hwang, F.R. Chen, K.C. Leou, C.H. Tsai, 2009 Synthesis of graphene on silicon carbide substrates at low temperature. Carbon 47 (11-2009) 2026–2031.

[19] A.A. Woodworth, C.D. Stinespring, 2010 Surface chemistry of Ni induced graphite formation on the 6H–SiC (0001) surface and its implications for graphene synthesis. Carbon 48 (7-2010) 1999–2003.

[20] T. Yoneda, M. Shibuya, K. Mitsuhara, A. Visikovskiy, Y. Hoshino, Y. Kido, 2010 Graphene on SiC(0001) and SiC(000(1)over-bar) surfaces grown via Ni-silicidation reactions. Surface Science 604 (17-18-2010) 1509–1515.

[21] K. Vassilevski, I.P. Nikitina, A.B. Horsfall, N.G. Wright, C.M. Johnson, 2010 Epitaxial graphene elaborated on 3C-SiC(111)/Si epilayers. Materials Science Forum 645–648, 589–592

[22] J. Hofrichter, B.N. Szafranek, M. Otto, T.J. Echtermeyer, M. Baus, A. Majeru, V. Geringer, M. Ramsteiner, H. Kurz, 2010 Synthesis of graphene on silicon dioxide by a solid carbon source. Nano Letters 10 (1-2010) 36–42.

[23] C. Li, D. Li, J. Yang, X. Zeng, W. Yuan, 2011 Preparation of single and few layer graphene sheets using Co depositing on SiC substrate. Journal of Nanomaterials 2011 (2011) 319624.

[24] B.V. Cunning, M. Ahmed, N. Mishra, A.R. Kermany, B. Wood, F. Iacopi, 2014 Graphitized silicon carbide microbeams: wafer-level, self-aligned graphene on silicon wafers. Nanotechnology 25 (32-2014) 325301.

[25] P. Machac, T. Fidler, S. Cichon, L. Miskova, 2012 Synthesis of graphene on SiC substrate via Ni-silicidation reactions. Thin Solid Films 520 (2012) 5215–5218.

[26] P. Machac, T. Hrebicek, 2016 Synthesis of graphene on Ni/SiC structure. Journal of Electrical Engineering 67 (2-2016) 147–149.

[27] M.A. Pimenta, G. Dresselhaus, M.S. Dresselhaus, L.G. Cancado, A. Jorio, R. Saito, 2007 Studying disorder in graphite-based systems by Raman spectroscopy. Physical Chemistry Chemical Physics 9 (11-2007) 1276–1291.

[28] Y. Hao, Y. Wang, L. Wang, Z. Ni, Z. Wang, R. Wang, C.K. Koo, Z. Shen, J.T.L. Thong, 2010 Probing layer number and stacking order of few-layer graphene by Raman spectroscopy. Small 6 (2-2010) 195–200.

[29] S. Tanuma, C.J. Powell, D.R. Penn, 1991 Calculations of electron inelastic mean free paths II. Data for 27 elements over the 50-2000 eV range. Surface and Interface Analysis 17 (13-1991), 911–926.

[30] Y. Wu, W. Ren, L. Gao, B. Liu, C. Jiang, H. Cheng, 2009 Synthesis of high-quality graphene with a pre-determined number of layers. Carbon 47 (2-2009) 493–499.

[31] P. Machac, T. Fidler, S. Cichon, V. Jurka, 2013 Synthesis of graphene on Co/SiC structure. Journal of Materials Science: Materials in Electronics 24 (10-2013) 3793–3799.

[32] P. Machac, S. Cichon, L. Miskova, M. Vondracek, 2014 Graphene preparation by annealing of Co/SiC structure. Applied Surface Science 320 (2014) 544–551.

[33] P. Machac, T. Fidler, S. Cichon, 2013 Graphene preparation by annealing of metal/SiC structure. Proceedings EDS'13 IMAPS CS International Conference Brno, Czech Republic, June 26–27, 2013, 11–14.

[34] A. Umair, H. Raza, 2012 Controlled synthesis of bi-layer graphene on nickel. Nanoscale Research Letters 7 (1-2012) 437–440.

[35] A. Galtayries, J. Grimblot, 1999 Formation and electronic properties of oxide and sulphide films of Co, Ni and Mo studied by XPS. Journal of Electron Spectroscopy and Related Phenomena 98–99 (1-1999) 267–275.

[36] J.G. Dillard, H. Glasbrenner, G. Pfennig, H. Klewe-Nebenius, H.J. Ache, 1990 Surface-analysis studies of Zr-Co alloy and Zr-Co alloy-films. Journal of the Less-Common Metals 166 (2-1990) 233–239.

[37] M.C. Biesinger, B.P. Payne, A.P. Grosvenor, L.W.M. Lau, A.R. Gerson, R.St.C. Smart, 2011 Resolving surface chemical states in XPS analysis of first row transition metals, oxides and hydroxides: Cr, Mn, Fe, Co and Ni. Applied Surface Science 257 (7-2011) 2717–2730.

[38] S.W. Park, Y.I. Kim, J.S. Kwak, H.K. Baik, 1997 Investigation of Co/SiC interface reaction. Journal of Electronic Materials 26 (3-1997) 172–177.

[39] T. Fujimura, S.I. Tanaka, 1999 In-situ high temperature X-ray diffraction study of Co/SiC interface reactions. Journal of Materials Science 34 (23-1999) 5743–5747.

[40] P.L. Tam, Y. Cao, L. Nyborg, 2012 XRD and XPS characterisation of transition metal silicide thin films. Surface Science 606 (3-4-2012) 329–336.

[41] U. Starke, C. Riedl, 2009 Epitaxial graphene on SiC(0001) and SiC(000$\bar{1}$): from surface reconstructions to carbon electronics. J. Phys: Condensed Matter. 21 (13-2009) 134016.

[42] E. Rollings, G.H. Gweon, S.Y. Zhou, B.S. Mun, J.L. McChesney, B.S. Hussain, A.V. Fedorov, P.N. First, W.A. de Heer, A. Lanzara, 2006 Synthesis and characterization of atomically thin graphite films on a silicon carbide substrate. Journal of Physics and Chemistry of Solids 67 (9-10-2006) 2172–2177.

[43] M. Ostler, F. Speck, M. Gick, T. Seyller, 2010 Automated preparation of high-quality epitaxial graphene on 6H-SiC(0001). Physica Status Solidi B – Basic Solid State Physics 247 (11-12-2010) 2924–2926.

[44] H. Ago, Y. Ito, N. Mizuta, K. Yoshida, B. Hu, C.M. Orofeo, M. Tsuji, K. Ikeda, S. Mizuno, 2010 Epitaxial chemical vapor deposition growth of single-layer graphene over cobalt film crystallized on sapphire. ACS Nano 4 (11-2010) 7407–7414.

[45] N. Liu, L. Fu, B. Dai, K. Yan, X. Liu, R. Zhao, Y. Zhang, Z. Liu, 2011 Universal segregation growth approach to wafer-size graphene from non-noble metals. Nano Letters 11 (1-2011) 297–303.

[46] A. Reina, X. Jia, J. Ho, D. Nezich, H. Son, V. Bulovic, M.S. Dresselhaus, J. Kong, 2009 Large area, few-layer graphene films on arbitrary substrates by chemical vapor deposition. Nano Letters 9 (1-2009) 30–35.

[47] A. Reina, S. Thiele, X. Jia, S. Bhaviripudi, M.S. Dressehaus, J.A. Schaefer, J. Kong, 2009 Growth of large-area single- and Bi-layer graphene by controlled carbon precipitation on polycrystalline Ni surfaces. Nano Research 2 (6-2009) 509–516.

[48] X. Li, W. Cai, J. An, S. Kim, J. Nah, D. Yang, R. Piner, A. Velamakanni, I. Jung, E. Tutuc, S.K. Banerjee, L. Colombo, R.S. Ruoff, 2009 Large-area synthesis of high-quality and uniform graphene films on copper foils. Science 324 (5932-2009) 1312–1314.

[49] S. Thiele, A. Reina, P. Healey, J. Kedziersky, P. Wyatt, P.L. Hsu, C. Keast, J. Schaefer, J. Kong, 2010 Engineering polycrystalline Ni films to improve thickness uniformity of the chemical-vapor-deposition-grown graphene films. Nanotechnology 21 (1-2010) 015601.

[50] J. Kang, D. Shin, S. Bae, B.H. Hong, 2012 Graphene transfer: key for applications. Nanoscale 4 (1-2012) 5527–5537.

[51] J.C. Schuster, 1993 Silicon carbide and transition metals: a critical evaluation of existing phase diagram data supplemented by new experimental results. International Journal of Refractory Metals and Hard Materials 12 (4-1993-1994) 173–177.

[52] H. Hiura, M.V. Lee, A.V. Tyurnina, K. Tsukagoshi, 2012 Liquid phase growth of graphene on silicon carbide. Carbon 50 (14-2012) 5076–5084.

[53] S.P. Cooil, F. Song, G.T. Williams, O.R. Roberts, D.P. Langstaff, B. Jorgensen, K. Hoydalsvik, D.W. Breiby, E. Wahlström, D.A. Evans, J.W. Wells, 2012 Iron-mediated growth of epitaxial graphene on SiC and diamond. Carbon 50 (14-2012) 5099–5105.

[54] T. Hrebicek: Graphene preparation by silicidation of the structure metal/SiC. Diploma Thesis, University of Chemistry and Technology, Prague, Czech Republic, 2016.

8

A Novel Grafting of Polymers onto the Surface of Graphene Oxide

Norio Tsubokawa, Takeshi Yamauchi,
Kazuhiro Fujiki and Shingo Tamesue

Abstract

A simple grafting of polymers onto graphene oxide (GO) was achieved by polymer radical trapping, ligand-exchange reaction, and surface initiated cationic and anionic graft polymerization. Grafting of poly(ethylene glycol) (PEG) onto GO was successfully achieved by trapping of PEG radicals formed by thermal decomposition of PEG macroazo initiator to give PEG-grafted GO. The grafting of copolymers containing vinyl ferrocene moieties onto GO surfaces was also successfully achieved by the ligand-exchange reaction between ferrocene moieties of these copolymers and GO. Carboxyl groups on GO have an ability to initiate the cationic polymerizations of vinyl monomers, such as *N*-vinylcarbazole and isobutyl vinyl ether. The corresponding vinyl polymers were grafted onto GO, during the cationic polymerization, based on the termination of growing polymer cation by counter anion (carboxylate) groups on GO. It was found that the anionic ring-opening alternating copolymerization of epoxides with cyclic acid anhydrides was successfully initiated by potassium carboxylate groups on GO, introduced by neutralization of carboxyl groups with KOH, to give the corresponding polyester-grafted GO. The dispersibility of GO in organic solvents was remarkably improved by the grafting of the above polymers onto GO. In addition, easy preparation of reduced GO-based conducting polyaniline composite organogel will be discussed.

Keywords: graphene, graphene oxide, grafting of polymers, surface-initiated polymerization, ligand-exchange reaction

1. Introduction

Recently, graphene and graphene oxide (GO), as well as carbon nanotube, nanodiamond, and fullerene, have emerged as excellent nanomaterials having electrical and thermal conductive, mechanical, gas barrier, optical, and antibacterial properties. Therefore, a polymer composite with GO has recently attracted much interest in the materials field due to its outstanding properties [1–8]. Especially, many researchers have reported the preparation and properties of GO/chitin and GO/biopolymer composites [9–15]. For example, Travlou et al. have reported the synthesis and applications of GO/chitosan and GO/polysaccharides nanocomposites [9, 14].

It is well known that in comparison with GO/chitin nanocomposite, we can obtain GO having an excellent dispersibility into organic solvents and various polymer matrices by grafting of conventional polymers onto GO and readily obtain thermosetting, thermoplastic, and thermoelastic polymer nanocomposite thin films having electro and thermal conductivity. The surface modifications of graphene oxide (GO) by grafting of polymers via atom transfer radical polymerization (ATRP) [4–6, 16, 17] and reverse addition fragmentation chain transfer polymerization (RAFT) have been reported by many researchers [7, 8, 18]. According to the above-mentioned method, polymer brush grafted onto GO can be obtained.

However, the above graft polymerization needs the complicated treatment for the introduction of surface-initiating groups. Therefore, it is desired to develop a simple and an easy method for the grafting of polymers onto GO without complicated procedures for the introduction of the initiating groups onto GO.

On the other hand, we have also achieved the grafting of various polymers onto various nanocarbons such as carbon black, carbon nanotubes, fullerene, and nanodiamond, by "grafting from" and "grafting onto" methods [19, 20]. We have designed a simple surface grafting of various polymers onto these nanocarbons by polymer radical trapping [21, 22], ligand-exchange reaction [23, 24], surface-initiated cationic [25], and anionic graft polymerization [26]. According to the above processes, they do not require complicated process for the introduction of initiating groups onto nanocarbons for the graft polymerization.

In this chapter, a novel and an easy grafting of polymers onto GO without complicated pretreatment by trapping of polymer radicals [27], ligand-exchange reaction of ferrocene-containing polymer with GO [27], and simple cationic and anionic graft polymerization initiated by carboxyl groups on GO [28] are reviewed. In addition, the dispersibility of various polymer-grafted GO in several organic solvents and easy preparation of conductive composite gel consisting of reduced GO and polyaniline will be discussed.

2. Grafting of polymers onto GO by polymer radical trapping

2.1. Radical reactivity against aromatic compounds and nanocarbons

It has been reported that reactivity of methyl radical to aromatic compounds increases with the increasing number of aromatic rings: the relative reactivity of methyl radical against to carbon black is ten million times of that of benzene [29]. Therefore, nanocarbons, such as carbon black, fullerene, and carbon nanotube, are known as a strong radical-trapping agent.

It is well known that radical polymerization of vinyl monomers is dramatically retarded (or inhibited) in the presence of nanocarbons, such as carbon black and carbon nanotubes.

These results indicate that during the radical polymerization in the presence of nanocarbons, initiator radicals and growing polymer radicals are readily trapped by nanocarbons [29].

2.2. Confirmation of radical trapping ability of GO

To make sure the radical trapping activity of GO, the effect of GO on thermally initiated radical polymerization of styrene (St) at 80°C was investigated. **Figure 1** shows the result of the thermally initiated radical polymerization of St in the absence (blank) and in the presence of GO at 80°C.

As shown in **Figure 1**, it is found that the thermal radical polymerization of St is remarkably retarded in the presence of GO. It was confirmed that during polymerization, a part of polySt formed was grafted onto GO to give polySt-grafted GO (GO-*g*-polySt), although the formation of ungrafted polySt preferentially proceeded: the grafting of polySt onto GO was confirmed by GC-MS: the percentage of polySt grafting was less than few percentage. Based on the above results, it is concluded that GO has a strong radical-trapping activity.

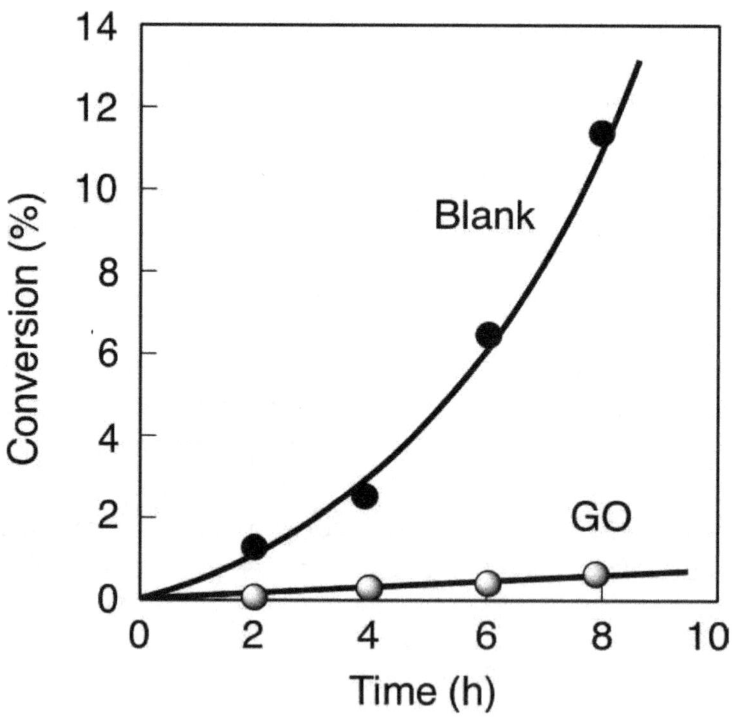

Figure 1. Thermally initiated radical polymerization of St in the absence (blank) and in the presence of GO (GO). GO, 0.10 g; St, 10.0 mL; Temp., 80°C.

2.3. Grafting of PEG by radical trapping

The grafting of poly(ethylene glycol) (PEG) onto the GO surface by trapping of PEG radicals produced by the thermal decomposition of PEG macroazo initiator (Azo-PEG) was investigated (**Scheme 1**). We used commercially available Azo-PEG [30].

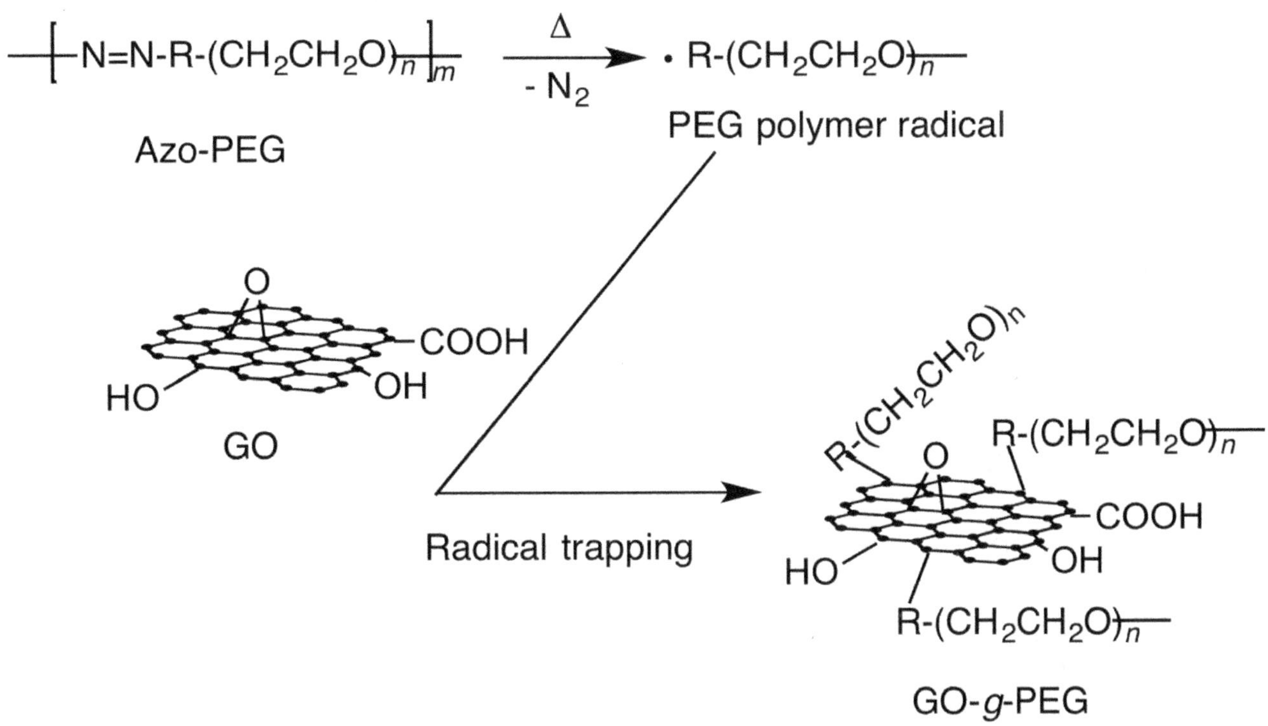

Scheme 1. Grafting of PEG onto GO by the trapping of PEG radicals formed by thermal decomposition of Azo-PEG.

Figure 2 shows the effect of polymerization time on the percentage of PEG grafting (percentage of grafted polymer on GO) onto the GO surface. It is found that the percentage of PEG grafting increased with the passage of the polymerization and reached over 15% after 24 h.

The grafting of PEG onto GO, on the contrary, was hardly observed when GO was reacted at room temperature, because of no decomposition of Azo-PEG for the generation of PEG radicals.

2.4. Identification of PEG grafting onto GO by GC-MS

Identification of PEG grafting onto GO was achieved by using gas chromatogram and mass spectra (GC-MS) of thermally decomposed gas of GO-*g*-PEG. The GC-MS of PEG, GO-*g*-PEG, and untreated GO is shown in **Figures 3** and **4**. **Figure 3** clearly shows that the GC of GO-*g*-PEG agreed with that of PEG.

Furthermore, as shown in **Figure 4**, the MS of thermally decomposed gas of GO-*g*-PEG at retention time 6.8 min was also in accord with that of PEG: the structures of fragment at 45, 59, 73, and 89 (m/z) estimated from MS database are shown in **Figure 4**.

The MS of thermally decomposed gas of GO-*g*-PEG at other retention time was also in accord with that of PEG. These results suggested that PEG radicals, formed by the thermal

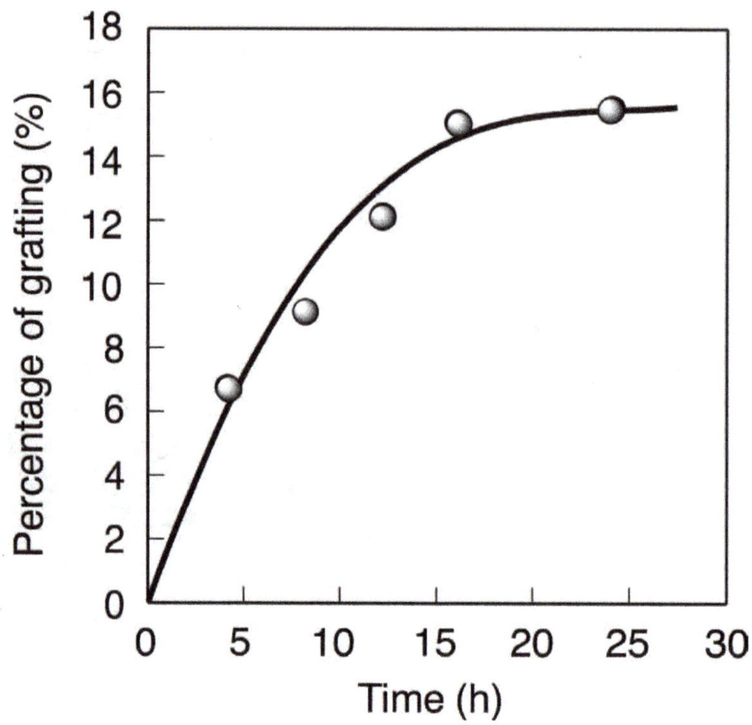

Figure 2. Grafting of PEG onto GO by the reaction of PEG with Azo-PEG. GO, 0.10 g; Azo-PEG, 2.00 g; toluene, 20.0 mL; Temp., 80°C.

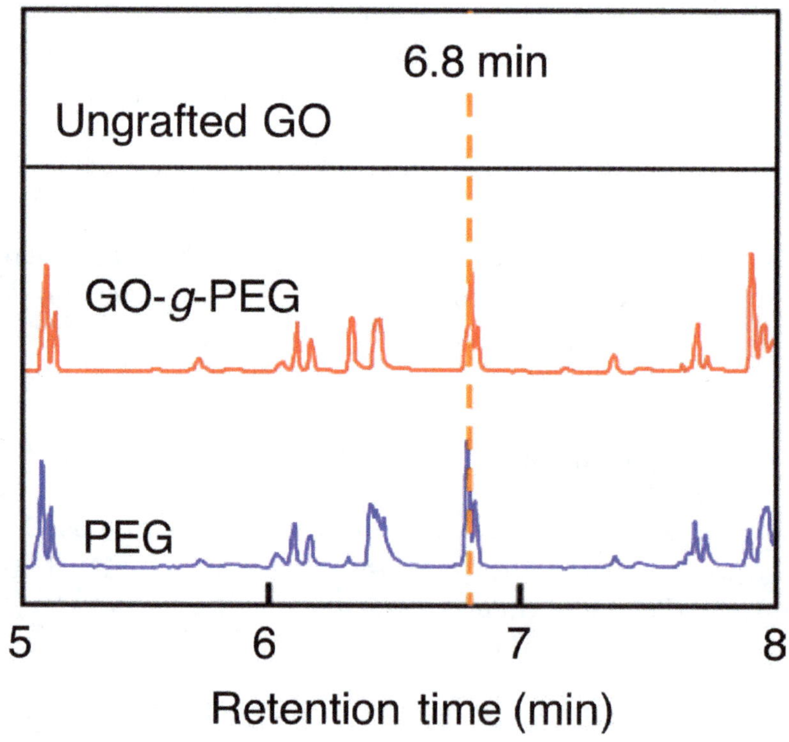

Figures 3. Thermal decomposition gas chromatograms of untreated GO, GO-*g*-PEG, and PEG.

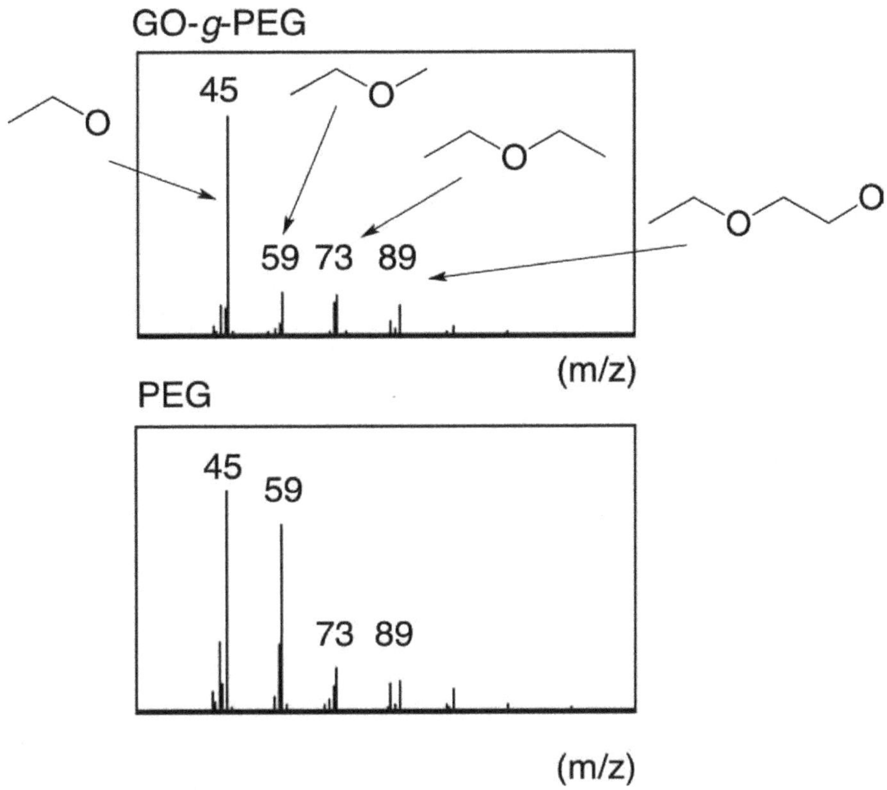

Figure 4. Mass spectra of decomposed gas of GO-*g*-PEG and PEG at retention time 6.8 min.

decomposition of Azo-PEG, are successfully captured by GO surface, and PEG is grafted (chemically bonded) onto GO.

3. Grafting of polymers onto GO by ligand-exchange reaction with ferrocene

3.1. Ligand-exchange reaction of ferrocene with graphene structure of nanocarbons

Morrison et al. reported that the h^6-benzene-h^5-cyclopentadienyliron cation could be readily prepared by the ligand-exchange reaction of ferrocene with benzene in the presence of $AlCl_3$ and Al powders as the catalysts [31, 32]. Furthermore, Miyake et al. reported that the ligand-exchange reaction is successfully applied for the introduction of functional groups to graphene structure of various carbon materials [32]. We have reported the grafting of polymers by ligand-exchange reaction of ferrocene moieties of polymers with graphene structure of carbon black, carbon fiber, and carbon nanofibers [23, 24]. Therefore, we designed the grafting of poly(Vf-*co*-MMA) onto GO surfaces by ligand-exchange reaction between ferrocene moieties of poly(Vf-*co*-MMA) and polycondensed aromatic rings of the GO surface, as shown in **Scheme 2**.

3.2. Preparation of poly(Vf-*co*-vinyl monomer) and characterization

Ferrocene containing copolymer of vinyl ferrocene (Vf) with methyl methacrylate (MMA) [poly(Vf-*co*-MMA)] and styrene (St) [poly(Vf-*co*-St)] was prepared by the copolymerization of the corresponding monomers, using 4,4'-azobisisobutylonitrile (AIBN) as an initiator.

Table 1 shows the number-average molecular weight (M_n) and molecular weight distribution (M_w/M_n), which are determined by SEC, Vf contents of poly(Vf-*co*-MMA), and poly(Vf-*co*-St), which are determined by elemental analysis, and are also shown in **Table 1**. As shown in **Table 1**, it is found that M_w/M_n of these copolymers prepared by the conventional radical copolymerization by using AIBN as an initiator is considerably narrow. The content of vinyl ferrocene (Vf) moieties of poly(Vf-*co*-MMA) and poly(Vf-*co*-St) is estimated to be 9.3 and 6.3%, respectively. ^{1}H-NMR and FT-IR are used to confirm the structures of poly(Vf-*co*-MMA) and poly(Vf-*co*-St).

3.3. Grafting of poly(Vf-*co*-MMA) and poly(Vf-*co*-St) onto GO by ligand-exchange reaction

The results of the grafting reaction of poly(Vf-*co*-MMA) with GO surface under several conditions are shown in **Table 2**. No grafting of the copolymer onto GO surface was hardly observed, even if, GO was reacted (heated) with poly(Vf-*co*-MMA) in 1,4dioxane in the absence of $AlCl_3$ and Al powders as a catalyst at 80°C for 24 h (Run 1). In addition, no grafting of the copolymer onto GO proceeded in the presence of Al powder alone (Run 2).

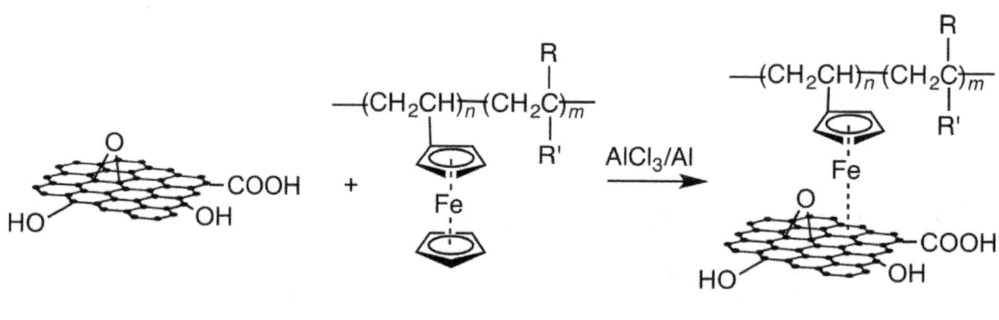

GO | Poly(Vf-*co*-MMA) / Poly(Vf-*co*-St) | GO-*g*-poly(Vf-*co*-MMA) / GO-*g*-poly(Vf-*co*-St)

Scheme 2. Grafting of poly(Vf-*co*-MMA) and poly(Vf-*co*-St) onto GO by ligand-exchange reaction of these copolymers with polycondensed aromatic rings of the surface.

Vf copolymer	Time (h)	*Mn* ×10⁴	*Mw*/*Mn*	Content of Vf (mol%)
Poly(Vf-*co*-MMA)	24	1.3	1.23	9.3
Poly(Vf-*co*-St)	48	2.1	1.38	6.3

Note: Molar ratio in feed, Vf: vinyl monomer = 1:9; Temp., 70°C.

Table 1. Molecular weight, molecular weight distribution, and Vf content of Vf copolymers.

On the other hand, the grafting reaction successfully proceeded in the presence of $AlCl_3$ alone (Run 3). In the coexistence of $AlCl_3$ and Al powders, the grafting of poly(Vf-*co*-MMA) successfully proceeded and the percentage of grafting reached 54.1% after 24 h (Run 4). These results indicate that $AlCl_3$ receives cyclopentadien of ferrocene, and after the grafting of the copolymers onto GO, iron in ferrocene moiety exists in reduced form, because it is reported that Al powder prevents ferrocene from being oxidized to ferrocenium cation [32].

Figure 5 shows the relationship between reaction time and percentage of grafting during the ligand-exchange reaction of GO with poly(Vf-*co*-MMA) at 80°C. The percentage of poly(Vf-*co*-MMA) grafting increased with progress of the reaction and reached 54.2% after 24 h, but no longer increased after 24 h. This may be due to the fact that GO surface was blocked by previously grafted poly(Vf-*co*-MMA) chains.

Run	Catalyst		Grafting (%)
	$AlCl_3$ (10^{-4} mol)	Al powder (10^{-4} mol)	
1	–	–	0
2	–	1.8	0
3	7.0	–	21.0
4	7.0	1.8	54.1

Note: GO, 0.05 g; poly(Vf-*co*-MMA), 0.05 g; 1,4-dioxane, 10.0 mL, Temp., 80°C; 24 h.

Table 2. Effect of catalyst on the grafting of poly(Vf-*co*-MMA) onto GO by ligand-exchange reaction.

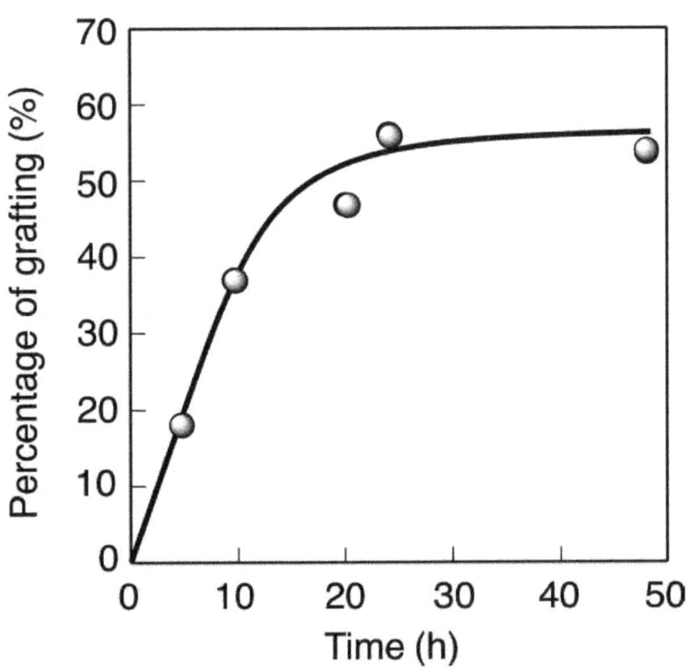

Figure 5. Grafting of poly(Vf-*co*-MMA) onto GO by ligand-exchange reaction of poly(Vf-*co*-MMA) with GO. GO, 0.05 g; poly(Vf-*co*-MMA), 0.05 g; $AlCl_3$, 0.70 mmol; Al powder, 0.18 mmol; 1,4-dioxane, 10.0 mL; Temp., 80°C.

The grafting of poly(Vf-*co*-St) onto GO was also achieved by the ligand-exchange reaction of the corresponding copolymer with GO: the percentage of poly(Vf-*co*-St) grafting was determined to be 61.1%.

The mole number of grafted poly(Vf-*co*-MMA) and poly(Vf-*co*-St) on GO was estimated to be 4.2×10^{-5} and 3.0×10^{-5} mol/g, respectively. This may be due to the fact that the ferrocene content of poly(Vf-*co*-St) is smaller than that of poly(Vf-*co*-MMA).

3.4. Identification of poly(Vf-*co*-MMA)-grafted GO by GC-MS

The GC of thermally decomposed gas of untreated GO, GO-*g*-poly(Vf-*co*-MMA) and poly(Vf-*co*-MMA) is shown in **Figure 6**. It was observed that the generation of same thermally decomposed gas was generated at retention time 5.8 min. However, the small difference in thermal decomposition GC between poly(Vf-*co*-MMA) and GO-*g*-poly(Vf-*co*-MMA) was observed. The same results were reported in the case of poly(Vf-*co*-MMA)-grafted vapor grown carbon fiber and carbon black [17]. It can be presumed that that poly(Vf-*co*-MMA) was grafted by coordinate bonds with graphene structure of GO, as shown in **Scheme 2**.

The MS of decomposed gas of GO-*g*-poly(Vf-*co*-MMA) and poly(Vf-*co*-MMA) at retention time 5.8 min is shown in **Figure 7**. As shown in **Figure 7**, the MS of decomposed gas of GO-*g*-poly(Vf-*co*-MMA) was also in accord with that of poly(Vf-*co*-MMA): the structure of fragments at 56, 65, and 91 (m/z) estimated from MS database are also shown in **Figure 7**. These results clearly indicate that poly(Vf-*co*-MMA) was successfully grafted onto GO, and poly(Vf-*co*-MMA) was immobilized by coordinate bonds between the graphene structure of GO and the ferrocene moiety of poly(Vf-*co*-MMA) as mentioned above.

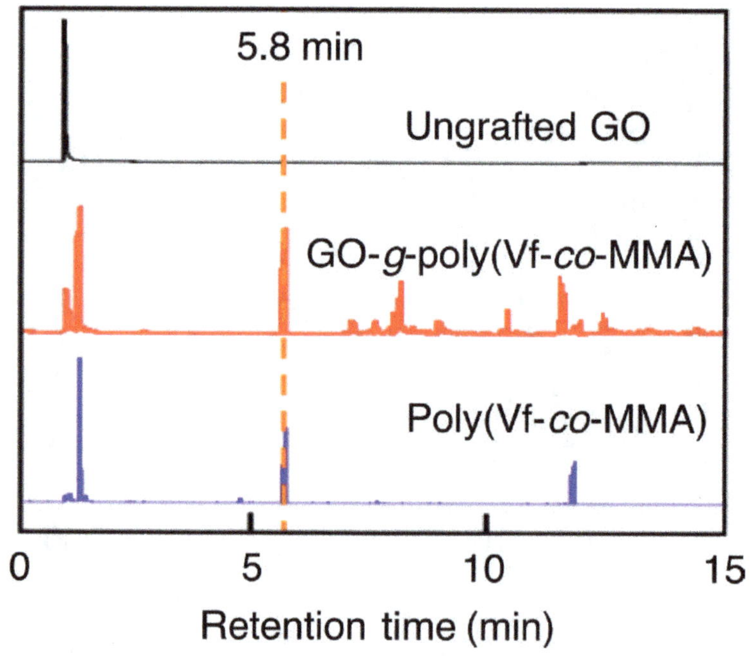

Figure 6. Thermal decomposition gas chromatograms of poly(Vf-*co*-MMA), GO-*g*-poly(Vf-*co*-MMA) and untreated GO.

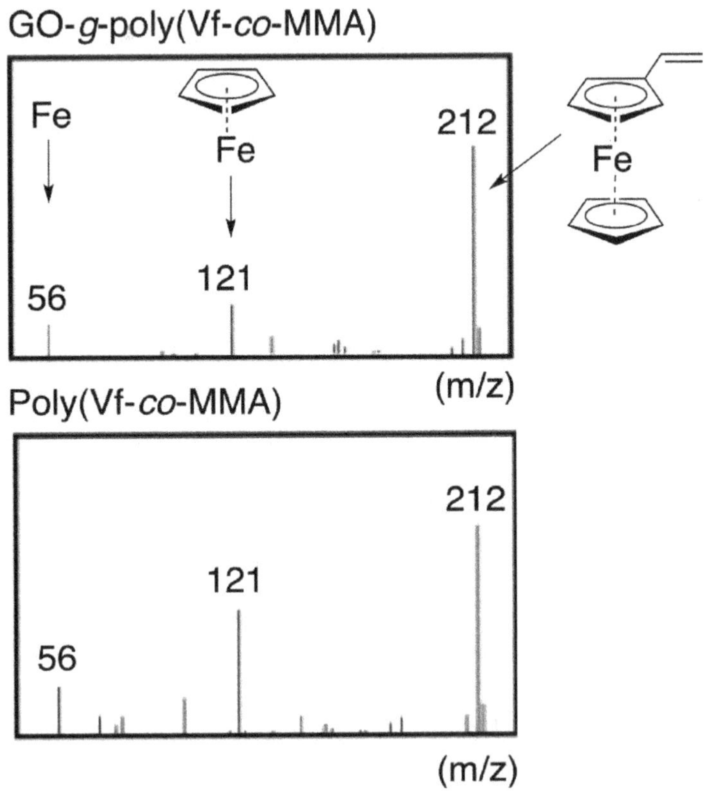

Figure 7. Mass spectra of decomposed gas of poly(Vf-*co*-MMA) and GO-*g*-poly(Vf-*co*-MMA) at retention time 5.8 min.

4. Cationic polymerization of vinyl monomers initiated by carboxyl groups on GO

4.1. Acidity of carboxyl groups on nanocarbons

We have pointed out that carboxyl (COOH) groups on nanocarbons, such as carbon black and vapor grown carbon fiber, have strong acidity, because of the effect of neighboring hydroxyl groups (*ortho*-effect) and have an ability to initiate the cationic polymerization of vinyl monomers to give the corresponding polymer-grafted nanocarbons.

For example, COOH groups on nanocarbons, such as carbon black and vapor grown carbon fiber, have strong acidity to initiate the cationic polymerization of vinyl monomers, such as vinyl ethers and *N*-vinylcarbazole (NVC) [25]. During the polymerization, the corresponding polymers were grafted onto these nanocarbon surfaces, based on the termination of growing polymer cation with surface carboxylate groups on GO as a counter anion.

4.2. COOH initiated cationic polymerization and grafting

The content of COOH and hydroxyl groups of GO used was determined to be 4.0 and 0.5 mmol/g, respectively. The content of COOH groups on GO used is much larger than those of acidic carbon black; for example, the content of COOH groups on channel black FW 200 (Deggusa AG.) is 0.61 mmol/g [28]. Therefore, it is expected that COOH groups on GO act as an effective initiator of the cationic polymerization of NVC, as shown in **Scheme 3**.

Figure 8 shows the relationship between conversion of NVC and reaction time during the polymerization of NVC in the presence of GO. As shown in **Figure 8**, the polymerization of NVC is successfully initiated even at 0°C and conversion of NVC increased with increasing reaction time and reached 95% after 12 h.

Figure 9 shows the FT-IR spectra of (A) polyNVC, (B) GO obtained from the cationic polymerization in the presence of GO, and (C) ungrafted GO. The FT-IR spectra of GO obtained from the polymerization in the presence of GO show characteristic absorptions of polyNVC. The result shows the grafting of polyNVC onto GO during the GO-initiated cationic polymerization of NVC.

Figure 10 shows the relationship between the percentage of polyNVC grafting onto GO and reaction time during the above-cationic polymerization. It was found that polyNVC was grafted onto GO, during the polymerization, and the percentage of grafting increased with increasing reaction time: the percentage of polyNVC grafting reached 14% after 12 h.

Figure 11 shows the relationship between conversion of isobutyl vinyl ether (IBVE) and reaction time during the GO-initiated cationic polymerization of IBVE. The conversion of IBVE increased with increasing reaction time and reached 78% after 1.5 h. The result indicates that GO also has an ability to initiate the cationic polymerization of IBVE.

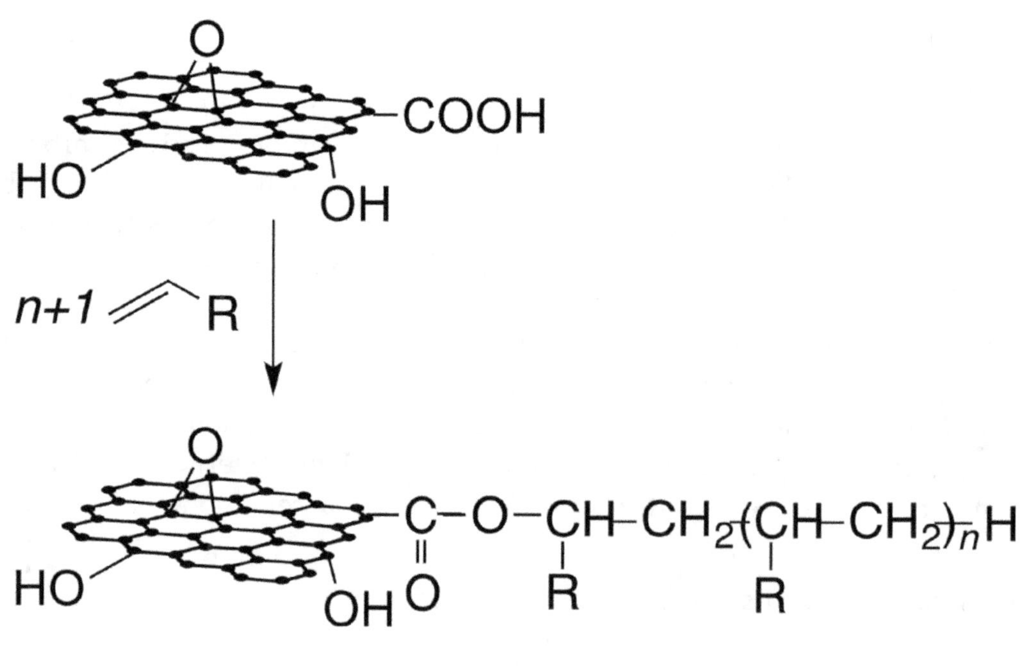

Scheme 3. Cationic polymerization of vinyl monomers initiated by COOH groups on GO.

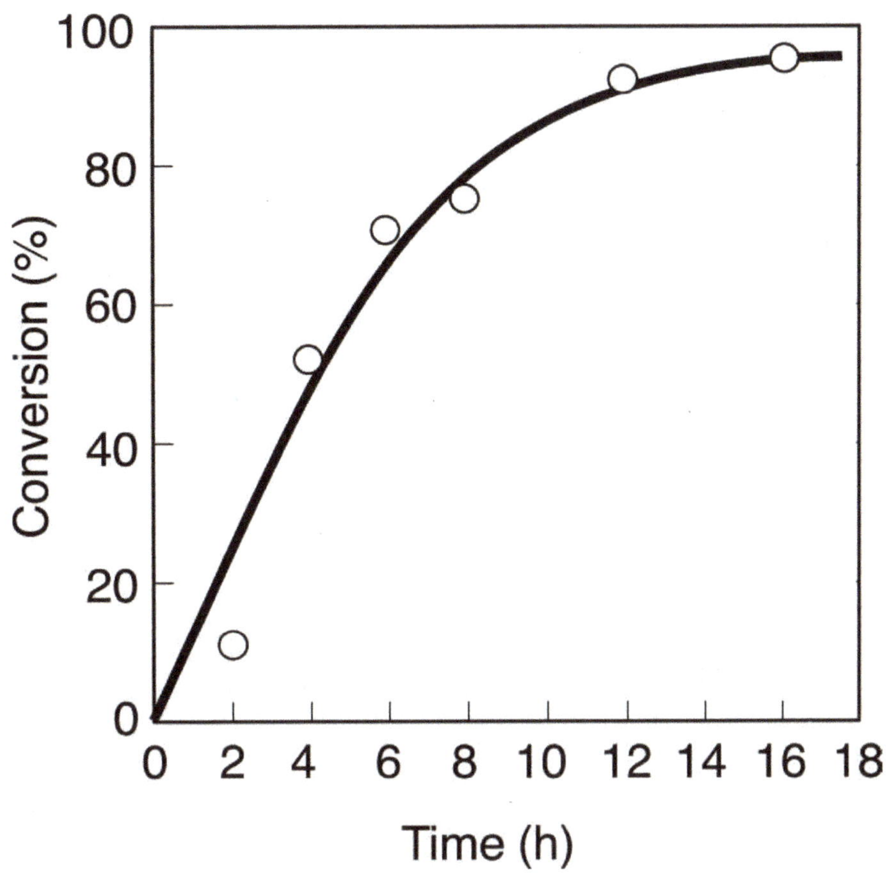

Figure 8. The relationship between conversion and polymerization time during the cationic polymerization of NVC initiated by COOH groups on GO. GO, 0.10 g; NVC, 1.0 g; toluene, 10 mL; Temp., 0°C.

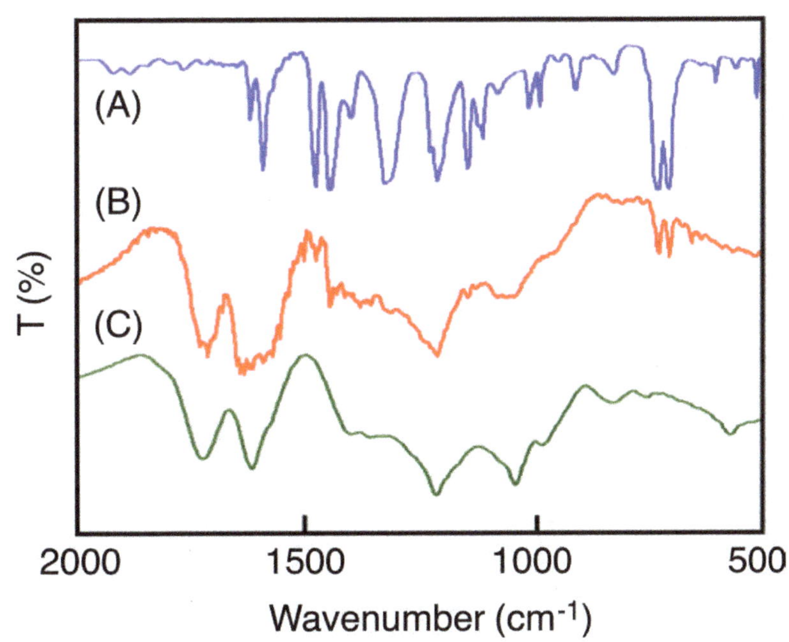

Figure 9. FT-IR spectra of (A) polyNVC obtained from the conventional cationic initiator, (B) GO obtained from the cationic polymerization of NVC in the presence of GO, and (C) ungrafted GO.

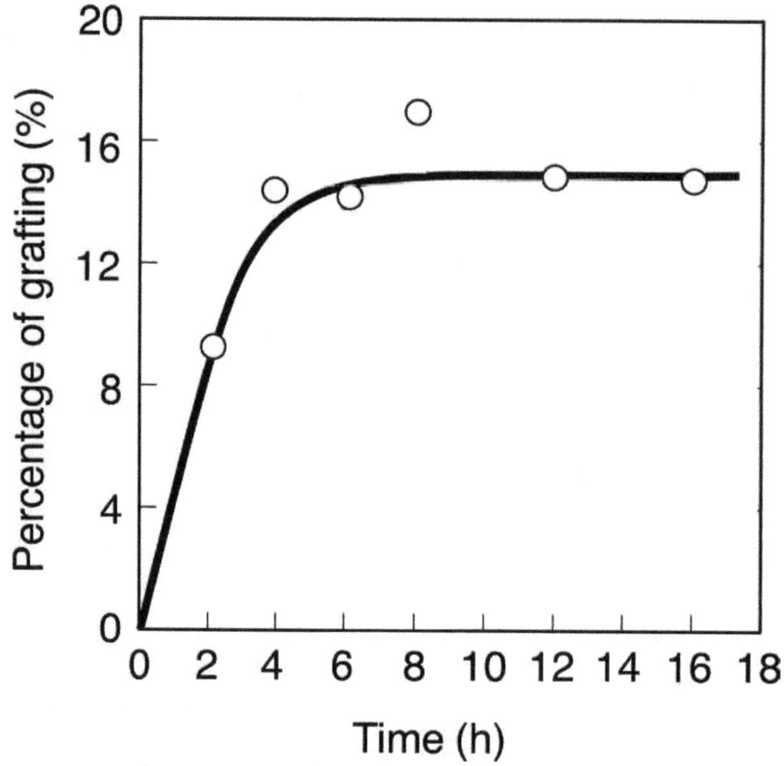

Figure 10. Relationship between percentage of polyNVC grafting onto GO and reaction time. Reaction conditions are shown in **Figure 8.**

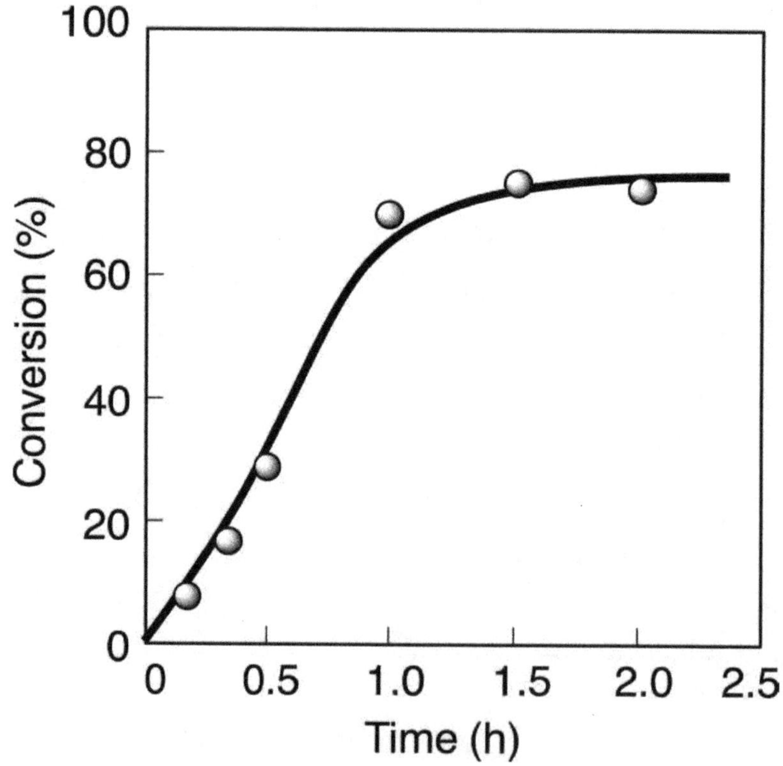

Figure 11. Cationic polymerization of IBVE initiated by COOH groups on GO. GO, 0.10 g; IBVE, 1.0 mL; toluene, 9.0 mL; Temp., 0°C.

4.3. Identification of polyIBVE grafting onto GO by GC-MS

The thermal decomposition GC of GO-*g*-polyIBVE, polyIBVE, and ungrafted GO is shown in **Figure 12(A)**. The thermal decomposition gas of GO-*g*-polyIBVE at retention time 1.2 min agreed with that of polyIBVE. On the other hand, the MS of thermally decomposed gas of polyIBVE and GO-*g*-polyIBVE at retention time 1.2 min is shown in **Figure 12(B)**. **Figure 12(B)** clearly shows that the MS of decomposed gas of GO-*g*-polyIBVE at retention time 1.2 min was in accord with that of polyIBVE: the parent peak at 74 (m/z) is considered to be isobutyl alcohol formed by the thermal decomposition of IBVE. The above results clearly indicate the grafting of polyIBVE onto GO during the GO-initiated cationic polymerization.

In addition, the GO-initiated polymerization of NVC and IBVE was completely inhibited by the addition of base, such as amines, indicating the initiation and propagation of the polymerization proceeded cationic polymerization mechanism.

4.4. Initiation and grafting mechanism

It is presumed that the cationic polymerization was initiated by proton addition of COOH groups on GO to vinyl monomer, as shown in **Scheme 4** (1) and then polymer chains propagated from carboxylate (COO^-) groups on GO as a counter anion, as shown in **Scheme 4** (2). The grafting of the corresponding polymer onto GO is considered to be termination (neutralization) grafting reaction of propagating polymer cation with carboxylate anion (counter ion) on GO, as shown in **Scheme 4** [25, 28].

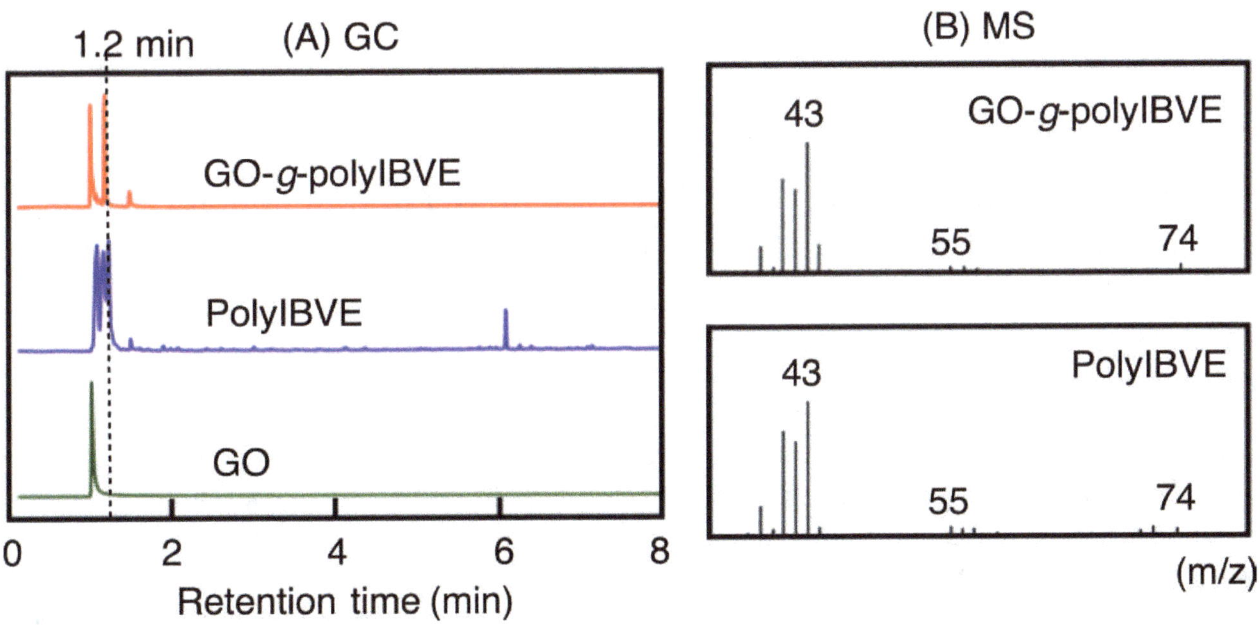

Figure 12. (A) Thermal decomposition GC of GO-*g*-polyIBVE, polyIBVE, and GO. (B) MS of thermal decomposition gas of GO-*g*-polyIBVE and polyIBVE at retention time 1.2 min.

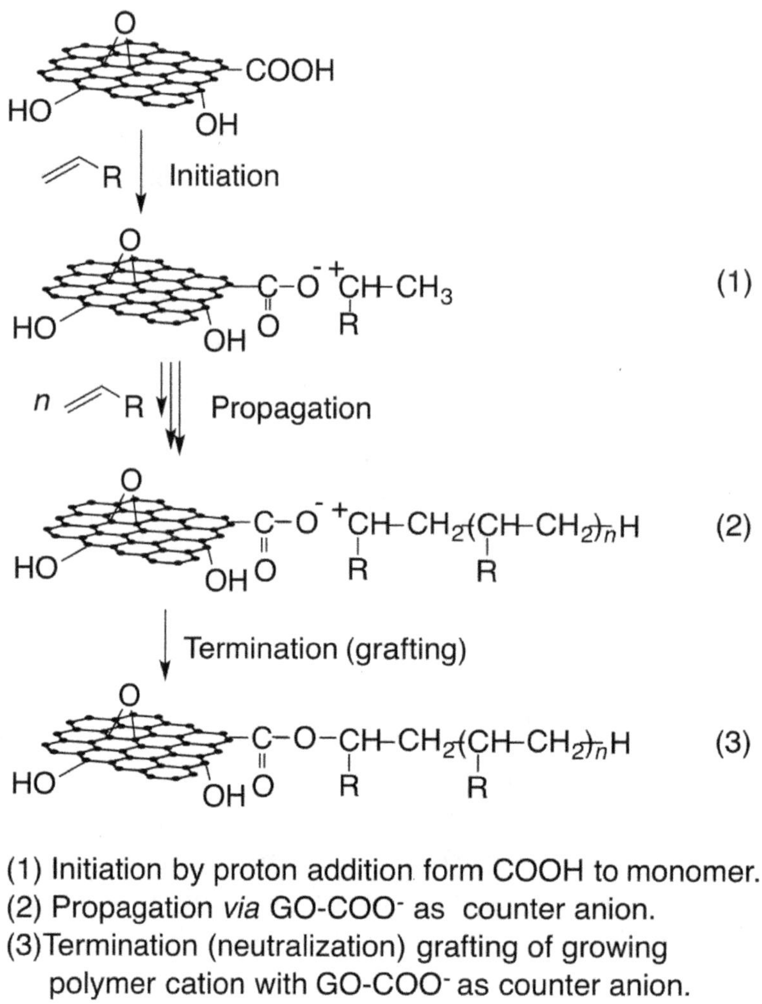

Scheme 4. Initiating and grafting mechanism of cationic polymerization of vinyl monomers initiated by COOH groups on GO.

5. Anionic ring-opening alternating copolymerization of epoxides with cyclic acid anhydrides initiated by potassium carboxylate groups on GO

5.1. Anionic ring-opening alternating copolymerization of epoxides with cyclic acid anhydrides

It is well known that alkali metal salts of aromatic carboxylic acid, such as potassium benzoate, have an ability to initiate the anionic ring-opening alternating copolymerization of epoxides with cyclic acid anhydrides to give the corresponding polyester. On the other hand, we have reported, in the previous paper, that potassium carboxylate (COOK) groups introduced onto the surface of nanocarbons, such as carbon black and vapor grown carbon fiber, can initiate the anionic ring-opening alternating copolymerization of epoxides with cyclic acid anhydrides and the corresponding polyester is readily grafted these nanocarbon surfaces [19, 20, 26, 28].

5.2. Anionic ring-opening alternating copolymerization initiated by COOK groups on GO

The anionic ring-opening alternating copolymerization of styrene oxide (SO) with phthalic anhydride (PAn) initiated by COOK groups on GO (GO-COOK) was examined (**Scheme 5**). GO-COOK can be readily prepared by the neutralization of COOH groups on GO with KOH. **Table 3** shows the results of the anionic ring-opening copolymerization of SO with PAn in the presence of GO-COOK under several conditions. In the polymerization, 18-crown-6 was added as an accelerator of the anionic copolymerization [19, 20, 26, 28].

As shown in **Table 3**, the ring-opening copolymerization of SO with PAn was hardly initiated in the presence of untreated GO (GO-COOH). In addition, no initiation of the anionic ring-opening polymerization of SO (or PAn) was observed even in the presence of GO-COOK. On the other hand, it was found that GO-COOK has an ability to initiate the anionic ring-opening alternating copolymerization of SO with PAn and the corresponding polyester, poly(SO-*alt*-PAn), was grafted onto GO.

The effect of polymerization time on the anionic ring-opening alternating copolymerization of SO with PAn [or maleic anhydride (MAn)] initiated by GO-COOK is shown in **Figure 13**. The copolymerization of SO with MAn was carried out in the presence of *N*-phenyl-naphthylamine (NPNA) in order to inhibit the radical cross-linking of formed unsaturated polyester, poly(SO-*alt*-MAn). It was found that the conversion increased with elapse of the polymerization time and the conversion of poly(SO-*alt*-PAn) and poly(SO-*alt*-MAn) exceeded 70 and 50% after 3 h at 120°C, respectively, suggesting the grafting of the corresponding polyester onto GO.

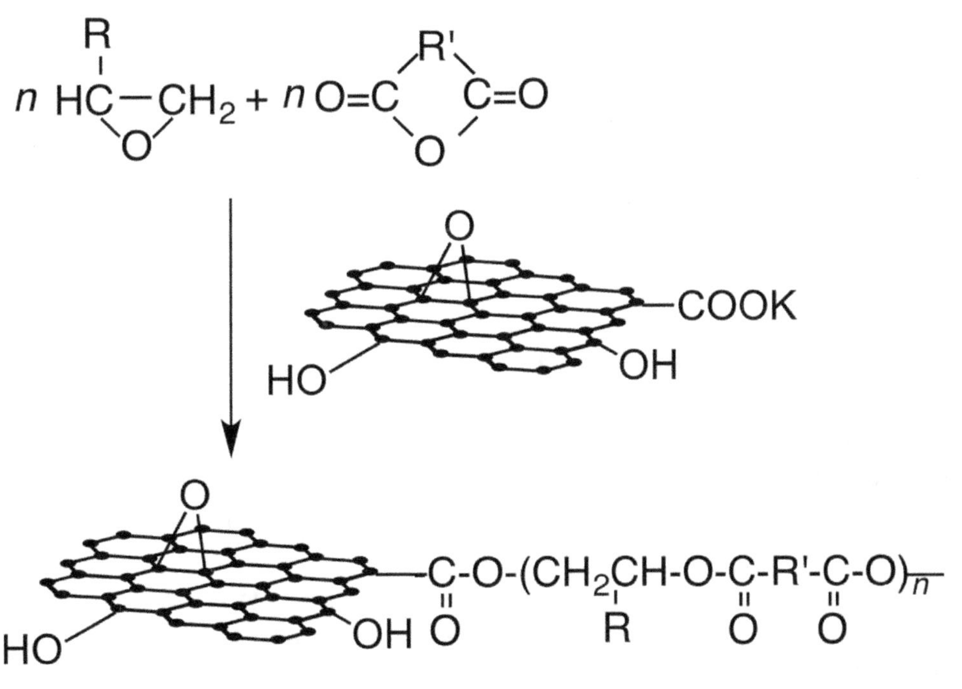

Scheme 5. Anionic ring-opening alternating copolymerization of epoxides with cyclic acid anhydrides initiated by COOK groups on GO.

GO	SO (mol)	PAn (mol)	Conversion (%)
GO-COOH	0.01	0.01	0
GO-COOK	0.01	–	0
GO-COOK	–	0.01	0
GO-COOH	0.01	–	72.0

Note: GO, 0.10 g; 18-crown-6, 0.02 g; Temp., 120°C; 8 h.

Table 3. Anionic ring-opening copolymerization of SO with PAn in the presence of GO-COOK and GO-COOH under several conditions.

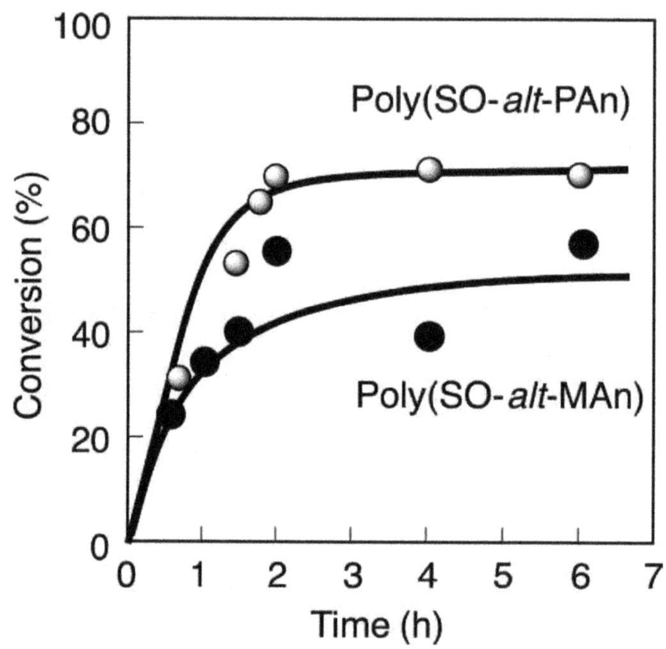

Figure 13. Effect of polymerization time on the anionic ring-opening alternating copolymerization of SO with PAn and MAn in the presence of GO-COOK. GO-COOK, 0.10 g; SO, 0.01 mol; cyclic acid anhydride, 0.01 mol; 18-crown-6, 0.02 mol; NPNA (in the case of MAn), 0.02 g, Temp., 120°C.

5.3. Identification of polyester-grafted GO by GC-MS

Figure 14(A) shows thermal decomposition GC of poly(SO-*alt*-PAn), which is obtained by using the conventional catalyst, GO-COOK, and GO-*g*-poly(SO-*alt*-PAn), obtained from the anionic ring-opening alternating copolymerization in the presence of GO-COOK. As shown in **Figure 14(A)**, the thermally decomposed gas of GO-*g*-poly(SO-*alt*-PAn) generated at retention time at 2.0 and 4.8 min was in accord with those of poly(SO-*alt*-PAn). In addition, the MS of thermally decomposed gas of poly(SO-*alt*-PAn) and GO-*g*-poly(SO-*alt*-PAn) at retention time 2.0 and 4.8 min, respectively, is shown in **Figure 14(B)** and **(C)**. It was found that the MS of thermally decomposed gas of poly(SO-*alt*-PAn)-grafted GO at retention time 2.0 min was in accord with that of poly(SO-*alt*-PAn), as shown in **Figure 14(B)**: the parent peak at 104 (m/z) was estimated to be the corresponding styrene generated by the thermal

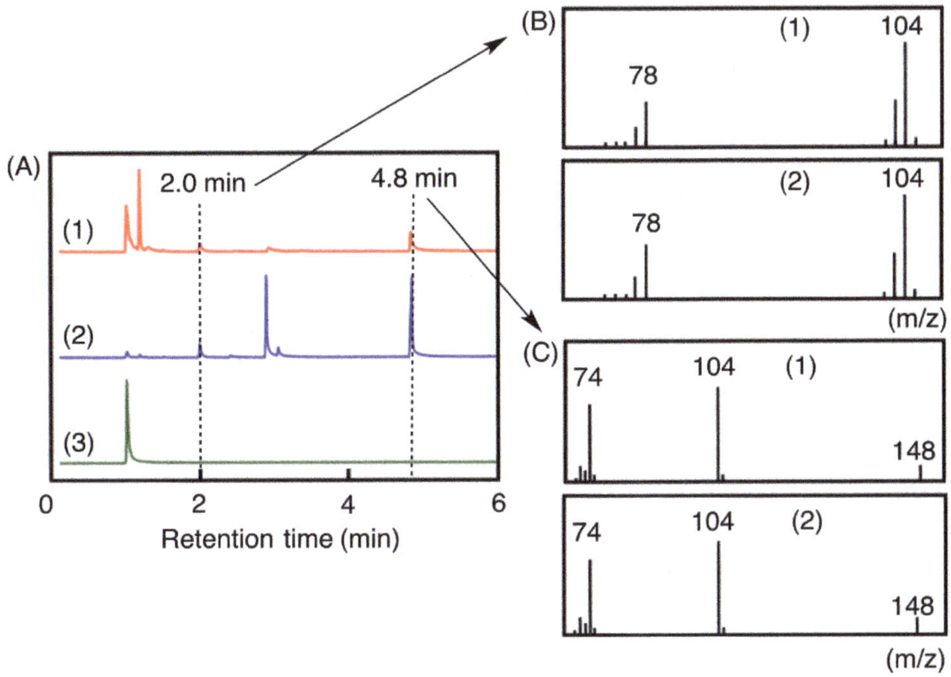

Figure 14. (A) Thermal decomposition gas chromatograms of (1) GO-*g*-poly(SO-*alt*-PAn), (2) poly(SO-*alt*-PAn), and (3) GO. (B) and (C) Mass spectra of thermal decomposition gas of GO-*g*-poly(SO-*alt*-PAn) and poly(SO-*alt*-PAn) at retention time 2.0 min and 4.8 min, respectively.

decomposition of poly(SO-*alt*-PAn). Furthermore, the MS of thermally decomposed gas of GO-*g*-poly(SO-*alt*-PAn) at retention time 4.8 min agreed with that of poly(SO-*alt*-PAn), as shown in **Figure 14(C)**; the parent peak at 148 (m/z) is estimated to be PAn generated by the thermal decomposition of poly(SO-*alt*-PAn). Based on the above results, it is concluded that poly(SO-*alt*-PAn) was successfully grafted onto GO during the anionic ring-opening copolymerization initiated by GO-COOK.

It was also confirmed by GC-MS that the grafting of poly(SO-*alt*-MAn) onto GO successfully achieved by the anionic ring-opening alternating copolymerization of SO with MAn initiated by GO-COOK.

6. Dispersibility of polymer-grafted GO in solvents

6.1. Dispersibility of GO obtained from radical trapping and ligand-exchange reaction

The untreated GO, GO-*g*-PEG, and GO-*g*-poly(Vf-*co*-MMA) were dispersed in good solvent of````` grafted polymer under irradiating ultrasonic wave and allowed to stand at room temperature. As a result, untreated GO precipitated within 15 min, but the dispersion of GO-*g*-PEG and GO-*g*-poly(Vf-*co*-MMA) in THF has excellent stability and precipitation of GO was scarcely observed even after 1 week at room temperature.

The stability of the GO dispersion in THF after the ultrasonic wave irradiation was also examined from the decrease of absorbance of GO dispersion at room temperature by use of UV-Vis spectrometer. The higher absorbance indicates no precipitation of GO, but lower absorbance indicates the precipitation of GO. As shown in **Figure 15**, untreated GO immediately precipitated, but the precipitation of GO-*g*-PEG and GO-*g*-poly(Vf-*co*-MMA) was scarcely observed, indicating the effect of grafting of polymers onto GO.

The results suggest that by the grafting of PEG and poly(Vf-*co*-MMA) onto GO surfaces, the aggregation of GO was successfully destroyed and grafted polymer chains onto the surface interfere with reaggregation of the GO in good solvents for grafted polymer, such as THF.

6.2. Dispersibility of GO obtained from surface-initiated cationic and anionic polymerization

Figure 16 shows the stability of dispersion of ungrafted GO, GO-*g*-polyNVC, GO-*g*-polyIBVE, and GO-*g*-poly(SO-*alt*-PAn) in THF at room temperature. Ungrafted GO precipitated completely within 3 h. On the contrary, polymer-grafted GOs gave stable dispersions in THF, a good solvent for grafted polymers.

Therefore, it is concluded that dispersibility of GO in THF was remarkably improved by grafting of polymers, such as polyNVC, polyIBVE, and polyesters.

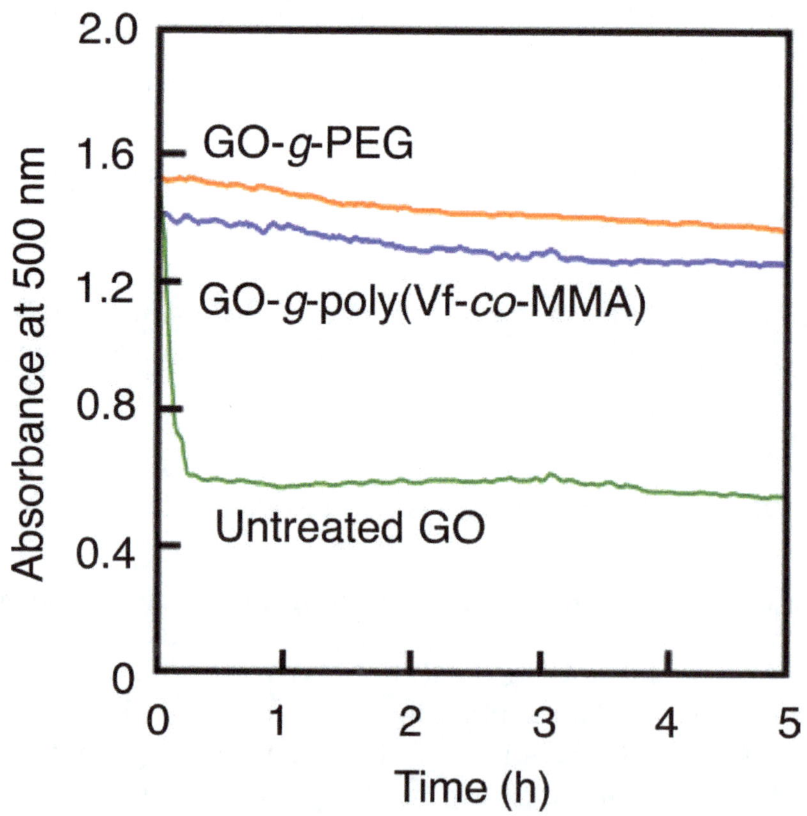

Figure 15. Dispersibility of GO-*g*-PEG and GO-*g*-poly(Vf-*co*-MMA) in THF at room temperature.

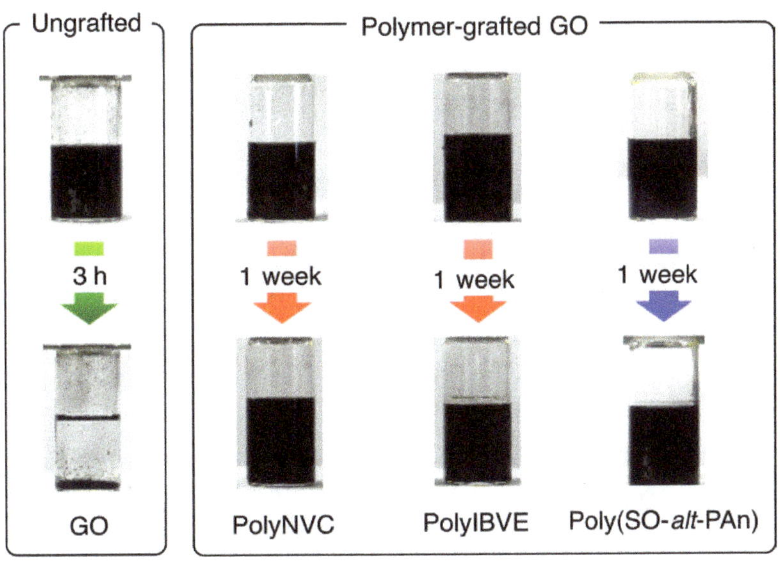

Figure 16. Dispersibility of ungrafted GO, GO-*g*-polyNVC, GO-*g*-polyIBVE, GO-*g*-poly(SO-*alt*-PAn) in THF at room temperature.

7. Easy preparation of reduced GO-based conducting polymer composite via organogel

Recently, we have reported that a highly electrically conducting graphene (reduced GO, rGO)-based polymer composite was successfully prepared via organogels using an organically dispersible electrically conductive polyaniline (polyANI) and a low molecular-weight organogelator consisting of cholesterol derivatives: organogelator used was complex salt of cholesterol hydrogen succinate with 1, 10-diaminodecane [33].

The plain gel was prepared by heat treatment of a mixture of toluene and the organogelator. PolyANI/rGO organogel was successfully obtained by the addition of polyANI and rGO to the plain gel and mixing them, as shown in **Figure 17**.

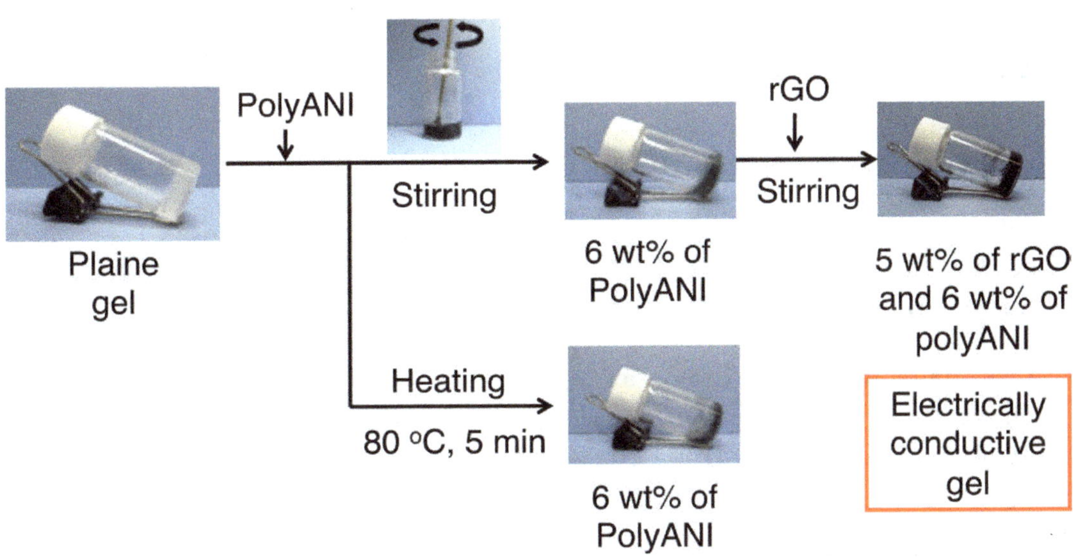

Figure 17. Easy preparation of polyANI/rGO organogel.

Similarly to the plain gel, polyANI/rGO organogels showed thixotropic behavior. It became apparent that these gels show a good recovery of viscoelasticity after the application of mechanical stress. Three-dimensional networks consisting of organogelators, polyANI aggregates, and rGO particles dispersed in the gel state were confirmed by the microscopic studies.

The value of electrical conductivity of the polyANI/rGO organogel is very low: it could be detected at 2.5×10^{-5} S/cm. The conductivity of dried polyANI/rGO composites synthesized via organogels was determined to be 3.2 S/cm. It is interesting to note that the contribution of the organogelator to form effective and complementary conducting pathways by polyANI and rGO.

A ballpoint pen was filled with the polyANI/rGO organogel and used to produce lines similar to the ones produced by commercially available ballpoint pen. It is interesting to note that the line showed electrical conductivity.

Acknowledgements

This work was supported by JSPS KAKENHI (grant number 24560836 and 24120008E).

Author details

Norio Tsubokawa[1*], Takeshi Yamauchi[1], Kazuhiro Fujiki[2] and Shingo Tamesue[1]

*Address all correspondence to: ntsuboka@eng.niigata-u.ac.jp

1 Graduate School of Science and Technology, Niigata University, Niigata, Japan

2 Niigata Institute of Technology, Kashiwazaki, Niigata, Japan

References

[1] Georgakilas V, editor. Functionalization of Graphene. Weinheim: Wiley-VCH; ISBN-13: 978-3527335510.

[2] Hu K, Kulkarni D. D, Choi I, Tsukruk V. V: Graphene-polymer nanocomposites for structural and functional applications. Prog. Polym. Sci. 2014; **39**: 1934–72.

[3] Song S, Zhai Y, Zhang Y: Bioinspired graphene oxide/polymer nanocomposite paper with high strength, toughness, and dielectric constant. ACS Appl. Mater. Interfaces. 2016; **8**: 31264–72.

[4] Rajender N, Suresh K. I, Surface-initiated atom transfer radical polymerization (SI-ATRP) from graphene oxide: effect of functionalized graphene sheet (FGS) on the synthesis and material properties of PMMA nanocomposites. Macromol. Mater. Eng. 2016; **301**: 81–92.

[5] Kumar A, Behera B, Thakre G. D, Ray S. S: Covalently grafted graphene oxide/poly(C-n-acrylate) nanocomposites by surface-initiated ATRP: an efficient antifriction, antiwear, and pour-point-depressant lubricating additive in oil media. Ind. Eng. Chem. Res. 2016; **55**: 8491–500.

[6] Sun Y, Zhou J. H, Cheng Q, Lin D. S, Jiang Q, Dong A. J, Liang Z. C, Deng L. D: Fabrication of mPEGylated graphene oxide/poly(dimethylaminoethyl mathacrylate) nanohybrids and their primary application for small interfering RNA delivery. J. Appl. Polym. Sci. 2016; **133**: 43303.

[7] Huynh V. T, Duc N, Such C. H, Hawkett B. S: Polymer coating of graphene oxide via reversible addition-fragmentation chain transfer mediated emulsion polymerization. J. Polym. Sci. Part A: Polym. Chem. 2016; **53**: 1423–21.

[8] Gu R, Xu W. Z, Charpentier P. A: Synthesis of graphene-polystyrene nanocomposites via RAFT polymerization. Polymer 2016; **55**: 5322–31.

[9] Travlou N. A, Kyzas, G. Z, Lazaridis N. K, Deliyanni A: Graphite oxide/chitosan composite for reactive dye removal. Chem. Eng. J. 2013; **217**: 256–65.

[10] He L, Wang H, Xia G, Sun J, Song R: Chitosan/graphene oxide nanocomposite films with enhanced interfacial interaction and their electrochemical applications. 2014; **314**: 510–15.

[11] Zuo P, Feng H, Xu Z, Zhang L, Zhang, Y, Xia W, Zhang W: Fabrication of biocompatible and mechanically reinforced graphene oxide-chitosan nanocomposite films. Chem. Central J. 2013; **7**: 39.

[12] Yuan H, Meng L, Park S: A review: synthesis and applications of graphene/chitosan nanocomposites. Carbon Lett. 2016; **17**: 11–17.

[13] Terzopoulou Z, Kyzas G. Z, Bikiaris D. N: Recent advances in nanocomposite materials of graphene derivatives with polysaccharides. Materials 2015; **8**: 625–83.

[14] Terzopoulou Z, Kyzas G. Z, Bikiaris D. N: Recent advances in nanocomposite materials of graphene derivatives with polysaccharides. Materials 2015; **8**; 652–83.

[15] Thakur V. K, Voicu S. I: Recent advances in cellulose and chitosan based membranes for water purification: a concise review. Carbohydrate Polym. 2016; **146**: 148–65.

[16] Fang M, Wang K, Lu H, Yang Y, Nutt S: Single-layer graphene nanosheets with controlled grafting of polymer chains. J. Mater. Chem. 2010; **20**: 1982–92.

[17] Ren L, Wang X, Guo S, Liu T: Functionalization of thermally reduced graphene by in situ tom transfer radical polymerization. J. Nanopart. Res. 2011; **13**: 6389–96.

[18] Cui L, Liu J, Wang R, Liu Z, Yang W: A facile "grafting from" method to prepare molecular-level dispersed graphene-polymer composites. J. Polym. Sci. Polym. Chem. 2012; **50**: 4423–4432.

[19] Tsubokawa N: Functionalization of carbon materials by surface grafting of polymers. Bull. Chem. Soc. Japan. 2002; **75**: 2115–36.

[20] Tsubokawa N: Preparation and properties of polymer-grafted carbon nanotubes and nanofibers. Polym. J. 2005; **37**: 637–55.

[21] Chen J, Wei G, Maekawa Y, Yoshida Y, Tsubokawa N: Grafting of poly(ethylene-block-ethylene oxide) onto a vapor grown carbon fiber surface by γ-ray radiation grafting. Polymer. 2003; **44**: 3201–7.

[22] Wakai H, Shinno T, Yamauchi T, Tsubokawa N: Grafting of poly(ethylene oxide) onto C60 fullerene using macroazo initiators. Polymer. 2007; **48**: 1972–80.

[23] Wei G, Saitoh S, Saitoh H, Fujiki K, Yamauchi T, Tsubokawa N: Grafting of vinyl polymers onto vapor grown carbon fiber surface by ligand-exchange reaction of ferrocene moieties of polymer with polycondensed aromatic rings of the wall-surface. Polymer. 2004; **45**: 8723–730.

[24] Cha I, Shirai K, Fujiki K, Yamauchi T, Tsubokawa N: Surface grafting of polymers onto nanodiamond by ligand-exchange reaction of ferrocene moieties of polymers with polycondednsed aromatic rings of the surface: Diamond Related Mater. 2011; **20**: 439–44.

[25] Tsubokawa N, Maruyama H, Sone Y: Carbon fiber as an initiator of cationic polymerization of N-vinylcarbazole. Polym. Bull. 1986; **15:** 209–14.

[26] Tsubokawa N, Hamada H, Sone Y: Grafting of polyesters from carbon fiber. Anionic ring-opening copolymerization of epoxides with cyclic acid anhydrides initiated by COOK groups on the surface of carbon fiber. Polym. Plast. Technol. Eng. 1989; **28**: 201–14.

[27] Cha I, Yagi Y, Kawahara T, Hashimoto K, Fujiki K, Tamesue S, Yamauchi T, Tsubokawa N: Grafting of polymers onto graphene oxide by trapping of polymer radicals and ligand-exchange reaction of polymers bearing ferrocene moieties. Colloid Surf. A: Physicochem. Eng. Aspects. 2014; **441**: 474–80.

[28] Nagata K, Kawahara T, Hashimoto K, Fujiki K, Tamesue S, Yamauchi T, Tsubokawa N: Grafting of polymers onto graphene oxide by cationic and anionic polymerization initiated by the surface-initiating groups. Composite Interfaces. 2015; **22**: 25–37.

[29] Ohkita K, Tsubokawa N, Saitoh E, Noda M: The free radical polymerization of vinyl monomers in the presence of carbon black. Carbon. 1975; **13**, 443–8.

[30] Wako Pure Chemical Ind. Ltd. Japan: Macro azo-initiator [Internet]. 2016. Available from: http://www.wako-chem.co.jp/kaseihin/macroazo/index.htm.

[31] Miyake M, Yasuda K, Kashihara T, Teranishi T: Chemical modification of carbon hexagonal plane by ligand-exchange reaction. Chem. Lett. 1999; **28**: 1037–8.

[32] Morrison Jr W. H, Ho E. Y, Hendrickson D. N: Reaction of ferrocene with polyaromatic molecules. π-Arene bis((π-cyclopentadienyl)iron) dications. J. Am. Chem. Soc. 1974; **96**: 3603–3608.

[33] Kuwahara R. Y, Oi T, Hashimoto K, Tamesue S, Yamauchi T, Tsubokawa N: Easy preparation of graphene-based conducting polymer composite via organogel. Colloid Polym. Sci. 2015; **293**: 1635–45.

9

Self-Assembly of Graphene Nanoribbons Induced by the Carbon Nanotube

Hui Li, Yifan Li and Wei Chen

Abstract

In this chapter, a series of molecular dynamics simulations have been carried out to explore the self-assembly of graphene nanoribbons (GNRs) induced by the single-walled carbon nanotubes (SWCNTs). Simulation results show that GNRs can insert and wrap SWCNTs spontaneously, forming helical configurations and maximizing the π-π stacking area between graphene and SWCNT. The helical configuration takes the least amount of energy and achieves the maximum occupancy. The size and function group of GNR and SWCNT should meet the required conditions to guarantee the self-assembly in insertion and wrapping processes. Several GNRs can spiral in an SWCNT simultaneously, and two formulas have come up in this study to estimate the quantity threshold for multiple GNR spiralling. The rolled GNRs can also spontaneously insert into SWCNTs, forming a DNA-like double helix, or collapsing to a linked double graphitic nanoribbon and wrapping in a helical manner around the tube.

Keywords: molecular dynamics simulation, graphene nanoribbon, structural evolution, helical configuration, self-assembly

1. Introduction

Carbon materials, especially the carbon nanotube and graphene, have attracted tremendous attention on the theoretical research and the potential applications because of their unique configurations and excellent performances. The cylindrical SWCNT possessing a large specific surface area and hollow interior could act as 'molecular straws' capable of absorbing dipolar molecules by capillary action [1]. Over the past decade, the self-assembled systems [2, 3] relied on the special hollow structure of cylindrical SWCNTs, are an attractive class of new bio-inspired nanomaterial for biologists and material scientists, because the self-assembled materials may not only be designed to be highly dynamic, displaying adaptive and self-healing properties, but could also help gain an understanding of the rules that govern biomolecule assembly processes [4]. In the self-assembly process, the hollow interior of SWCNT can serve as nanometre-sized moulds and templates [5] to control the configuration of other materials, or as a protective layer [6] to prevent the filler from oxidation and shape fragmentation. The physical and chemistry properties of the heterostructure are expected to be modified due to the interaction between SWCNT and exotic materials [7]. Thus, filling SWCNTs with chosen fillers can produce one-dimensional nanostructures with exciting new applications [8]. A wide array of fillers, including various metal atoms [9], halides [10], C60 [11, 12] and polyacetylene [13], were found to be filled into the SWCNT, with novel configurations and properties. Strano et al. [14] found that water molecules displayed different phase transition temperatures when confined in the nanotube. C60 [15] or other spherical metallofullerenes [16] were distributed evenly along the axial direction after self-assembling into the SWCNT just as beans distributed in pod, while those linear fillers such as DNA [17] and polyacetylene [13] were easy to insert into or wrap the SWCNT spontaneously with an interesting helical configuration.

Recently, our studies [18, 19] indicated that GNRs, the material with unique optical, magnetic and electrical properties [20–23], have been succeeded in spontaneously and spirally wrapping and inserting into the SWCNT. GNRs possess intriguing electronic structures ranging from semiconducting to half-metallic, depending on their geometry and dimensions. Thus, the helical composite structures formed by two novel carbon materials are expected to have excellent properties, such as high carrier mobility and high mechanical strength, to be used in microcircuit interconnectors, nanoelectronic devices and nanosensors [24–26]. And a convincing model on the interaction between the GNR and SWCNT may inspire great efforts in theoretical studies, synthesis and chemical modifications which focus on their electronic, biological, chemical and even magnetic properties. In this chapter, the self-assembly behaviour of GNRs, especially multiple GNRs, induced by SWCNT is studied in detail, while the mechanism and influencing factors of self-assembly are also described, providing an opportunity for a comprehensive and satisfactory understanding of how to control the composite structure of this SWCNT-GNR nanomaterial hybrid. This discovery is of great significance in the exploration of the properties of the GNR-SWCNT system and may expand the applications of GNR and SWCNT in extensive fields involving medicine, chemistry, biology and even fuel cells. The perfect GNR helix in SWCNT may be used to control the band gap of various GNT-based nanodevices, which will pave the way for the progress of nanoelectronics.

2. Simulation method

Molecular dynamics method is used to study the self-assembly behaviour of GNRs induced by SWCNTs. The force field of condensed-phase optimized molecular potentials for atomistic simulation studies (COMPASS) [27] is applied to model the atomic interaction. COMPASS is an ab initio force field which has been parameterized and validated using condensed-phase properties, various ab initio calculations and experimental data, with a functional form that includes covalent terms, van der Waals interactions and electrostatic forces. The aim of the force field is to achieve high accuracy in predicting the properties of complex mixtures [28] and it has been widely used due to its potential to obtain reasonable results in terms of the mechanical properties of CNTs [29, 30]. The van der Waals energy is described by LJ-96 function [28] in COMPASS force field, the functional forms of which is listed as: $E = D_0\left[2\left(\frac{R_0}{R}\right)^9 - 3\left(\frac{R_0}{R}\right)^6\right]$, where D_0 = 0.064 Kal/mol and R_0 = 4.01Å [31, 32]. The temperature in this paper is chosen as 298 K in the NVT canonical ensemble (number of particles, volume and temperature are constant). Andersen thermostat [33] is employed to control the thermodynamic temperature and generate the correct statistical ensemble by allowing the simulated system to exchange energy with a 'heating bath'. The speed of atoms follows the Maxwell-Boltzmann distribution and the time integration is undertaken using the velocity Verlet algorithm. The simulation time step is 1.0 fs. Each system is simulated for sufficient time to reach equilibrium. Trajectory is recorded every 5.0 ps for further analysis. The GNR with the opening edge is prepared by cutting the parallel sheet of block graphite.

3. Helical insertion of GNRs

Direct simulations in **Figure 1(a)** show the self-assembly of six GNRs into the fixed SWCNT. The length of SWCNTs is 98.38Å and six GNRs have the same size of 5.681Å in width and 147.600Å in length. When one end of the GNR is captured by the SWCNT, the GNR adheres onto the tube wall parallel and moves towards the tube. At the beginning, six GNRs fill into the cavity of SWCNT synchronously and spontaneously under the action of van der Waals coupling. When the simulation time is up to 2.3 ns, a clear and symmetrical multiple helical configuration with uniform pitch is formed in SWCNT, like an upgrade of the double helix of DNA [34] or the triple helix of collagen [35] in biology, suggesting nature's preference for helical structure. While the processes of encapsulating and spiralling are terminated, no more axial movement of the helical GNRs has been observed due to the van der Waals potential well, but rotation is continued in their circumferential direction, resulting the GNR tails outside SWCNT keeps a new subtle spiral morphology by movement inertia. In addition, for the helical GNRs in SWCNT, the essential corrugation [36] is suppressed by the enhanced interaction between GNRs and SWCNT, inspiring an efficient way to straighten the twisted and corrugated GNRs. The filling snapshots of six GNRs into the unfixed SWCNT are established in **Figure 1(b)** where a slight deformation of carbon nanotube appears followed by the formation of a perfect single helix of GNRs at 0.95 ns. The unfixed SWCNT-GNR system takes shorter time to attain equilibrium, demonstrating that the deformation and movement of carbon nanotubes do not hinder the self-assembly of GNRs, but improve their filling speed.

To further research the helical configuration of the multiple GNRs self-assembled into SWCNT, different numbers of annular and multilayer GNRs filling into SWCNT are shown in **Figure 2**. For convenience, a new and canonical definition for $(a_1, a_2 \ldots a_i \ldots a_N)$ and $[a_1, a_2 \ldots a_i \ldots a_N]$ is set, where a_i means the number of GNRs in *i*-layer and parentheses and square brackets refer to the simulated model and result, respectively. **Figure 2** illustrates that different numbers of GNRs can self-assemble into SWCNTs to form different layers of helical structures. As shown in **Figure 2(a)**, seven GNRs self-assemble and form a monolayer helical configuration in SWCNT, while for the system of eight GNRs, a GNR deviates from the first helix into the second one in **Figure 2(b)**, suggesting that there is a quantity threshold for every GNR helix. The helical angle of inner GNRs is easily affected by the outer one to ensure the maximum overlap between GNRs in different helices when few GNRs in the inner helix are located, as the royal blue GNR in **Figure 2(b)**. But when several GNRs locate on the inner helix (**Figure 2(c)**), their helical angle does not easily affect each other. This phenomenon might be useful when adjusting or controlling the self-assembly morphology of GNRs in SWCNT. Another noteworthy phenomenon is that the outer GNRs can enter the inner layer spontaneously in **Figure 2(d)** and **(e)**, which is closely related to the self-assembly mechanism of GNRs and is explained in detail in Section 4. When encapsulating the (7, 4, 2) GNRs into SWCNT in **Figure 2(f)**, the GNRs arrange densely to obtain a perfect multiple spiral structure, and no GNR transfers into the inner helix. When setting more GNRs in SWCNT like in **Figure 2(g)** and **(h)**, the self-assembly process also exists but the helical configuration becomes imperfect due to their undistinguished layer. In addition, the well-organized spiral configuration of (7, 4, 2) consists of an array of helicoids formed concentrically (shown in **Figure 2(j)**), similar to the stranded wire in cable (shown in **Figure 2(k)**), so the GNR@SWCNT composite structure could be expected to have excellent performance in mechanics and electrics like stranded wire.

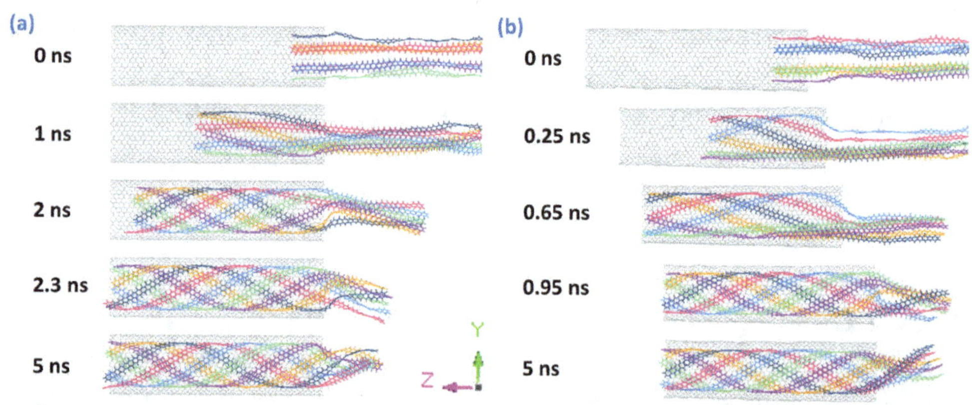

Figure 1. The insertion and helix-forming process of six GNRs into SWCNT: (a) snapshots of GNRs encapsulated in the fixed SWCNT; (b) snapshots of GNRs encapsulated in the unfixed SWCNT.

The rolled GNR, which is obtained by connecting the two ends of straight GNR, can also self-assemble into SWCNT to form a double-helical configuration. **Figure 3(a)** provides the representative snapshots of a rolled GNR inserted helically into the SWCNT. At the beginning, a rolled GNR is placed near the SWCNT separated by 5Å. As time continues, the rolled GNR tends to be collapsed. Due to the Van der Waals force [37] between the rolled GNR and SWCNT, the rolled GNR stretches its cross section and is close to the SWCNT at 0.1 ns. Then, the portion of the rolled GNR near the entrance is captured by the SWCNT hollow space and starts to move forward along the inner wall of the SWCNT. After an initial correlation time, the rolled GNR begins to curl in the SWCNT (at t = 3.5 ns). When the simulation time reaches 4.8 ns, the rolled GNR shows a clear helical conformation with a large pitch. Then, the helix becomes denser to occupy the entirety of the tubes. Eventually, a perfect DNA-like double-helix configuration with remarkably constant pitches is formed in the SWCNT. The result of two rolled GNRs self-assembling into the SWCNT is shown in **Figure 3(b)**. The two rolled GNRs with the same length are placed at the two ends of the SWCNT. When the simulation begins, the two rolled GNRs are captured by the inner hollow of the SWCNT simultaneously. After they contact with each other, one of them is pushed down to a two-layer graphitic nanoribbon, and then spirals like an unrolled GNR. As the simulation time reaches 2.25 ns, helices arise at both of the rolled GNRs. At the simulation time of 5.0 ns, the two rolled GNRs display a clear helical conformation.

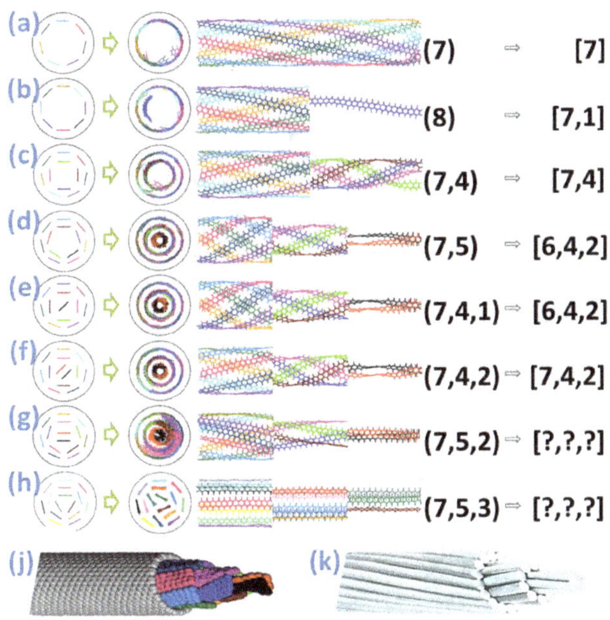

Figure 2. The initial arrangement and equilibrium configuration of the annular and multilayer GNRs in SWCNT. From (a) to (h), the number of GNRs is 7, 8, 11, 12, 12, 13, 14 and 15, respectively; (j) Corey-Pauling-Koltun configuration of the (7, 4, 2) GNRs; and (k) configuration of the stranded conductor in cable.

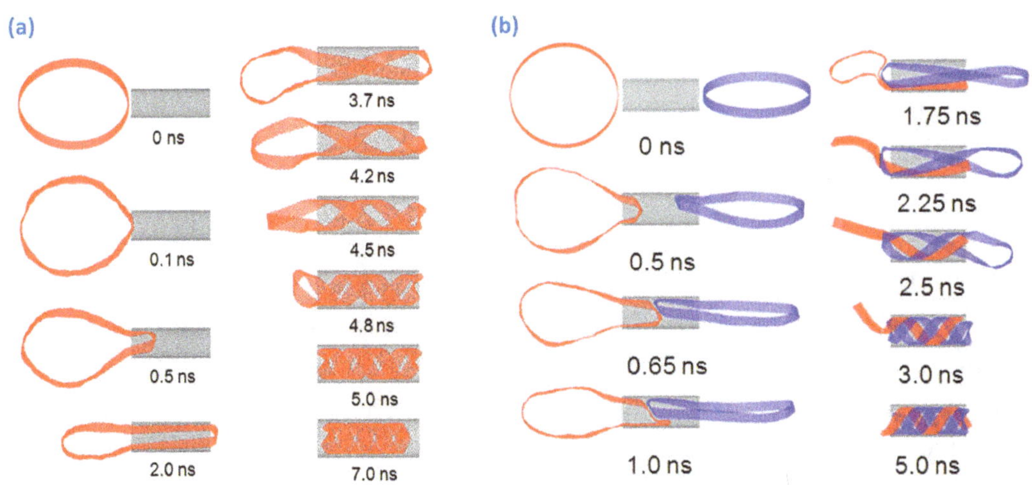

Figure 3. Representative snapshots of (a) a rolled GNR and (b) two rolled GNRs self-assembling into the SWCNT to form helical configurations. The length of the GNR is 46.40Å.

4. Interaction mechanism

Why the GNR could self-assemble into the SWCNT and develop a helical manner? To further study its interaction mechanism, the final geometric configuration and energy evolution of this self-assembly system are displayed in this section.

Figure 4 shows the concentration distribution profile of the multiple-layered GNR@SWCNT system to help characterize the parameters of the multiple helical configurations. Although the peak of concentration distribution profiles is split near the centre of SWCNT due to the rigidity of C-C bond and the larger curvature in the inner helix, it can be measured that the separation of the adjacent layers is 3.5Å, similar to the basal plane separation in graphite or the wall thickness of the multi-walled carbon nanotubes (3.4Å). That is to say, GNRs are arranged very closely in the limited space of SWCNTs. In the crowded environment, long molecular chains frequently adopt ordered, helical conformations, which has been confirmed by Snir and Kamien [38]. In their simulation system, a solid, impenetrable but flexible tube was immersed into a solution of hard spheres. It is amazing that the flexible tube will take on a regular helical configuration which takes up the least space. The enlarged environment entropy is considered to be the driving force of helix formation.

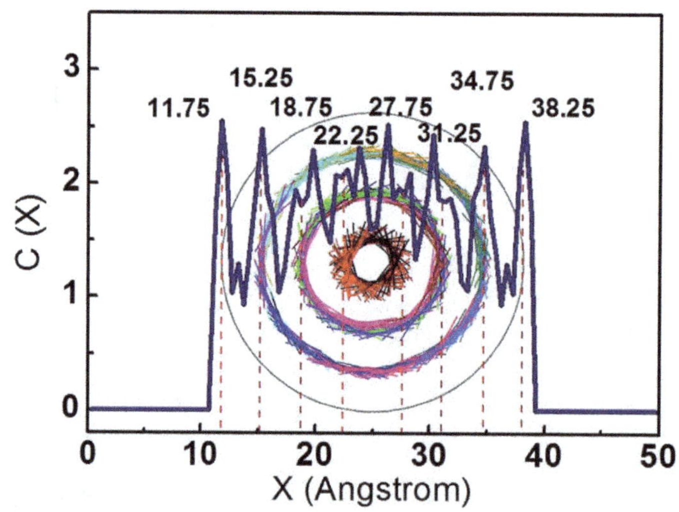

Figure 4. Concentration distribution profiles of the GNR and SWCNT in the (7, 4, 2) system in the X-direction.

On the other hand, the spontaneous filling and intertangling of GNRs are also driven by the increasing π-π stacking area between GNRs and SWCNT, because the system energy is released when carbon six-membered rings are stacked. In order to confirm this view, a series of simulations about different SWCNTs are studied. **Figure 5(a)** shows that similar multiple helical configurations are taken in cylindrical SWCNT shell to maximize their π-π stacking area, while the flat GNRs are obtained in the special cubical fixed SWCNT because this configuration can guarantee the largest stacking area with the straight and stationary tube wall. The partial enlarged view of the GNR@SWCNT helical configuration is displayed in **Figure 5(b)**, demonstrating that the GNR ring is offset from SWCNT ring slightly and their centroid-centroid distance is 3.858Å, consistent with the typical aromatic-aromatic offset π-π stacking [39].

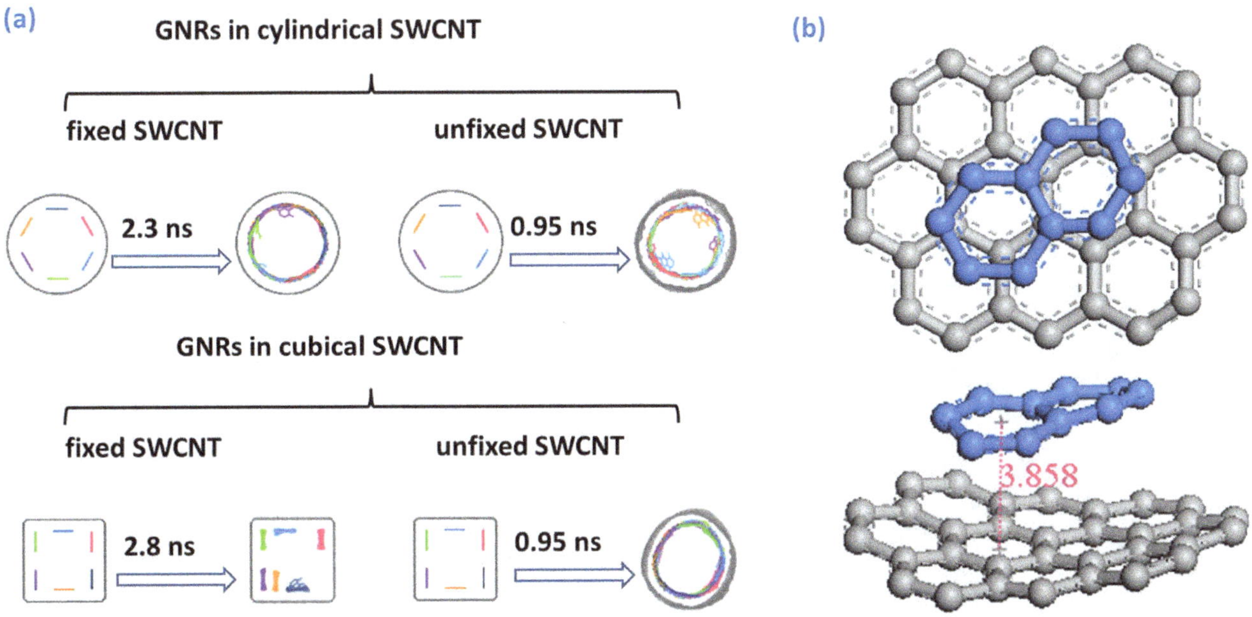

Figure 5. (a) The effect of the SWCNT configurations on the self-assembly process and (b) partial enlarged detail of the GNR@SWCNT helical configuration where SWCNT is grey and GNR is blue.

Figure 6 illustrates the potential energy (E_p), van der Waals interaction energy (E_{vdw}) and π-π stacking area for the GNR@SWCNT system as functions of simulation time in the self-assembly process. The evolution of π-π stacking area and system energy is identical, demonstrating that the energy decreasing is caused by the π-π stacking area expanding and the van der Waals interaction between the GNRs and SWCNT plays a dominant role in driving the formation of the helical configuration. The decline of potential energy indicates that the self-assembly course is spontaneous. Taking the self-assembly of (7, 4, 1) GNRs as an example, a clear change of energy gradient is shown in **Figure 6(a)**. The initial large energy gradient is contributed by the rapid expansion of π-π stacking areas as multiple GNRs filling into SWCNT. Then, the energy gradient decreases when the GNRs reach the other end of the carbon nanotube, because the tail of GNRs inserts into the SWCNT slowly with GNRs spiralling and configuration optimizing. As for the (7, 4, 1) GNRs in **Figure 6(b)**, both energy curves are divided into two stages, corresponding to two processes of structural changing. In the first stage from 0 to 1.8 ns, multiple GNRs encapsulate into carbon nanotube along their original directions and remain stable for some time. Then in the second stage (2.3–2.6 ns), two GNRs transfer into inner helical structure individually from the first and the second helix to obtain the largest π-π stacking area for the system, so the system configuration is changed from the (7, 4, 1) to (6, 4, 2) with a rapid decrease of potential energy, suggesting that this process can improve their stability. In this process, the van der Waals energy drops 0.7 Mcal/mol and the potential energy drops just 0.5 Mcal/mol, indicating that the disappeared energy is stored efficiently by the bending of new inner GNRs.

In principle, the helical self-assembly is determined by the competition between the van der Waals energy and the bending strain energy of GNRs [40]. The former provides an attractive force to spiral GNRs, and the latter usually tends to maintain the GNRs in their original form [41]. During the self-assembly process, the van der Waals interaction helps the GNRs overcome their energy barrier caused by the bending strain energy, and drives the GNRs to be trapped in the SWCNT and undergo self-scrolling. Throughout the helix-forming course, the van der Waals energy has partially converted into the internal energy of GNRs for their mechanical deformation and partially converted into kinetic energy.

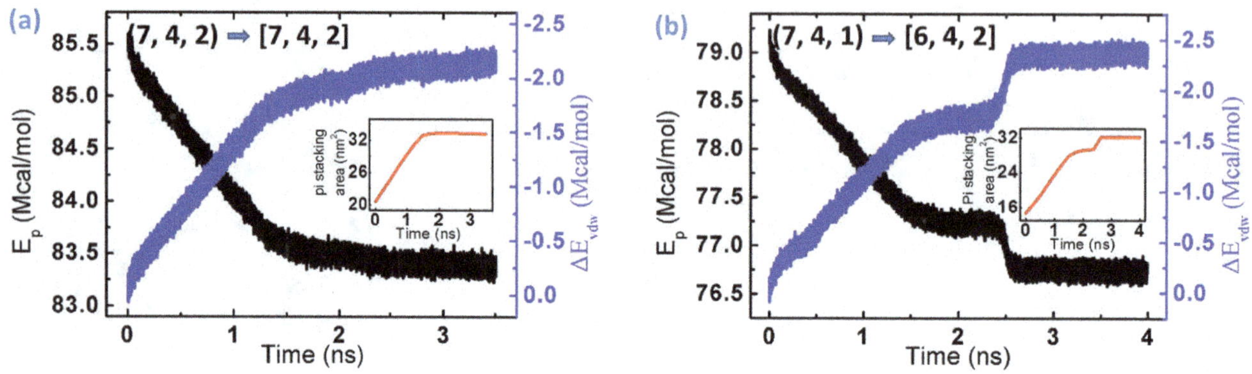

Figure 6. The potential energy (E_p) of GNR@SWCNT system and the van der Waals interaction energy (E_{vdw}) between the GNR and SWCNT as functions of simulation time in the helical encapsulation of (a)13 GNRs and (b)12 GNRs, and the inserted graph is the corresponding evolution of π-π stacking area.

5. Dependence of size, function group and initial arrangement of GNRs

In order to further exploit the effect of system size on self-assembly, a series of simulations and parameter analyses have been completed. **Figure 7** reveals that all GNRs of different widths in the SWCNT of different diameters are able to form the perfect helical structures, with varying screw pitches. The helical angle increases with GNR width but have no clear monotone change tendency with SWCNT diameter. In the self-assembly process, more time is needed to form a large helical angle because of the sluggish motion and coordination of multiple GNRs. Besides, the relation between system size and threshold value is also shown in **Figure 7**. It can be seen that the simulated threshold of GNRs in the first perfect helix is inverse with their width due to the limited space inside SWCNT but has a positive correlation with SWCNT diameter.

A simple mathematical model (**Figure 8**) is built to calculate the quantity threshold for the helical GNRs self-assembled in SWCNT. Three formulas are given as follows.

The threshold of GNRs in *i*-layer is

$$V_i = \left\lfloor \frac{(D-2iA)\pi}{A'+W} \right\rfloor \quad (1)$$

where D is the SWCNT diameter, i is the sequence of the GNR helix, A is the fixed distance between GNRs and SWCNT, A' is the fixed distance between adjacent turns of GNRs and W is the GNR width. The specific square brackets mean that the calculated value should be rounded down to the nearest whole unit. All distances are in units of Angstroms.

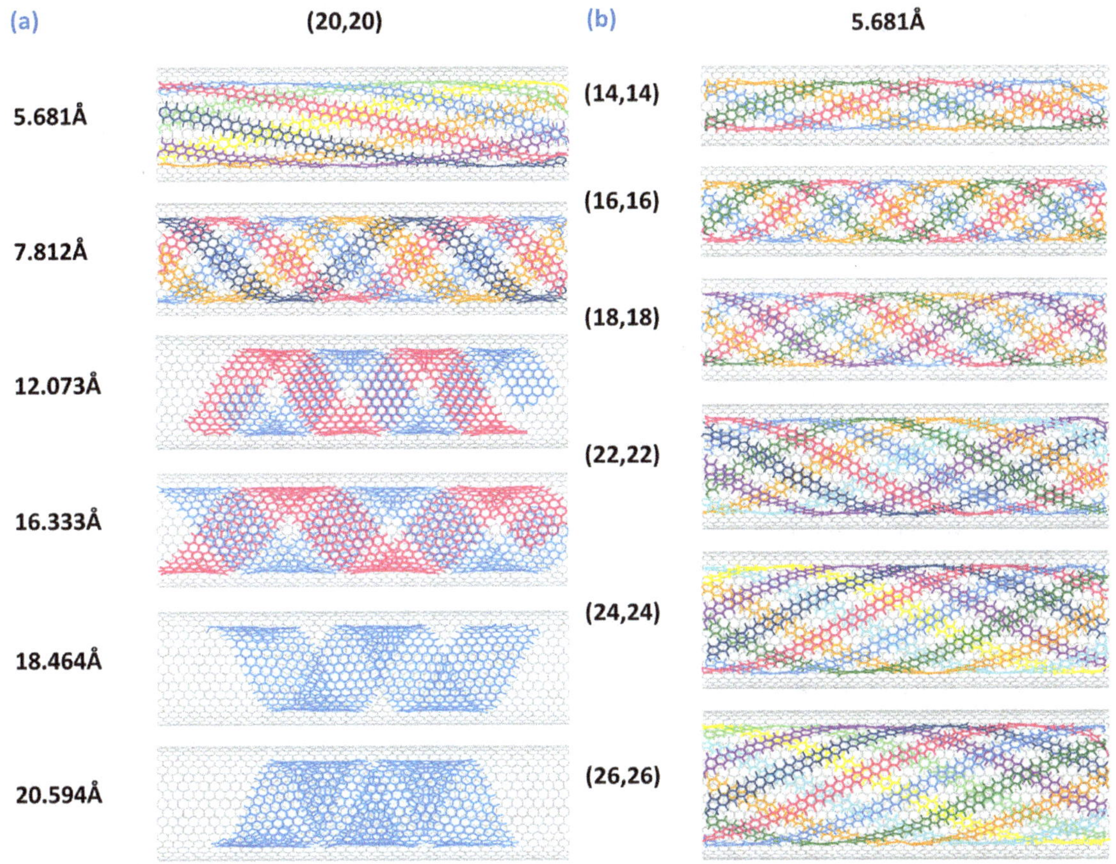

Figure 7. The dependence of the diameter of SWCNT and the width of GNRs in the evolution: (a) helical encapsulation of GNRs with different width into the SWCNT (20, 20); (b) helical encapsulation of GNR into the SWCNT with different diameter.

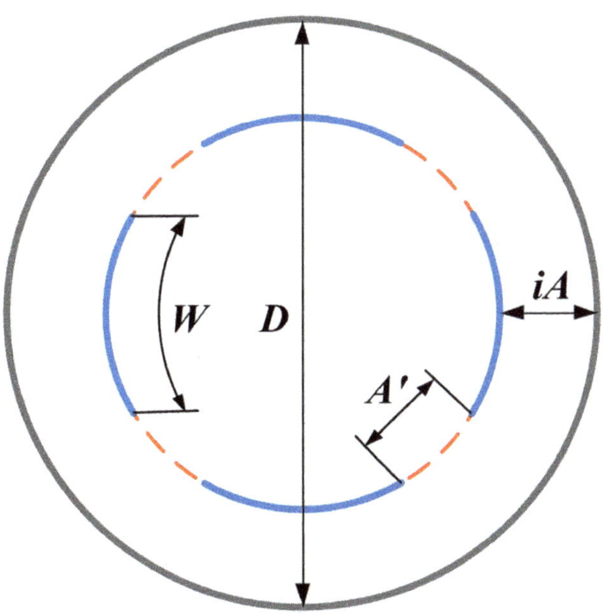

Figure 8. The ideal model of the cross section of GNR@SWCNT helical structure. SWCNT and GNR are referred to as grey circle and blue arc, respectively.

The whole threshold of the self-assembly system is

$$V_{\text{total}} = \sum_{i=1}^{N} V_i \tag{2}$$

where N is the number of GNR helix in SWCNT, which is described by

$$N = \left\lfloor \frac{D}{2A} \right\rfloor \tag{3}$$

The above formulas are used to calculate the theoretical threshold for 5.681Å GNR in (20, 20) SWCNT, and the result is consistent with the acquired thresholds of the first, second GNR helix and whole system very well. So the formulas are proved applicable to estimate the quantity threshold for the helical GNR in SWCNT.

Then, the effect of system size on the self-assembly of rolled GNRs is further studied. **Figure 9(a)** shows the diameter threshold of SWCNT which could be successfully encapsulated by the rolled GNRs with certain widths. It can be observed that the SWCNT diameter should be larger than the threshold to ensure the collapse and insertion of the rolled GNR. When the SWCNT diameter is too small, the van der Waals interaction cannot overcome the large energy barrier in system and the distance between GNR and SWCNT cannot maintain 3.5Å due to the confinement. Consequently, the SWCNT size should be chosen appropriately according to the GNR width to facilitate the self-assembly. In addition, **Figure 9(b)** demonstrates that a perfect helix inside the nanotube can be obtained when the space inside the nanotube is sufficient, which is independent of the diameter of the SWCNT. But what would happen if the length of the SWCNT is too short? **Figure 9(c)** shows the snapshots of the rolled GNR self-assembling into the short SWCNT. The rolled GNR inserts into the SWCNT successfully with the trapped portion to form a helix. After the helical configuration in SWCNT is closely packed, the front helical GNR is pushed out and the GNR tail continuously enters the SWCNT in a helical manner. Finally, the rolled GNR attempts to be symmetric to the tube, with the middle GNR in the short SWCNT forming a perfect helical configuration.

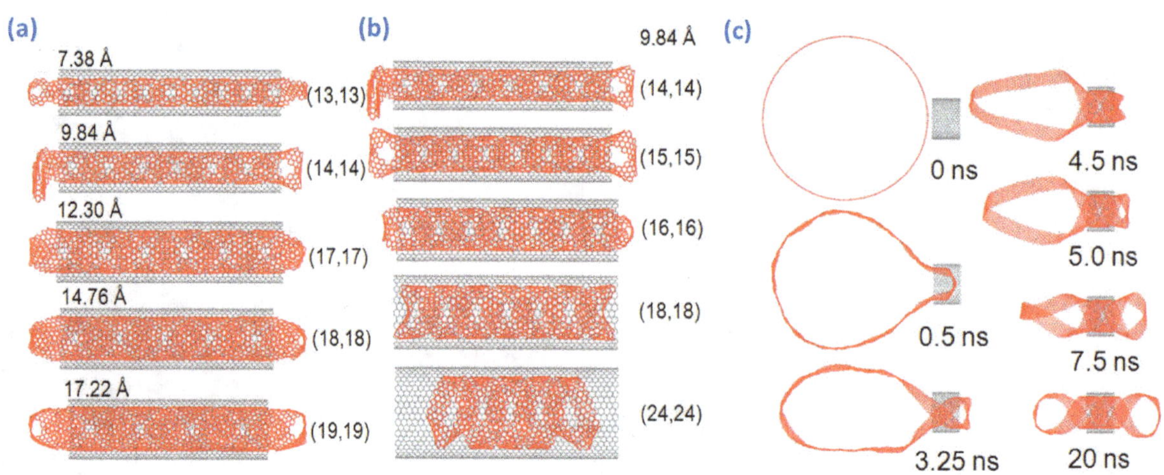

Figure 9. (a) The critical diameters of the SWCNTs to guarantee the insertion of the rolled GNR with different widths successfully: GNR(7.38Å)@SWCNT (13,13), GNR(9.84Å)@SWCNT (14,14), GNR(12.30Å)@SWCNT (17,17), GNR(14.76Å)@SWCNT (18,18), GNR(17.22Å)@SWCNT (19,19), respectively. (b) The helical insertion of GNR with the width of 9.84Å into the SWCNT with different diameters. (c) Representative snapshots of a rolled GNR inserting into the SWCNT (25, 25) with the length of 24.60Å to form a helical configuration.

The self-assembly of GNRs into SWCNT can be utilized to deliver substances into the nanoscale-confined space without any other external force. In order to explore its possible applications, different functional groups are modified on the edge of the rolled GNR. As shown in **Figure 10(a)**, the rolled GNR modified with 133 –OH can self-assemble into the SWCNT spontaneously to form a helix, and the interval between neighbouring segments is proved to be influenced by the chemical groups –OH. Then, a series of simulations are performed on the rolled GNR that is modified with different number of –OH, as shown in **Figure 10(b)**. The results demonstrate that the rolled GNR can form a helix in the SWCNT when one end of the GNR is modified with 50,100 and 133 –OH. By contrast, the rolled GNR cannot form a helix when fully modified with 200 –OH at one end, and even cannot be trapped by the hollow interior of the SWCNT when modified with 400 –OH. Another functional group, the –H, is also chemically attached to the rolled GNR uniformly. When the dangling σ-orbitals on the GNR are all saturated by hydrogen atoms, the rolled GNR can still self-assemble into the SWCNT and form a helix, which is quite different from the result of the GNR modified with –OH, indicating that the functional groups on the GNRs have a significant influence on their self-assembly. In addition, the spiral of the GNR without any dangling σ-orbitals also reveals that the van der Waals interaction is the main driving force of the helix self-assembly, rather than the dangling σ-orbitals at the edge of GNRs [42]. At the same time, the stability of the final configuration in different systems has been verified. **Figure 10(c)** shows that the potential energy of the 200 –OH system decreases to a certain value and then maintains at this level for 7 ns, indicating that the non-helical configuration is comparably stable. The potential energy in the system of GNR modified with 133 –OH is very different (**Figure 10(d)**) as it is divided into several periods: two 'rapid drop' periods, a 'platform' period and an equilibrium state. So it may be suspected that the modification of the GNR with different functional groups can influence the final configuration of the GNR@SWCNT system.

In addition, we further simulate the influence of the temperature on the final structure of the GNR@SWCNT system. It is surprising that the final composite structure of the GNR@SWCNT system can maintain stability even when the temperature is 1000 K.

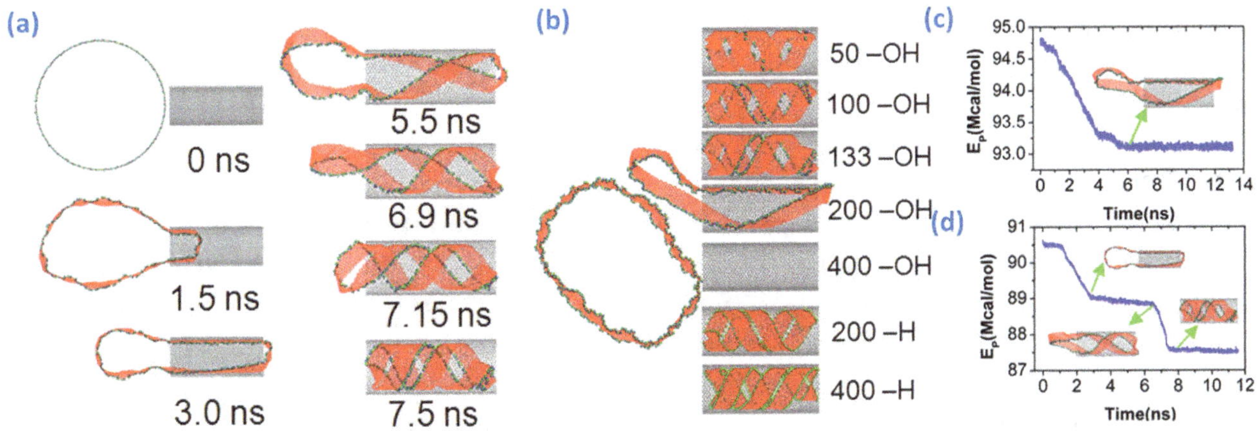

Figure 10. (a) Self-assembly snapshots of the GNR modified with –OH entering with the SWCNT (30, 30). The length of the GNR is 135.6Å, the width of the GNR is 14.76Å and the number of the –OH is 133. (b) Final configuration of this GNR modified with functional groups encapsulated into the SWCNT (30, 30). Total potential energy (E_p) of the GNR-SWCNT systems as functions of time: (c) the GNR modified with 200 –OH and (d) the GNR modified with 133 –OH.

6. Helical wrapping of GNRs

The above studies mainly focus on the insertion of GNRs into the hollow SWCNTs, but there is another form of self-assembly for the GNR/SWCNT system. This section investigates the collapse and wrapping characteristic of the rolled GNRs around the SWCNT. **Figure 11(a)** shows the typical snapshots of the rolled GNR spontaneously wrapping on the SWCNT (15, 15). Initially, the circular rolled GNR is positioned in the centre of the SWCNT, with the axis parallel to the SWCNT and the offset distance of about 5Å above the SWCNT. As shown in **Figure 11(a)**, the rolled GNR approaches the SWCNT rapidly due to the strong attractive force between them. During the approaching process, the rolled GNR stretches its cross section from a circle to an oval gradually because the carbon atoms close to the SWCNT endure stronger van der Waals force than the uppers. When the lower atoms in rolled GNR come into contact with the SWCNT, the upper atoms continue approaching the SWCNT, inducing the oval cross-section collapses to a triangle one [43] that could cover the upper surface of the SWCNT completely. Then the GNR rotates and its cross section changes into a bow shape in 5.5 ns. During the above process, the rolled GNR collapses rather slowly because the van der Waals interaction on the top semi-circular tube is so weak. But when th'e simulation reaches 9 ns, the collapse of the entire rolled GNR is further accelerated due to the increasing contact area [40] between SWCNT and GNR. The sequential collapse of these rings manufactures a linked double-layered GNR and then results in a double-walled helix wrapping around the SWCNT with a large helical pitch. Regardless of the angle between the two axes of the rolled GNR and SWCNT, the GNR can wrap around the SWCNT helically. The GNR helix out of the SWCNT wall is similar to the tendril of a morning glory climbing around a tree trunk where growth pattern helps the plant to maximize the sunlight absorption in the minimum space and grow strong. **Figure 11(b)** and **(c)** shows the concentration distribution profiles of the final configuration in the X-direction and the total potential energy of the GNR/SWCNT system in the wrapping process. The former figure reveals that the distance between two layers of the CNR outside the tube is also about 3.5Å, and the latter energy curve has a decreasing tendency with the simulation time and finally reaches the minimum at equilibrium, similar to the insertion of GNRs into SWCNT, meaning that the wrapping self-assembly of GNR/SWCNT has the same mechanism as the filling self-assembly.

The different final configurations of the rolled GNRs adhering on the SWCNTs are summarized in **Figure 12**. The length of all SWCNTs (15, 15) is 73.79Å, and the width of all rolled GNRs is 14.76Å. All rolled GNRs are circular initially. As illustrated in **Figure 12**, the rolled GNRs with a length larger than the threshold of 38.33Å are collapsed. However, the rolled GNRs with diameters between 157.61 and 230.03Å cannot fully collapse, but simply deform from cylindrical symmetry to bow shapes, which is because the van der Waals interaction in these system cannot overcome the energy required for the fully mechanical deformation of the round GNRs, leading to the mechanically bistable configuration [41]. When the GNR length continues to increase more than 234.30Å, the perfect helix outside of SWCNT can be guaranteed.

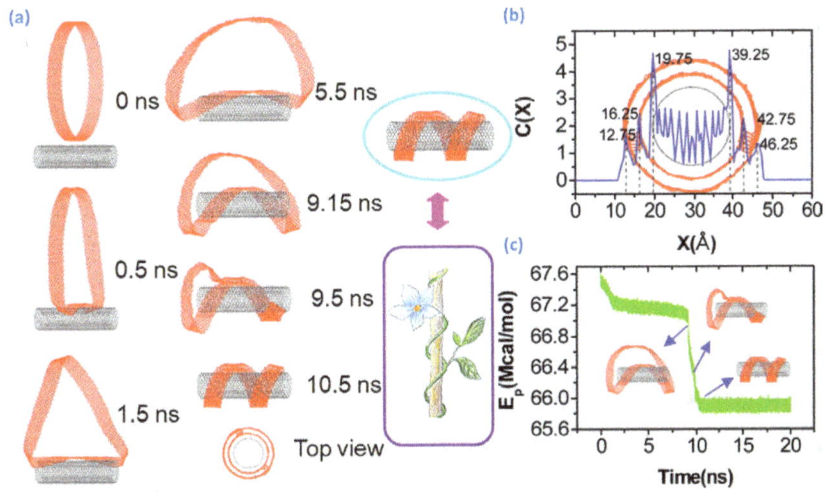

Figure 11. (a) Representative snapshots of a rolled GNR wrapping on the SWCNT (15,15) to form a spiral configuration, which resembles the tendril climbing around a trunk (hand drawn sketches by Wei Chen). The width of the GNR is 14.76Å and the length of the SWCNT is 73.79Å. (b) Concentration distribution profiles of the GNR and SWCNT in the system in the X-direction. (c) Total potential energy (E_P) of the GNR-SWCNT systems as a function of time.

Length of GNR	Final configuration	Representive snapshots	Conclusion (L: length)
25.57	●	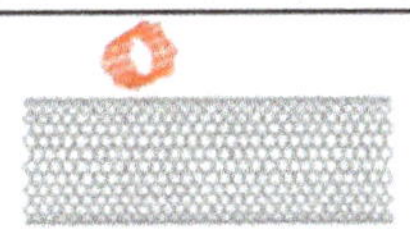	Uncollapsed: L≤34.09Å
29.81	●		
34.09	●		
38.33	▲	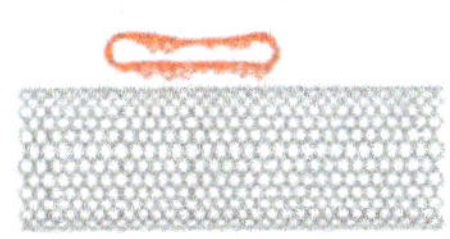	Full-collapsed: 34.09Å≤ L≤153.37Å
46.87	▲		
63.90	▲		
127.80	▲		
136.31	▲		
153.37	▲		
157.61	★	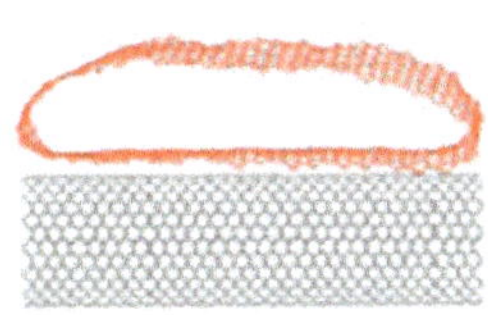	Semi-collapsed: 153.37Å≤ L≤230.03Å
170.40	★		
191.70	★		
213.00	★		
225.79	★		
230.03	★		
234.30	■	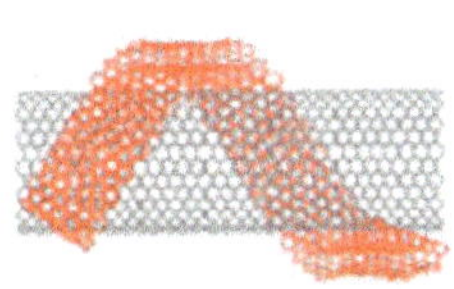	Helical configuration: L≥230.03Å
247.09	■		
255.60	■		
340.80	■		

●stands for circle configuration without collapse;
▲stands for the full-collapse parallel to the axe of the nanotube;
★stands for the semi-collapse;
■stands for the full-collapse to form a helical configuration.

Figure 12. The final configuration of rolled GNRs with different lengths adhering on the SWCNT (15, 15). The width of the GNRs is 14.76Å and the length of SWCNTs is 73.79Å.

7. Conclusion

In summary, this work proves that the multiple and rolled GNRs can self-assemble into or wrap around the cylindrical SWCNT to form a well-organized multiple helical configuration which takes the least amount of energy and takes up the least space. The self-assembly process is driven by the van der Waals interaction between the GNRs and the SWCNT, because this energy is released when carbon six-membered rings are stacked. The final self-assembly configuration of GNRs is responsible for the combined effect of system size and functional groups. The sizes of the SWCNTs should exceed a certain value to ensure the insertion of GNRs with certain widths, while they have negligible effects on the formation of the perfect helix after GNR encapsulation. The quantity threshold of GNRs inside SWCNT increases with the SWCNT diameter and decreases with the GNR width, and it can be estimated by the empirical formulas shown in this paper. The modification on GNR edge can influence the final configuration in the self-assembly process. In addition, the results of this study also reveal that the size of GNRs and SWCNTs should meet some required conditions to guarantee the wrapping self-assembly outside of the SWCNT in a helical form.

The helical configuration of the GNRs is similar to the ordered, spiral structure of long molecular chains such as DNA and protein in the confined cell. In fact, the spiral phenomenon is common but interesting in the nature, such as the spiral arrangement of florets in sunflower, the spiral nucleation of SiC crystal in a limited space and the spiral air flow in the centre of a tropical cyclone. What's more, Chinese old Tai Chi totem also emphasizes the philosophy of helical configuration, which implies that the universe derives from a series of helix. Thus, the findings in this paper are of great importance towards the better understanding of the helix, and of significance in predicting and controlling the self-assembly configuration. Moreover, the unique phenomenon of self-assembly in the GNRs/SWCNT system has potential advantages in the application as nanocontainer for drug delivery and molecular transportation. And this composite structure can also be used to fabricate functional nanodevices, such as sensor, nanoelectronic components, integrated circuits and optoelectronic devices.

Acknowledgements

The authors would like to acknowledge the support from the National Natural Science Foundation of China (Grant No.51671114). This work is also supported by the Special Funding in the Project of the Taishan Scholar Construction Engineering and National Key Research Program of China (Grant No. 2016YFB0300501).

Author details

Hui Li*, Yifan Li and Wei Chen

*Address all correspondence to: lihuilmy@hotmail.com

Key Laboratory for Liquid-Solid Structural Evolution and Processing of Materials, Ministry of Education, Shandong University, Jinan, People's Republic of China

References

[1] Pederson MR, Broughton JQ. Nanocapillarity in fullerene tubules. Physical Review Letters. 1992;**69**(18):2689–92. doi:10.1103/PhysRevLett.69.2689

[2] Serpell CJ, Rutte RN, Geraki K, Pach E, Martincic M, Kierkowicz M, et al. Carbon nanotubes allow capture of krypton, barium and lead for multichannel biological X-ray fluorescence imaging. Nature Communications. 2016;**7**:13118. doi:10.1038/ncomms13118

[3] You Y, Wei R, Yang R, Yang W, Hua X, Liu X. Crystallization behaviors of polyarylene ether nitrile filled in multi-walled carbon nanotubes. RSC Advances. 2016;**6**(75):70877–83. doi:10.1039/C6RA11783J

[4] Sadownik JW, Ulijn RV. Dynamic covalent chemistry in aid of peptide self-assembly. Current Opinion in Biotechnology. 2010;**21**(4):401–11. doi:10.1016/j.copbio.2010.05.010

[5] Nakanishi R, Kitaura R, Warner JH, Yamamoto Y, Arai S, Miyata Y, et al. Thin single-wall BN-nanotubes formed inside carbon nanotubes. Scientific Reports. 2013;**3**:1385. doi:10.1038/srep01385

[6] Choi WY, Kang JW, Hwang HJ. Structures of ultrathin copper nanowires encapsulated in carbon nanotubes. Physical Review B. 2003;**68**(19):193405. doi:10.1103/PhysRevB.68.193405

[7] Shao J, Yang C, Zhu X, Lu X. Melting and freezing of Au nanoparticles confined in armchair single-walled carbon nanotubes. The Journal of Physical Chemistry C. 2010;**114**(7):2896–902. doi:10.1021/jp910289c

[8] Gately RD, in het Panhuis M. Filling of carbon nanotubes and nanofibres. Beilstein Journal of Nanotechnology. 2015;**6**:508–16. doi:10.3762/bjnano.6.53

[9] Sun FW, Li H, Liew KM. Compressive mechanical properties of carbon nanotubes encapsulating helical copper nanowires. Carbon. 2010;**48**(5):1586–91. doi:10.1016/j.carbon.2009.12.056

[10] Nie C, Galibert A-M, Soula B, Flahaut E, Sloan J, Monthioux M. A new insight on the mechanisms of filling closed carbon nanotubes with molten metal iodides. Carbon. 2016;**110**:48–50. doi:10.1016/j.carbon.2016.09.001

[11] Hirahara K, Bandow S, Suenaga K, Kato H, Okazaki T, Shinohara H, et al. Electron diffraction study of one-dimensional crystals of fullerenes. Physical Review B. 2001;**64**(11):115420. doi:10.1103/PhysRevB.64.115420

[12] Farimani AB, Heiranian M, Aluru NR. Nano-electro-mechanical pump: Giant pumping of water in carbon nanotubes. Scientific Reports. 2016;**6**:26211. doi:10.1038/srep26211

[13] Lv C, Xue Q, Shan M, Jing N, Ling C, Zhou X, et al. Self-assembly of double helical nanostructures inside carbon nanotubes. Nanoscale. 2013;**5**(10):4191–9. doi:10.1039/c2nr33157h

[14] Agrawal KV, Shimizu S, Drahushuk LW, Kilcoyne D, Strano MS. Observation of extreme phase transition temperatures of water confined inside isolated carbon nanotubes. Nature Nanotechnology. Forthcoming. doi:10.1038/nnano.2016.254

[15] Smith BW, Luzzi DE. Formation mechanism of fullerene peapods and coaxial tubes: A path to large scale synthesis. Chemical Physics Letters. 2000;**321**(1–2):169–74. doi:10.1016/S0009-2614(00)00307-9

[16] Sato Y, Suenaga K, Okubo S, Okazaki T, Iijima S. Structures of D5d-C80 and Ih-Er3N@C80 fullerenes and their rotation inside carbon nanotubes demonstrated by aberration-corrected electron microscopy. Nano Letters. 2007;**7**(12):3704–8. doi:10.1021/nl0720152

[17] Caoduro C, Hervouet E, Girard-Thernier C, Gharbi T, Boulahdour H, Delage-Mourroux R, et al. Carbon nanotubes as gene carriers: Focus on internalization pathways related to functionalization and properties. Acta Biomaterialia. 2017;**49**:36-44. doi:10.1016/j.actbio.2016.11.013

[18] Li Y, Chen W, Ren H, Zhou X, Li H. Multiple helical configuration and quantity threshold of graphene nanoribbons inside a single-walled carbon nanotube. Scientific Reports. 2015;**5**:13741. doi:10.1038/srep13741

[19] Chen W, Li H. How does carbon nanoring deform to spiral induced by carbon nanotube? Scientific Reports. 2014;**4**:3865. doi:10.1038/srep03865

[20] Dutta S, Pati SK. Novel properties of graphene nanoribbons: A review. Journal of Materials Chemistry. 2010;**20**(38):8207–23. doi:10.1039/C0JM00261E

[21] Bai J, Cheng R, Xiu F, Liao L, Wang M, Shailos A, et al. Very large magnetoresistance in graphene nanoribbons. Nature Nanotechnology. 2010;**5**(9):655–9. doi:10.1038/nnano.2010.154

[22] Yang L, Cohen ML, Louie SG. Excitonic effects in the optical spectra of graphene nanoribbons. Nano Letters. 2007;**7**(10):3112–5. doi:10.1021/nl0716404

[23] Ni Y, Yao K, Fu H, Gao G, Zhu S, Wang S. Spin seebeck effect and thermal colossal magnetoresistance in graphene nanoribbon heterojunction. Scientific Reports. 2013;**3**:1380. doi:10.1038/srep01380

[24] Xia D, Xue Q, Xie J, Chen H, Lv C, Besenbacher F, et al. Fabrication of carbon nanoscrolls from monolayer graphene. Small. 2010;**6**(18):2010–9. doi:10.1002/smll.201000646

[25] Wang L, Zhang HW, Zhang ZQ, Zheng YG, Wang JB. Buckling behaviors of single-walled carbon nanotubes filled with metal atoms. Applied Physics Letters. 2007;**91**(5):051122. doi:10.1063/1.2767249

[26] Cicero G, Grossman JC, Schwegler E, Gygi F, Galli G. Water confined in nanotubes and between graphene sheets: A first principle study. Journal of the American Chemical Society. 2008;**130**(6):1871–8. doi:10.1021/ja074418+

[27] Sun H. COMPASS: An ab initio force-field optimized for condensed-phase applications overview with details on alkane and benzene compounds. The Journal of Physical Chemistry B. 1998;**102**(38):7338–64. doi:10.1021/jp980939v

[28] Bunte SW, Sun H. Molecular modeling of energetic materials: The parameterization and validation of nitrate esters in the COMPASS force field. The Journal of Physical Chemistry B. 2000;**104**(11):2477–89. doi:10.1021/jp991786u

[29] Wang Q, Duan WH, Liew KM, He XQ. Inelastic buckling of carbon nanotubes. Applied Physics Letters. 2007;**90**(3):033110. doi:10.1063/1.2432235

[30] Wang Q. Atomic transportation via carbon nanotubes. Nano Letters. 2009;**9**(1):245–9. doi:10.1021/nl802829z

[31] Sun H, Mumby SJ, Maple JR, Hagler AT. An ab initio CFF93 all-atom force field for polycarbonates. Journal of the American Chemical Society. 1994;**116**(7):2978–87. doi:10.1021/ja00086a030

[32] Rigby D, Sun H, Eichinger BE. Computer simulations of poly(ethylene oxide): Force field, pvt diagram and cyclization behaviour. Polymer International. 1997;**44**(3):311–30. doi:10.1002/(SICI)1097-0126(199711)44:3<311::AID-PI880>3.0.CO;2-H

[33] Andersen HC. Molecular dynamics simulations at constant pressure and/or temperature. The Journal of Chemical Physics. 1980;**72**(4):2384–93. doi:10.1063/1.439486

[34] Wittung P, Nielsen PE, Buchardt O, Egholm M, Nordén B. DNA-like double helix formed by peptide nucleic acid. Nature. 1994;**368**(6471):561–3. doi:10.1038/368561a0

[35] Brodsky B, Ramshaw JAM. The collagen triple-helix structure. Matrix Biology. 1997;**15**(8–9):545–54. doi:10.1016/S0945-053X(97)90030-5

[36] Bets KV, Yakobson BI. Spontaneous twist and intrinsic instabilities of pristine graphene nanoribbons. Nano Research. 2009;**2**(2):161–6. doi:10.1007/s12274-009-9015-x

[37] Ruoff RS, Tersoff J, Lorents DC, Subramoney S, Chan B. Radial deformation of carbon nanotubes by van der Waals forces. Nature. 1993;**364**(6437):514–6. doi:10.1038/364514a0

[38] Snir Y, Kamien RD. Entropically driven helix formation. Science. 2005;**307**(5712):1067. doi:10.1126/science.1106243

[39] Janiak C. A critical account on [small pi]-[small pi] stacking in metal complexes with aromatic nitrogen-containing ligands. Journal of the Chemical Society, Dalton Transactions. 2000(21):3885–96. doi:10.1039/B003010O

[40] Yan K, Xue Q, Xia D, Chen H, Xie J, Dong M. The core/shell composite nanowires produced by self-scrolling carbon nanotubes onto copper nanowires. ACS Nano. 2009;**3**(8):2235–40. doi:10.1021/nn9005818

[41] Lu W, Chou T-W, Kim B-S. Radial deformation and its related energy variations of single-walled carbon nanotubes. Physical Review B. 2011;**83**(13):134113. doi:10.1103/PhysRevB.83.134113

[42] Li Y, Sun F, Li H. Helical wrapping and insertion of graphene nanoribbon to single-walled carbon nanotube. The Journal of Physical Chemistry C. 2011;**115**(38):18459–67. doi:10.1021/jp205210x

[43] Yan K, Xue Q, Zheng Q, Xia D, Chen H, Xie J. Radial collapse of single-walled carbon nanotubes induced by the Cu_2O surface. The Journal of Physical Chemistry C. 2009;**113**(8):3120–6. doi:10.1021/jp808264d

10

Growth Study and Characterization of Single-Layer Graphene Structures Deposited on Copper Substrate by Chemical Vapour Deposition

Stefanos Chaitoglou, Enric Bertran and Jose Luis Andujar

Abstract

Currently, the graphene industry is moving forward to the import of graphene in a number of novel applications. To take full advantage of the excellent properties of the material, the standardization of the growth process is an emergency. The suitable growth technique should ensure the high yield, accompanied by high quality of single-layer graphene sheets. Chemical vapour deposition is the technology that fulfils the above requirements, promoting the growth of largescale graphene films through automatized processes. In the present chapter, we present the latest advances in this field, summarizing the most recent publication activity of the authors. The results outline how the control in the growth process over parameters like the gases flow, growth temperature and pressure can affect the nucleation density of graphene domains, the growth rate and percent coverage. Growth of graphene domains with different morphologies depends on the crystallographic orientation of the copper lattice. At the same time, the formation of ripples occurs in the graphene surface as a result of the copper foil compression during the cooling step. These ripples are responsible for the appearance of a compressive stress in the graphene sheets. We demonstrate the control over such stress through the variation in the hydrogen flow during the growth.

Keywords: graphene films, chemical vapour deposition, ripples formation, hydrogen effect

1. Introduction

Bottom-up synthesis methods like chemical vapour deposition (CVD) are those that better ensure the growth of continuous graphene films [1]. Evidence of the potential to move to industrial scale synthesis has been proven [2, 3]. When the various graphene nuclei reach the coalescence phase, they join to each other, forming a continuous layer. In this growth model, the grain boundaries are proven defects points affecting the quality of the graphene layer, both concerning mechanical and electrical properties. Thus, it is an emergency to proceed with the growth of wafer-scale single crystal graphene domains [4–10]. Efficient growth approaches should ensure a low nucleation density of graphene domains, followed by high growth rates. Parameters that affect the synthesis process are the gas mixture and growth pressure, temperature and time [11]. Copper foil is one of the preferred substrates for the CVD growth as it permits the production of large-scale, single-layer graphene domains [12]. Carbon atoms show a low solubility into the copper. More carbon species remain in the surface of the foil participating in the formation of the graphene sheet [1]. A conventional CVD growth method consists of a continuous flow of carbon gas precursor/H_2/Ar mixture in order to generate the graphene growth [13]. In all experiments described in this chapter, methane is used as carbon precursor. As described in the work of Li et al. [14], the steps for the graphene CVD growth are the following:

1. Exposure of Cu to methane, argon and hydrogen.
2. Catalytic decomposition of methane on Cu to mainly form CH_y species.
3. As a result of the temperature, methane pressure, methane and hydrogen flow and partial pressure, the Cu surface appears undersaturated, saturated or supersaturated with CH_y species.
4. Formation of nuclei as a result of local supersaturation of CH_y.
5. Nuclei grow to form graphene islands.
6. Full Cu surface coverage by graphene under specific temperature flow rates and pressure conditions.

The hydrogen presence has proved to play a critical role in the process, as it affects different mechanisms. It reduces the native copper oxide layer and activates the graphene growth. At the same time, it can apply an etching on the material. Thus, it is fundamental to fully understand the above mechanisms. It has been investigated before how hydrogen can affect thickness, shape, size, edge configuration and crystalline quality of graphene islands/domains as well the control of the nucleation density [15–21], which is very important, taking advantage of the low solubility of carbon into copper [21–27]. Polishing of the copper surface or even enclosure of the substrate in special 'pockets', enables the control of the gases mixture, has proven to help in the control of the nuclei density [12]. The use of high growth temperatures favours the copper recrystallization, permitting the growth of high-quality graphene [28]. Searching for means to further reduce the nucleation, it has been proposed the oxidation of the copper surface [29, 30] resulting in very low domain density with size up to 1 cm^2 [31]. In order to reduce the copper surface, thermal annealing processes are usually applied, which require long periods of baking [32].

In this chapter, we present new experiments and results evidencing the effect of the partial pressure of H_2 at a given CH_4/H_2 gas flow ratio on the growth of bi-dimensional crystals of graphene and on their morphology.

To provide a better understanding of the physicochemical mechanisms that govern film formation, we include them in the framework of atomic level phenomena such as adsorption, diffusion and nucleation, providing the possibility to have a full control on the growth process. We study the CVD growth of graphene in the range between 970 and 1070°C, examining the temperature effect in the graphene growth ratio and nucleation density. By fitting the graphene coverage ratio as a function of growth time, we estimate an activation energy of 3.01 eV.

Finally, we study the formation of ripples in the surface of the graphene sheet. Ripples in the graphene surface are proven to affect its electronic structure [33], carrier transport[34] and chemical properties.[35] Therefore, the capability to control the formation of this kind of corrugation is urgent.

In other recent works, Park et al. have demonstrated the capability to control the density and height of such ripples through control of the cooling ratio, which follows the growth step [36, 37]. Slow cooling permits the formation of molecular hydrogen, which can result in the suspension of the graphene film. Moreover, they have demonstrated the importance of such ripple morphology in the mobility properties of the graphene. Graphene with fewer and smaller ripples demonstrates better transport properties when it is embodied in electronic devices.

2. Experimental part

The whole process was carried out in a suitable CVD oven (schematically drawn in **Figure 1**). The reactor consists of a furnace, where the CVD process takes place, coupled to a spherical chamber with a magnetron sputtering system. The turbomolecular pump can achieve a vacuum in the order of ~10^{-4} Pa, providing a very pure atmosphere for the graphene growth. This allows us to deposit different thin films (copper-nickel on silicon) in ultrapure conditions (without oxygen) and to perform directly the graphene growth without exposing the substrate to the ambient. By this method, all the treatments, since the cleaning process and reduction/annealing of the copper foil or the deposition of the metal catalyst by sputtering, can be performed '*in situ*' by a single run. For the evaluation of the obtained graphene, we used Raman spectroscopy as well as scanning electron and atomic force microscopy. The scanning electron microscopy (SEM) images were taken with a JEOL JSM7100F microscope in 5 keV. The Raman spectroscopy was performed with a Jobin-Yvon LabRam HR 800 system. A green laser with a 2 μm spot diameter has been used. Atomic force microscopy (AFM) image has been obtained with an AFM Multimode 8, electronica Nanoscope V (Bruker), and operated by the NOVA software.

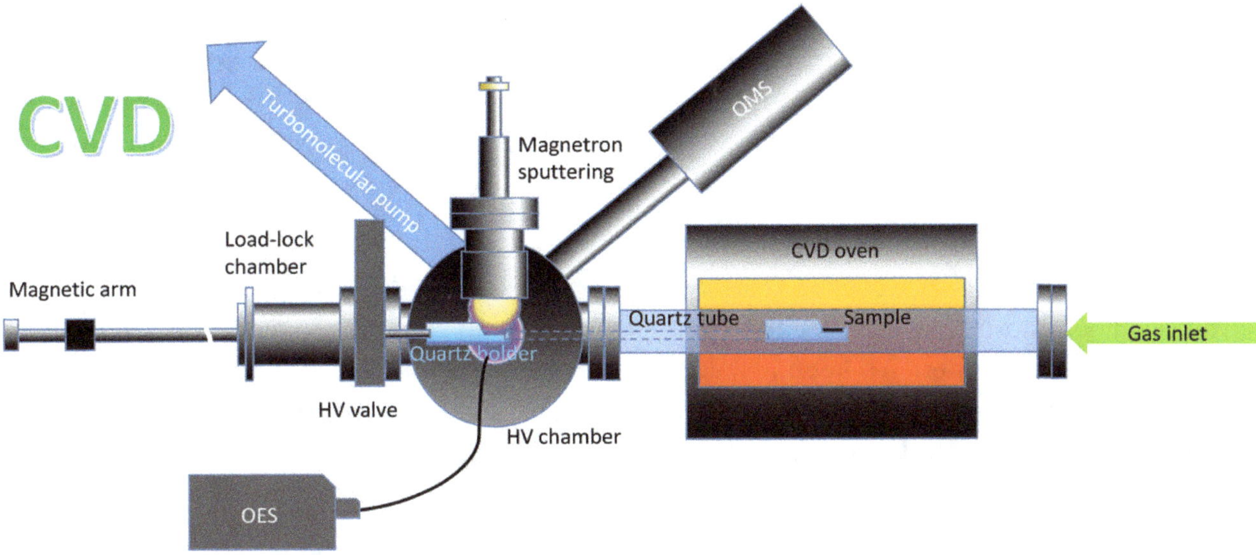

Figure 1. Illustration of the reactor. The system consists of two parts, a chamber with a magnetron sputtering head for thin films deposition and a quartz tube where the CVD processes take place.

2.1. Pretreatment of copper foil

Polycrystalline copper foil 75 μm thick and 99% pure was cut in pieces of 2 × 2 cm^2. The foil piece is cut in pieces of 2 × 2 cm^2. It is pre-cleaned with acetone and isopropanol immersed in an ultrasonic bath for 10 min. Then it is being introduced into the CVD reactor. In other works, the application of argon/hydrogen of helium/hydrogen plasma for the effective pre-treatment of the copper foil is being introduced. Plasma pre-treatment is, moreover, found to improve the crystallinity of the formed graphene [38]. Here we present the performance of hydrogen plasma for the same purpose. From previous studies made in our group [39], we have been able to optimize the parameters necessary to use in order to reduce the native copper oxide layer. According to these results, first, a background pressure of 7×10^{-4} Pa is applied with a turbomolecular pump to secure that the whole process is taking place under high purity conditions. Radio frequency (RF) hydrogen plasma is then applied to chemically reduce the copper substrate. To generate the plasma, we introduce hydrogen with 20 sccm flow rate and applied 100WRF power at 20 Pa. The hydrogen radicals react with the copper oxide .

$$CuO + 2H \rightarrow Cu + H_2O \quad (1)$$

2.2. Graphene CVD growth

The authors have investigated in various works [40, 41] the growth mechanisms of graphene and the effect of parameters like the hydrogen concentration, growth pressure, time and temperature. The hydrogen flow appears to perform an etching in the graphene domains, which affects their morphology and uniformity, when its concentration exceeds some limits. The results reveal that a sufficient graphene growth is possible when we optimize the switching of the carbon precursor/hydrogen flow ratio during the process. This reduces the etching effect that the hydrogen is performing, allowing the growth of graphene and the full cover of the substrate by it.

We first perform experiments varying the gas mixture ratio, hydrogen/methane and the total pressure. By analysing images obtained by scanning electron microscopy, under stable methane/hydrogen flow rate and varying the total pressure, we study the variation in the nucleation density and size of the graphene domains. Under stable pressure and varying the hydrogen flow, we observe the anisotropical etching that the second performs when its concentration exceed an equilibrium amount

Samples grown under different conditions have been observed by SEM, using the Image J software. The amount of samples analysed can be seen in the size distribution histograms presented in this chapter. We keep a stable methane/hydrogen ratio, 5/20 sccm, and control the total pressure in the chamber by rotating the conductance valve which is placed in the turbomolecular pump. In **Figure 2**, we have the series of samples that were grown at 20 min, in a range of pressures between 12.5 and 20 Pascal. To control the hydrogen effect, we perform growths of the same duration, under stable pressure and methane flow and varying the hydrogen flow in the range between 10 and 20 sccm (**Figure 3**). All the experimental conditions are presented in **Table 1**. To study the size of the graphene domains, we use the length of the lobes.

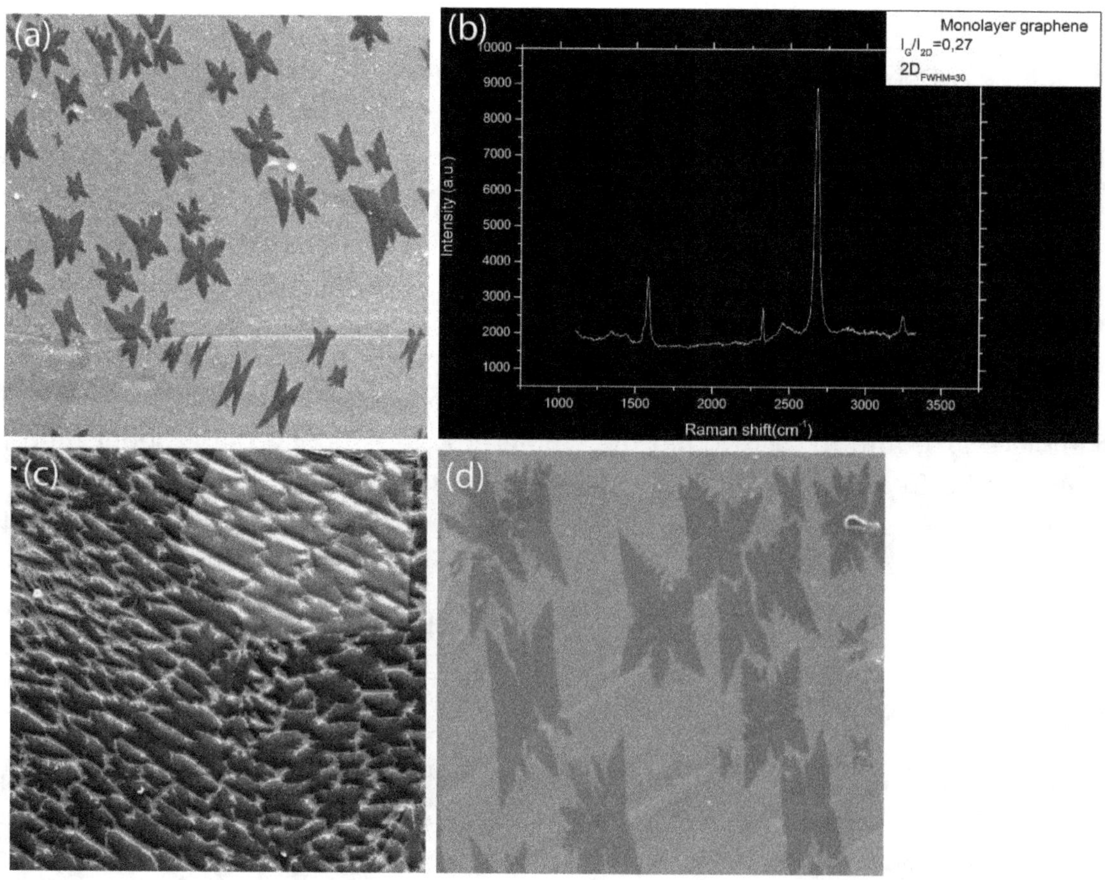

Figure 2. Scanning electron emission images of graphene grown in stable methane/hydrogen (5/20 sccm) flows under different total pressures. Figure (a) 12.5 Pa total pressure, (b) 15 Pa, (c) to 17.5 Pa, (d) to 20 Pa. The size of the crystals is dependent on the growth total pressure in a not linear way. In panel (b) is being provided a Raman spectrum of the continuous graphene film after its transfer on top of SiO_2. The length of the X direction is 200 μm in all figures.

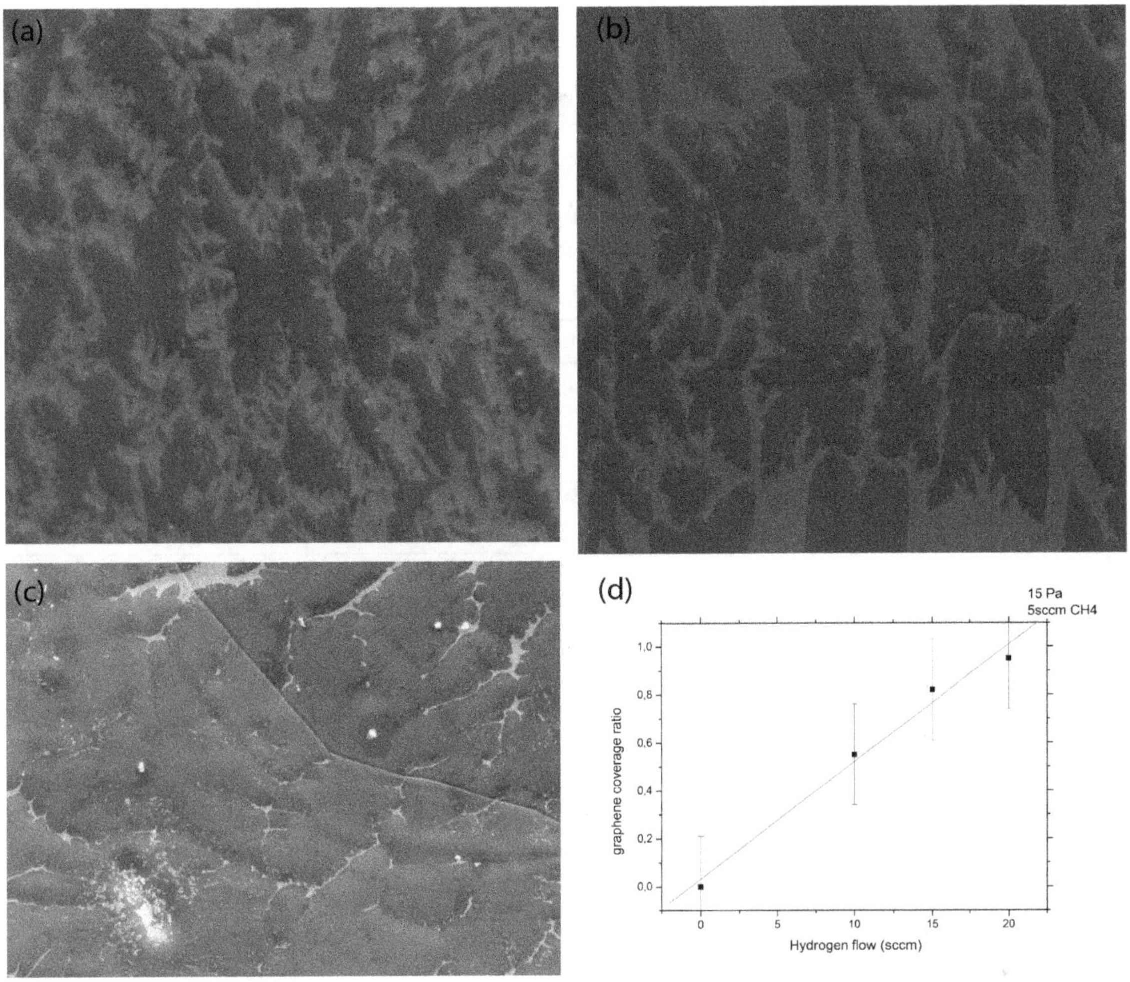

Figure 3. Scanning electron microscopy images of graphene grown under 5 sccm of methane, at 15 Pa total pressure, in 1040°C, during 20 min. Varying the hydrogen flow, at (a) 10, (b) 15 and (c) 20 sccm affects the morphology of the film. The scale bar is in all the images 10 μm. In panel (d) the plot demonstrates the percent graphene covered area. The highest coverage, ~95% corresponds to the highest hydrogen flow.

Sample (corresponding image)	Total pressure (Pa)	Hydrogen flow (sccm)	Crystal's lobe length (μm)	Nucleation density (nucleus/1000 μm^2)	% Surface cover
Image 2a	12.5	20	13	0.8	15
Image 2b	15	20	27	1.2 (full cover)	100
Image 2c	17.5	20	14	3.7	82
Image 2d	20	20	32	0.5	40
Image 3a	15	10	30.5	2.6	55
Image 3b	15	15	30.1	0.8	82
Image 3c	15	20	27	1.3	95

Table 1. Experimental conditions of the samples presented in **Figures 2** and **3**. The methane flow is the same in all experiments, 5 sccm, as well as the growth time (20 min) and temperature is 1040°C. In the table, the effect of growth total pressure and hydrogen flow to the crystal size as well as the nucleation density is presented.

2.3. Temperature effect

We investigated the growth of graphene in the range between 970 and 1070°C. Below that temperature graphene nucleation was not observed, while the upper limit was chosen in order to be as close as possible to copper melting temperature (1084°C). The rest of the growth process is followed as explained in the previous chapter. The pressure is stable at 15 Pa and the $CH_4/H_2 = 5/20$. The only varying factor is the growth temperature (**Table 2**).

Sample	Methane flow (sccm)	Hydrogen flow (sccm)	Pressure (Pa)	Temperature (°C)
1	5	20	15	970
2	5	20	15	990
3	5	20	15	1030
4	5	20	15	1070

Table 2. (a) Graph image exhibiting the various steps of the growth process and (b) growth conditions of the different samples.

2.4. Strain control via the H_2 flow

To study the effect of hydrogen in the formation of ripples in the graphene surface, we have performed growths where only the flow of hydrogen is varying. The rest of the parameters are stable. Growth temperature is always 1040°C and the growth time is set in 20 min. The flow of methane is 5 sccm. We performed growth from 10 to 35 sccm H_2 (**Table 3**). The growth time was always 20 min. After that time, the furnace was opened and cooling to 500°C took about 10 s.

Sample	Hydrogen flow (sccm)	Ripples density (number/μm)	2D FWHM (cm^{-1})	2D position (cm^{-1})	Strain ε (%)
A	15	2.6	32.6	2719	–0.002
B	20	3.6	26.6	2701.6	–0.0011
C	25	3.8	24.75	2694.75	–7.3E-4
D	30	4.3	30.4	2680.5	–2.6E-5
E	35	4.3	30.75	2696	–8.4E-4

Table 3. Experimental data of the ripples density and the 2D peak characteristics (position and FWHM). The strain is calculated through the 2D peak shift.

3. Results and discussion

3.1. Copper oxide reduction

During the hydrogen plasma treatment the hydrogen radicals react with the copper oxide reducing it to metallic copper [23]. We use optical emission spectroscopy to determine and evaluate the reduction of the copper surface. The outcoming light from the discharge was collected by a spectrophotometer (Stellarnet EPP2000C), which operated in the range 300–850 nm. The evolution of the OH radical integral spectral intensity in the spectral range of 305–330 nm is a sufficient tool to evaluate the removal of the oxide [42]. Results show that the intensity of the peaks corresponding to OH radicals decrease after 5 min of plasma treatment. In particular, the first peak at 305 nm decreases from 22.6 to 5.9 a.u., as we can see in **Figure 4**. The reduction in the OH concentration reveals the reduction of the oxygen radicals (as a result of the oxide reduction). Nevertheless, as the copper oxide cannot be completely removed, the peak does not decrease more with further treatment [43]. **Figure 5(a)** shows the copper foil used as a substrate before (up and down sample) and after (medium sample) the hydrogen etching and the graphene growth. Graphene protects copper from oxidation which explains that even after the passage of days the reflected colour of the substrate does not change [44]. Copper foil before etching appears with a higher pitch of red because of the copper oxide layer on the top. After the treatment it appears brighter and large crystalline domains can be distinguished at naked eyes. This process is much more time sufficient, compared to the usual pre-annealing required for the Cu substrates taking up to several hours [32]. As an additional tool to characterize the copper surface, we used electron backscattered diffraction (EBSD). It

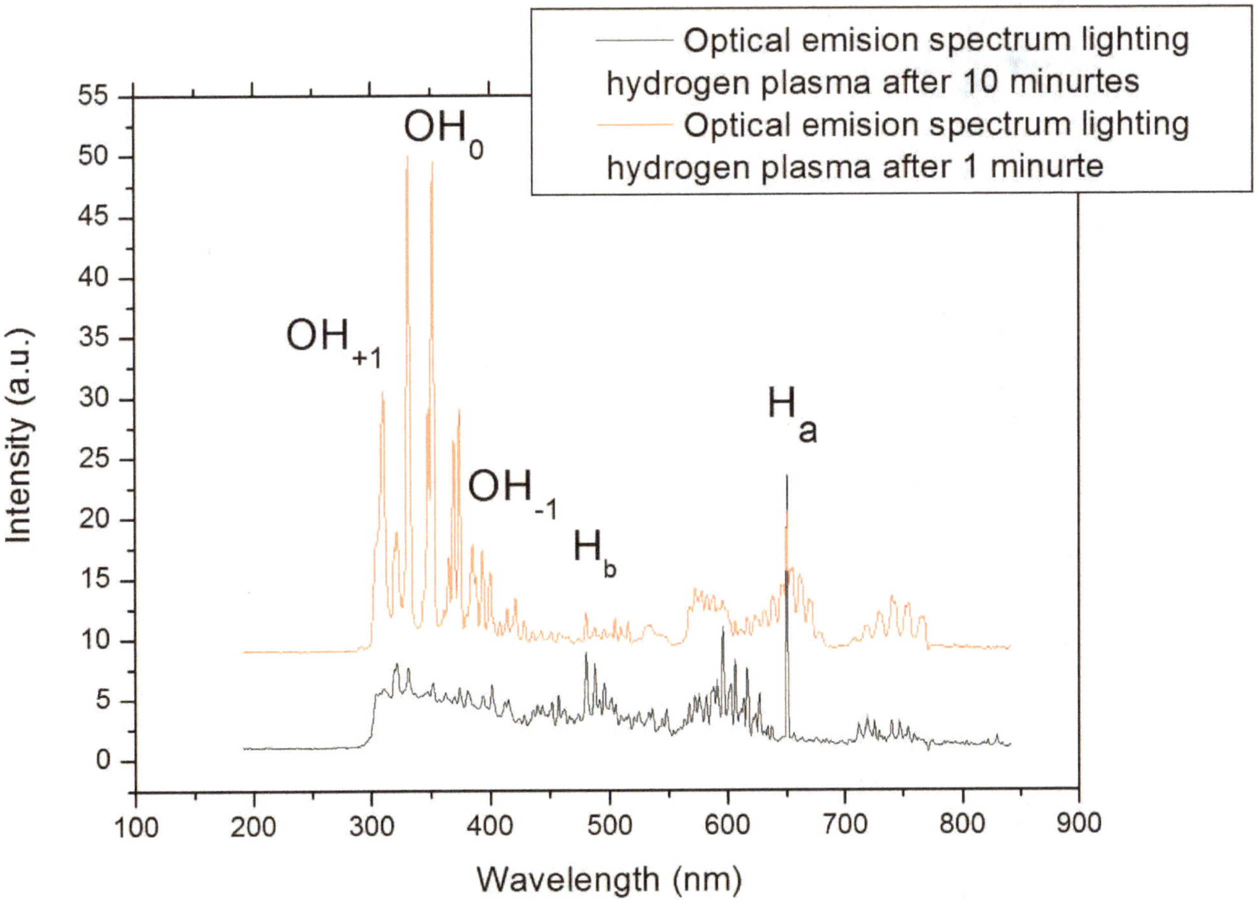

Figure 4. Optical emission spectroscopy of hydrogen plasma during reduction of copper: after 1 min (red line) and after 5 min (black line). The reduction of the OH radical peak intensity in the spectral range of 305–330 nm reveals the removal of the oxide layer.

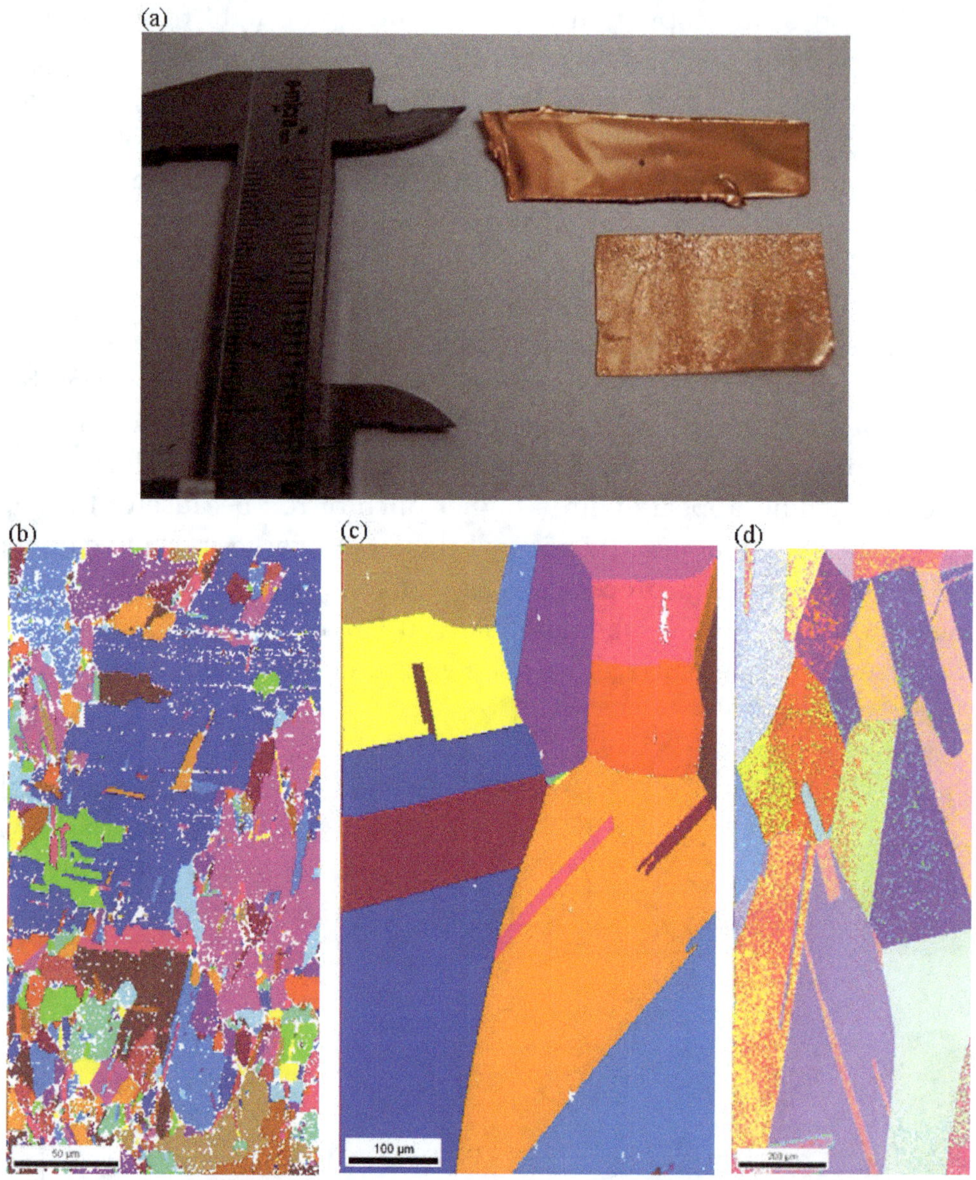

Figure 5. (a) Polycrystalline copper foil before (up piece) and after (down piece) the hydrogen etching. The difference in the surface colour indicates the deoxidation of it. The red line indicates 4 cm. EBSD maps of (b) the as-received polycrystalline copper foil (scale bar 50 μm), (c) the foil after plasma annealing and 20 min graphene growth at 1040°C (scale bar 100 μm), and (d) the foil after plasma annealing and 50 min graphene growth (scale bar 200 μm).

is a large area imaging method which detects the different index copper facets. Here, EBSD is used in order to demonstrate the growth of the copper grains after the H_2 plasma and the growth process, with respect to the as-received copper foil. In **Figure 5**, we provide the EBSD mapping of untreated polycrystalline copper foil of 75 μm (**Figure 5(b)**), copper foil after 20 min growth (**Figure 5(c)**) and copper foil after 40 min growth (**Figure 5(d)**). As we see, the grains of the untreated copper have various sizes with most of them not overcoming ~30 μm^2. The boundaries between the grains are randomly oriented. During the heating and the growth the necessary energy is provided to the grains in order to increase in size. In our experiments, the heating is taking place under no presence of gas, in very high vacuum, and in the order of 10^{-4} Pa. The growth is taking place under the presence of hydrogen and methane, in a pressure of ~20 Pa and in a temperature of ~1040°C. As we see in **Figure 5(c)** and **(d)**, with the longer growth, the grains grow larger. We see gains of sizes up to ~2000 μm² for the longer growth. The edges between the grains now are much straighter. These factors affect the quality of the grown graphene later. Graphene needs smooth and large grains to grow over. Usually, the boundaries are defect sites for the graphene, so the less possible boundaries mean the less possible defects.

3.2. Lobe's length and nucleation density of graphene bi-dimensional islands

Graphene crystals are grown forming planar dendritic geometries. The carbon species are being attached in the border of the islands. Varying the pressure results in formation of different graphene, considering both the size and the nucleation density of the domains. The nucleation density varies 8, 12, 37, 5 nuclei/10.000 μm², respectively, for 12.5, 15, 17.5 and 20 Pa. The different shapes of the domains that can be seen in **Figure 2a** are a result of the different copper orientation. As it has been explained [45] domains with four cusps grow over Cu (221) and domains with six cusps over Cu (310). The Raman spectra in the inset of **Figure 2b** ensures us about the single-layer nature of the graphene film. We can also observe some second-layer nucleation in some spots, below the first layer [46, 47]. We are interested in growing graphene with the lowest nucleation density possible, in order to decrease the amount of grain boundaries in the film, which usually introduce defects. For this reason, samples grown under 12.5 and 20 Pa are more adequate. We use a residual gas analyser to record the methane and hydrogen pressure during the growth. The $\delta P^{H}/\delta P^{CH4}$ ratio increases exponentially with the increase of the total pressure. This is probably due to the insufficiency of the turbomolecular pump to evacuate the low-mass atomic hydrogen. This results in higher concentration of hydrogen in the chamber.

Making some calculations, we observe that if the surface coverage is normalized with the quadratic lobe´s length , we can extract the linear dependence between it and the nucleation density. In **Figure 6(a)**, we see the agreement between our experimental data and this model.

$$\frac{C}{L^2} = \frac{3\sqrt{3}}{2}n \tag{2}$$

The value of the slope, $b = 1.5 \times 10^{-4}$, of the fitted line in **Figure 6(a)** is the constant of the surface growth rate, dS/dt as a function of the lobes growth rate dL/dt.

In **Figure 6b–d**, we present the histograms of the results of the lobe's length size distribution measurements (independently of the copper facet) for the growth processes taking place at 12.5, 17.5 and 20 Pa. The two-peak distribution in histogram of **Figure 6d** is a result of the different growth velocity for the orientations of graphene cusps grown on the (310) copper crystalline facet. At 15 Pa, we have formation of a continuous layer, therefore we cannot estimate the size of the single domains.

3.3. Nucleation density and copper facet

From the SEM figures, we also observe a relation between the nucleation density and the copper orientation. Similar observations have been reported elsewhere [48]. According to the

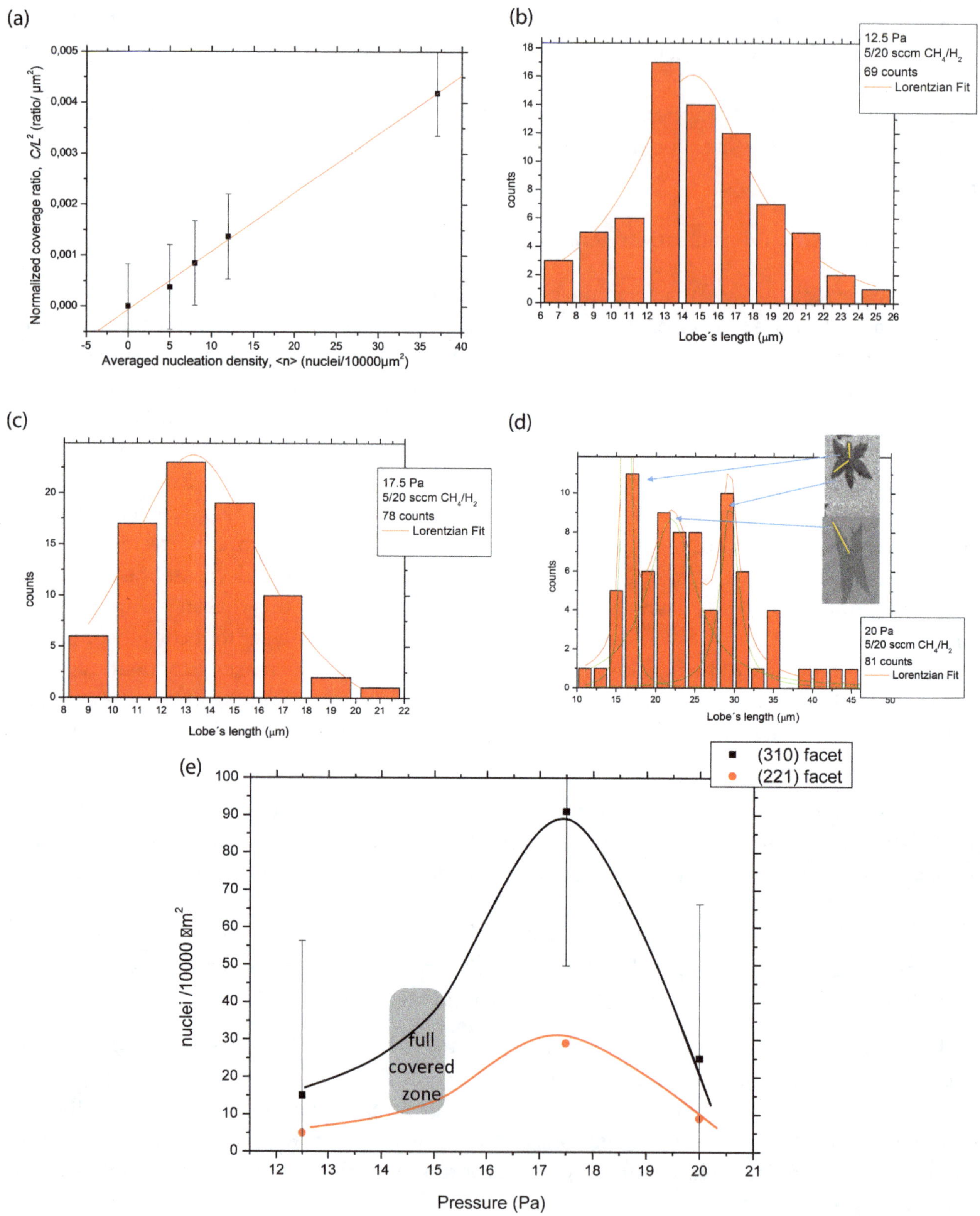

Figure 6. (a) Diagram demonstrating the linear dependence between the C/L^2 and the nucleation density. (b) Histogram of the lobe's length distribution measurements for the growth taking place in 12.5 Pa total pressure (c) and in 17.5 Pa. The tree peaks here is a result of the anisotropic growth of the lobe's in the (310) facet as well as the lobe's length in the (221) facet. (d) Same for 20 Pa total pressure. (e) Diagram presenting the variation in the nucleation density for each pressure and in relation to the copper facet. The copper facet (310) favours the nucleation.

shape of the graphene domains, in our samples, we observe nucleation over the copper facets (221) and (310). We observe that in all pressures, nucleation density is higher in the (310) facet. The results are demonstrated in **Figure 6e**. Thus, we can conclude that the facet (221) is more suitable for the growth of films with fewer boundaries. This is interesting to know as it establishes the possibility of growing graphene in single crystallinity copper surfaces.

3.4. Graphene coverage

Using the Image J software, we can know the percent surface coverage with graphene. We see that for the lowest nucleation, in 12.5 and 20 Pa, the coverage is lower that for the higher ones (**Figure 7**). The highest coverage appears in 15 Pa growth pressure. We observe that coverage plot follows a similar fluctuation as the nucleation density plot. On the right axis, we presented the $\langle P_{H2}\rangle/\langle P_{CH4}\rangle$ ratio of the average relative pressure values as a function of time, for the different total pressures, which follow a similar fluctuation as well, revealing a connection between the two. From the above, we underline the critical role of hydrogen. When its concentration is not sufficient, it cannot catalyze the graphene growth through the formation of active surface-bound carbon species (C_yH_x)s. When the concentration is too high, the etching effect is reinforced. Graphene nucleation, the growth rate and the termination size of grains are affected by competition of these two processes.

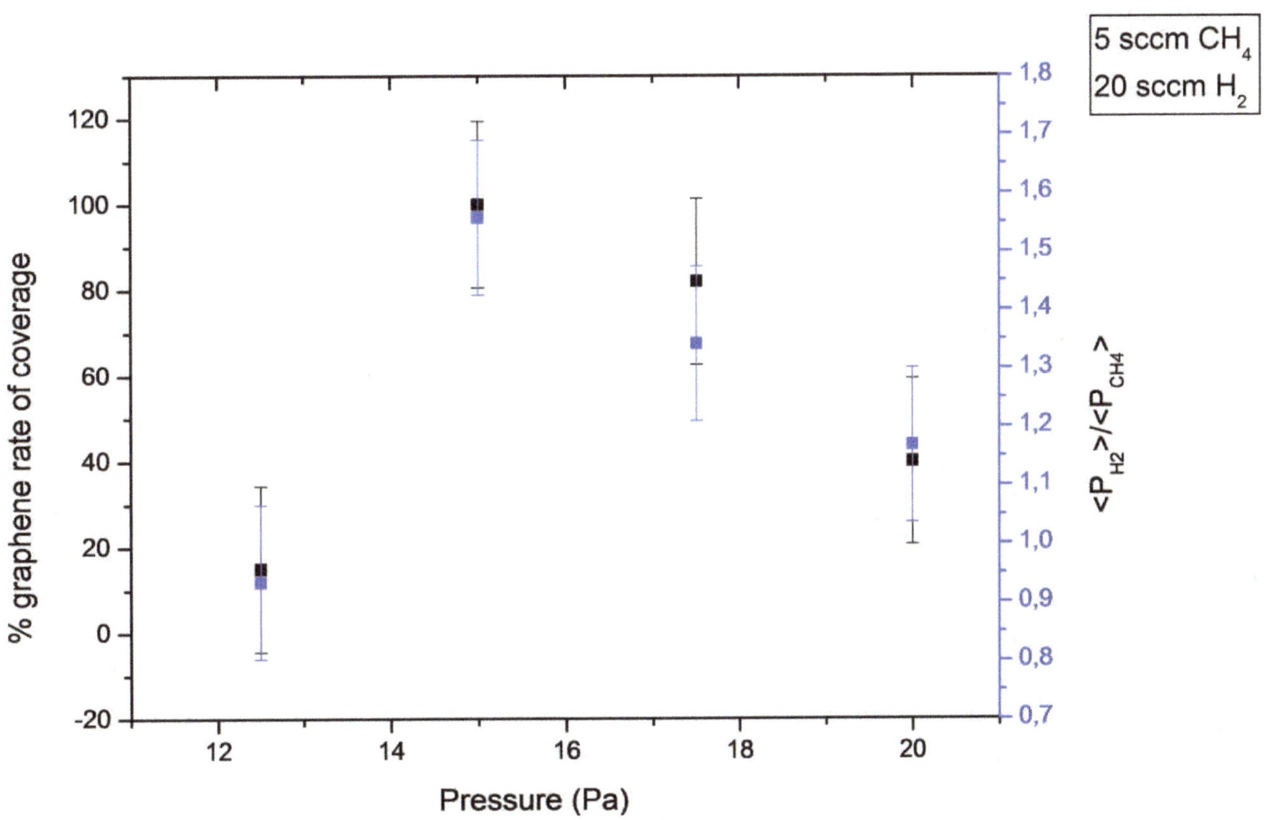

Figure 7. On the left Y axis we see thegraphene percent coverage of the surface as a function of the total pressure. On the right Y axis we see the $\langle P_{H2}\rangle/\langle P_{CH4}\rangle$ ratio of the average relative pressure values as a function of time, for the different total pressures. The match in the variation of the values between the two graphs is evidence of the importance of hydrogen in the graphene growth.

3.5. Effect of the growing time

When we study the effect of growth time in the formation of graphene film, we can observe clearly the etching effect (**Figure 8**). We obtain a continuous film in 20 min. If we prolong the growth time, we observed the etching of the film and its damage as a result of the high hydrogen concentration in the chamber. To overcome this problem, techniques have been proposed which involve performing a two-step growth [14]. In the first step, a lower flow of methane is introduced in order to achieve low nucleation. In the second step, the methane flow and pressure increase in order to rapidly fill the gaps in the surface of the film.

3.6. Growth of graphene on large copper surface

We perform growth in larger copper pieces, 12 cm^2 with respect to 4 cm^2 that were used before. We do an extensive Raman analysis to study if graphene of same quality grows in the whole surface. The results indicate that single-layer graphene grows in the whole area, from the centre up to the edges. The spectra are obtained with a green laser. The G position appears at 1578 ± 3 cm^{-1} and the 2D at 2671 cm^{-1}, confirming the single-layer nature of the film [49]. The I_G/I_{2D} ratio is 0.27 ± 0.03 with a FWHM of 30 cm^{-1}. All data are presented in **Figure 9a**. Graphene of same quality is grown in the back side of the foil (**Figure 9b**). Then, the graphene is transferred onto different substrates, quartz and SiO_2, to study possible effects. The transfer is made by electrochemical delamination of the graphene [32]. By using this method, we avoid introducing any impurities from the metal etchant. G and 2D peak appear shifted by very few cm^{-1}. Such shifts can be due to edges, dislocations, cracks or vacancies in the sample that cause the so-called self-doping of the graphene [50].

3.7. Temperature effect

We are interested to study how the growth temperature can affect the nuclei formation and growth velocity of graphene. In addition, we study the formation of ripples in the graphene layer. The ability to control the formation of such ripples is of high interest if we want to obtain graphene films with enhanced mobility properties [37].

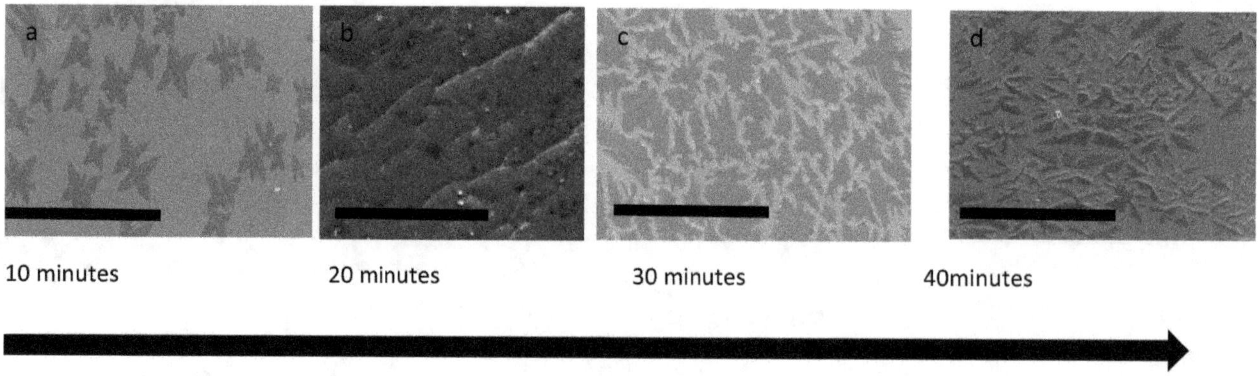

Figure 8. Variation of growth time. (a) At 10 min growth, 20 μm lobe's length crystals are obtained. (b) At 20 min growth, the film appears continuous. (c) At 30 min growth, the film appears partially damaged. (d) At 40 min growth, most of the graphene have been etched away from the catalyst substrate. The scale bar is always 50 μm.

(a)

sample	2D position (1/cm)	G position(1/cm)	I_G/I_{2D}	2D FWHM (1/cm)
Cu foil	2671	1582	0,38	59
Cu foil	2668	1575	0,40	64
Cu foil	2674	1578	0,42	65
Cu foil	2671	1575	0,45	60
Cu foil	2668	1575	0,44	63
After transfer on glass	2679	1575	0,64	67
Back side	2663	1578	0,43	65
After Transfe on SiO2	2676	1582	0,40	35
After transfer on SiO2	2676	1582	0,48	40
After transfer on SiO2	2676	1582	0,03	37
After transfer on SiO2	2671	1575	0,27	30

(b)

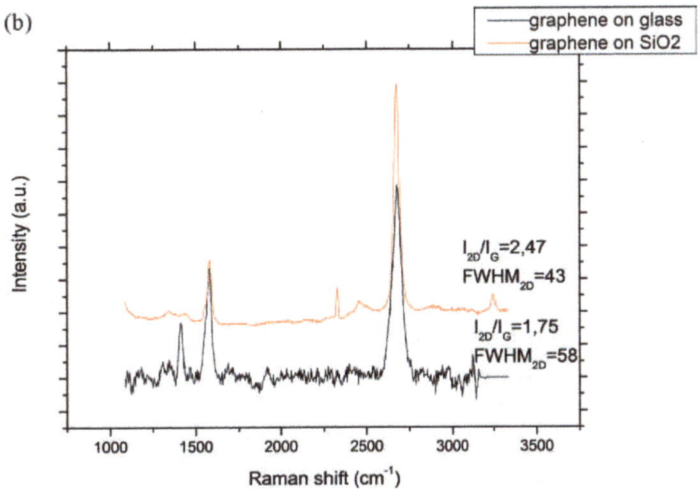

Figure 9. (a) Table presenting the Raman spectroscopy results. (b) Raman spectra after the transfer of the graphene film to glass (with thermally grown quartz on top, black line) and on SiO_2 (red line).

Considering the application of the hydrogen plasma, its efficiency in removing the copper oxide layer has been discussed in previous works of the authors [40] . In the same work, we provided back scattered electron microscopy figures where the evolution of the copper domains is shown. During the heating, the copper grains grow up, forming domains with size up to 10^4 μm^2.

In **Figure 10**, we present the various scanning electron microscopy (SEM) images of graphene grown in different temperatures. **Figure 10a** corresponds to growth taking place at 970, (b) 990, (c) 1030°C and (d) 1070°C. We observe that nucleation of the graphene only occurs in the initial instants. This explains the high homogeneity in the graphene domains' size. In addition, higher temperatures lead to lower nucleation density and higher growth rates. No second layer nucleation below the first layer is observed in any of the figures.

In these figures, with the exception of **Figure 10d**, the domains do not reach the coalescence phase. Further growth times should be required to obtain a continuous graphene layer. Before, we have studied in depth the growth mechanisms of such continuous layers. We underline the dual role of the hydrogen in this process. Control of the growth is also performed through the growth pressure and its influence on the residence time of the carbon atoms in the surface. Careful selection of the above two parameters resulted in the formation of large continuous layers. Further extension of the growth time resulted in the anisotropic etching of the graphene.

Here we focus in the temperature effect and we study the kinetics of the graphene growth based on it. The study of the SEM figures provides information considering the nucleation density of the graphene crystals, growth velocity and surface coverage. Kinetics of two-dimensional graphene nucleation over copper has been studied both theoretically and experimentally by Kim et al. [51] before. When methane reaches the copper surface, it breaks down. The concentration of carbon active species increases until it reaches a critical supersaturation level. Nucleation starts upon this point. Nucleation and growth of the supersaturated nuclei deplete the adsorbed carbon species surrounding them and the nucleation rate becomes negligible. The growth of the nuclei continues until supersaturated amount of surface carbon species is consumed and equilibrium

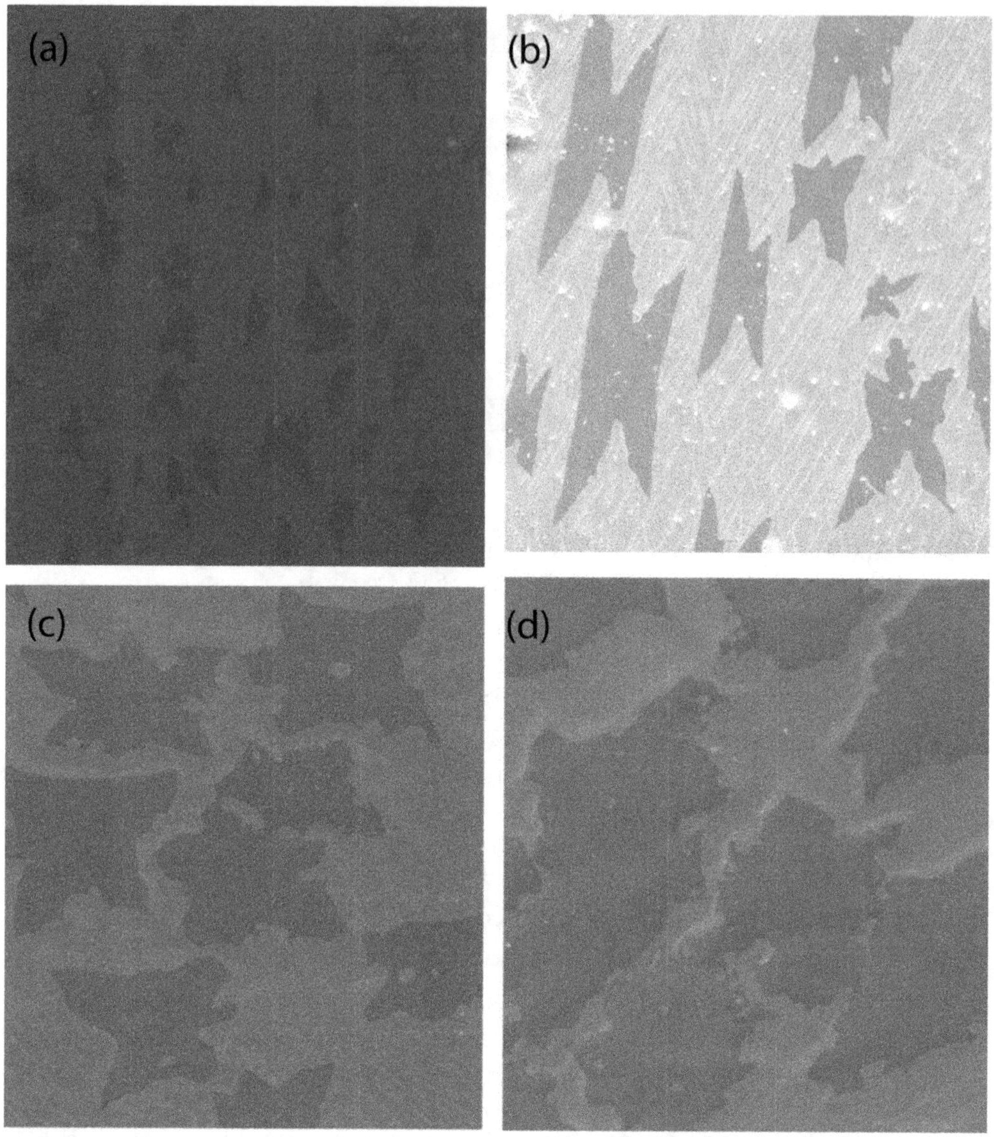

Figure 10. SEM images of graphene domains grown in different temperatures: (a) at 970, (b) 990, (c) 1030 and (d) 1070°C. The *X* axis is always 100 μm.

between graphene, surface carbon and CH_4/H_2 is reached. Depending on the amount of available carbon, graphene domains will either coalesce to form a continuous layer or stop growing to reach a saturated, final incomplete coverage [52].

To understand the kinetics of the existing nucleation model, we study the nucleation density and growth rate with respect to the growth temperature. In **Figure 11a**, we show the Arrhenius plot of the temperature-dependent density of graphene nuclei. According to Robinson model [53], there are two nucleation regimes which are a result of the competition between the processes of adatom capture, surface diffusion and re-evapouration. In the high temperature regime ($T > 870$°C), the desorption of the carbon adatoms is significant compared to their mobility, so their lifetime and nucleation rate is desorption-controlled. The decrease in the nucleation density for increasing temperatures is explained due to the increase of the desorption rate, reducing the probability of further nucleation, as discussed by Kim et al. [51].

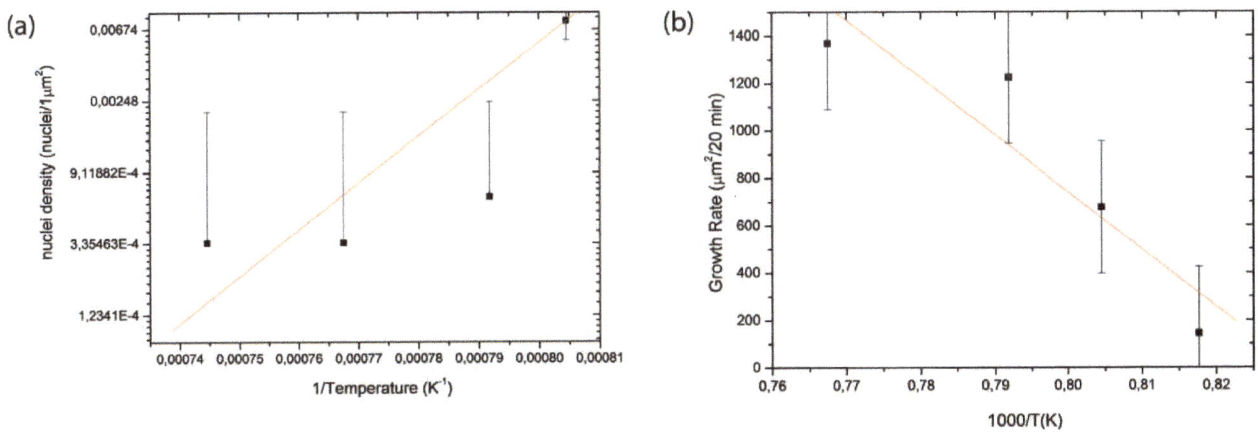

Figure 11. (a) Analysis of graphene nucleation behaviour. Natural logarithm of density of graphene nuclei versus $1/T$ from SEM analysis. (b) Linear plot of the graphene nucleus growth rate versus $1000/T$.

In **Figure 11b**, we observe how the growth rate increases with the increase of the growth temperature. In this plot, we assume that the growth is linear for the growth time. Higher growth rates occur thanks to the lower nuclei density. Previous studies [51] have shown no significant differences of the growth rates on three main crystalline orientations [Cu(111), Cu(100) and Cu(103)] of the polycrystalline Cu.

To calculate the activation energy, we will apply the kinetic model proposed by Xing et al. [54]. According to this model, growth rate is proportional to the uncovered copper surface.

$$\frac{d\text{coverage}}{dt} = a(1 - \text{coverage}) \tag{3}$$

According to the above equation, the rate is the highest in the initial moments and decreases during the graphene growth. By integrating it, we derive the coverage equation coverage $= -e^{at}+1$ where t is the time and a is a constant defined by the growth temperature. As seen in **Figure 12b**, coverage increases with the increase in temperature, result which can be explained taking into account the previous observations considering the nucleation density and growth rate behaviour. On the same time, a increases with the temperature as well. To calculate the activation energy E of the graphene growth, we use the Arrhenius equation $a \sim e^{-E/kT}$, where k is the the Boltzmann's constant and T is the absolute temperature. From the Arrhenius plot of **Figure 12a**, we can extract the activation energy from the slope of the linear fit. The activation energy is calculated to be 3.01 eV. In previous studies, a wide range of activation energies (1–3) have been proposed for graphene growth over copper. The energy barrier depends on the growth temperature and on the dominant nucleation mechanism. In lower temperatures, the adsorption energies of carbon monomers depend on the crystalline orientation of copper [55]. In higher temperatures, where desorption is the principle responsible nucleation mechanism, the differences in activations energies are minimal with respect to the copper orientation, both for low and atmospheric pressure CVD [11].

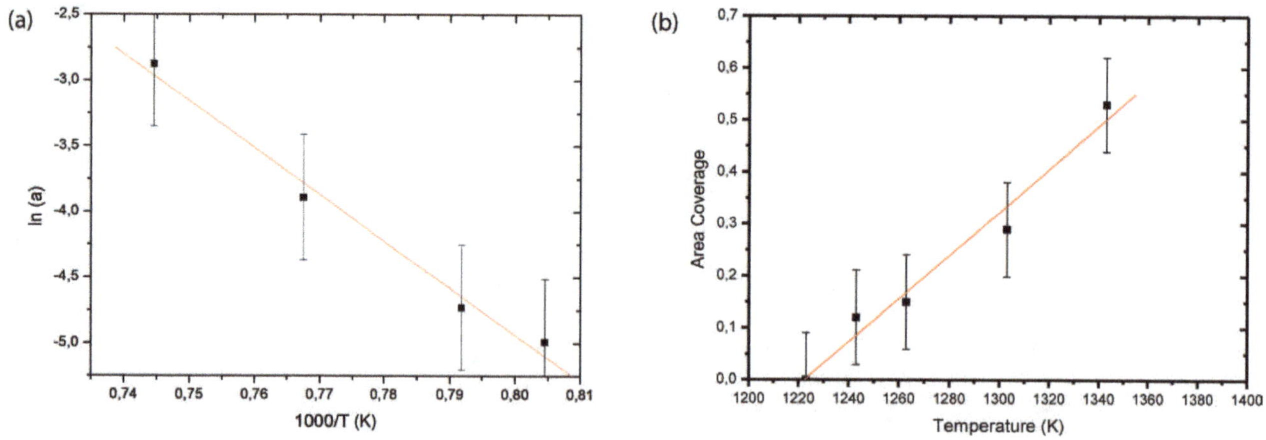

Figure 12. (a) Arrhenius plot of the ln (a) as a function of 1000/T. From the slope of the linear fitting we obtain the activation energy. (b) Graphene coverage on Cu as a function of different growth temperatures.

3.8. Strain control via the H_2 flow

By the SEM images, we could determine the density of the ripples in each sample. **Figure 13** shows the SEM images for the different hydrogen flows in the studied range.

Figure 14a provides the Raman spectrum corresponding to samples grown at different hydrogen flows. The copper photoluminescence background has been removed from all the spectra. In previous works, the copper luminescence has been attributed to the different quenching effect of the deposited graphene owing to probably irregular thickness distribution. Another explanation can be the quenching effect of graphene as a function of distance between Cu surface and graphene. The suspended graphene can result from the ripple's formation in the copper surface during the cooling step [37]. In **Figure 14b**, we can see an enlarged image of the above spectra where the shift in the 2D peak becomes evidenced by the evolution of the hydrogen flow.

The absence of a D peak in the spectra presented in **Figure 14a** proves the highly crystalline nature and the absence of defects in the graphene films. The ration I_D/I_G is proportional to the mean distance between two defect points in a graphene film. The absence of the D peak corresponds to pristine graphene [56]. A small D peak can be distinguished in the spectra corresponding to graphene growth under the higher hydrogen flow rate (35 sccm) probably owing to the hydrogen etching on the graphene surface. The defects concentration of the graphene surface can be obtained by Eq. (4) [57]

$$n_D(\mathrm{cm}^{-2}) = \frac{10^{14}}{\pi^2\left[C_A\left(r_A^2 - r_S^2\right) + C_S r_S^2\right]} \frac{I_D}{I_G} \tag{4}$$

$$n_D = 1.02 \times 10^{11}\ \mathrm{cm}^{-2} \tag{5}$$

Where r_A and r_S are the radius of two circular areas measured from the defect site. The first length, r_S, is the radius of the structurally disordered area around the defect. r_A defines the disk where the D peak scattering takes place and it defines the activated area. C_A depends only on the Raman mode, being roughly given by the ratio of the electron-phonon coupling

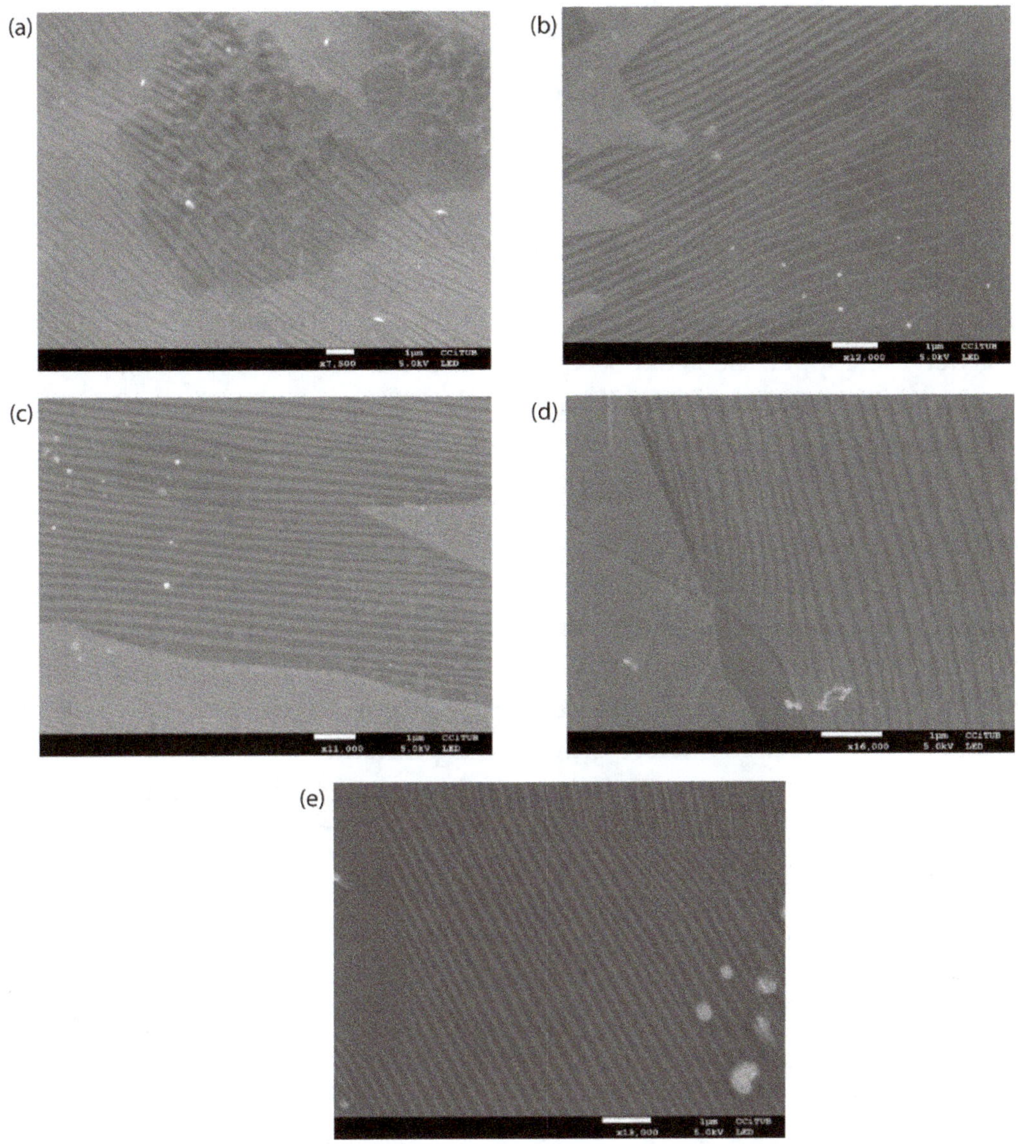

Figure 13. SEM images of graphene/copper surface: (a–e) graphene films growth at different hydrogen flows, resulting in an increase in the ripple's density. (a) 15, (b) 20, (c) 25, (d) 30 and (e) 35 sccm hydrogen flow. Scale bar is always 1 μm.

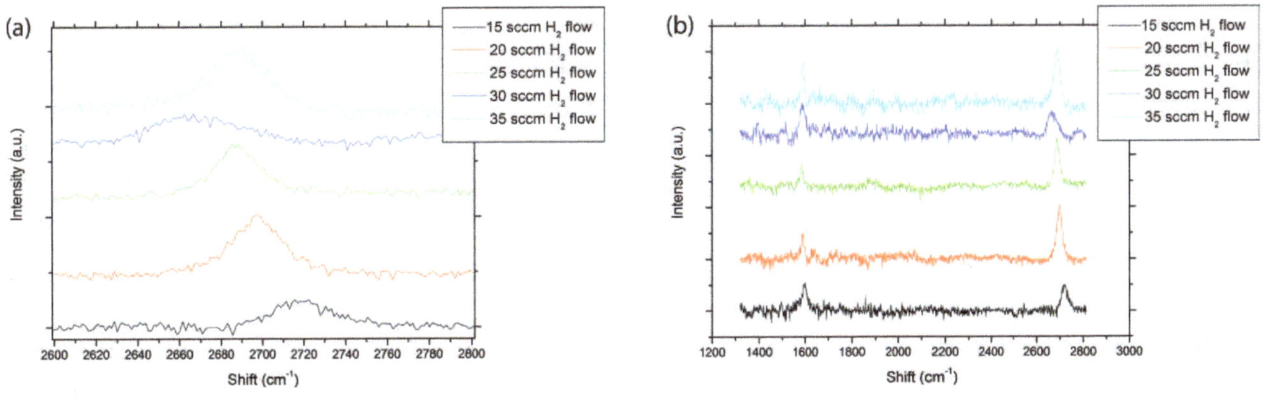

Figure 14. (a) Raman spectra of the different samples grown at several hydrogen flow rates. (b) Enlarged figure of the 2D peak of the same samples.

between the two phonons considered. $C_{S,x}$ represents the Raman cross section of $I(x)/I(G)$ associated with the distortion of the crystal lattice after defect introduction per unit of damaged area. The following parameters were reported for intensity measured as height: $r_A \sim 3$ nm, $r_S \sim 1$ nm, $C_A = 4.2$, $C_S = 0.87$ [57]. For samples grown at lower hydrogen flow rate, the presence of defects from Raman analysis is negligible.

In previous works [58], the formation of ripples in graphene during CVD growth has been studied. During CVD growth, graphene tends to replicate the morphology of the substrate. When graphene is grown over copper foil, the formation of these ripples is a common phenomenon. When the copper foil is exposed at the same heating and cooling process (as CVD graphene growth) but in the absence of any hydrocarbon, no ripples are formed and the formation of dendrite-like structures or 'snowflakes' is observed. Therefore, the presence of hydrocarbon seems a necessary condition for the formation of ripples.

One possible mechanism describing the ripple's formation is the following. At high temperatures, above 1000°C, a massive near-surface movement of the copper atoms takes place underneath the growing graphene. This gives rise to the surface reconstruction and to the formation of nano-ripples. While cooling down to room temperature, the copper substrate contracts, whereas the graphene which is on top tries to expand according to its negative thermal expansion coefficient [59]. In the areas where graphene is pinned to the copper, it is displayed to a strong compressive stress, whereas in the areas where the graphene is suspended above the nano-ripples it is allowed to relax some amount of the compressive strain through out-of-plane deformation (rippling) [60–61].

In the present work, when we measure the ripples in the samples corresponding at different hydrogen flows, a sigmoid increase of the density of ripples at higher hydrogen flows takes place (**Figure 15**). More specifically, the ripple density increases from 2.6 to 4.3 ripples/μm while the hydrogen flow increases from 15 to 30 sccm (with a step of 5 sccm). Finally, at 35 sccm of hydrogen flow the ripple density is 4.3 ripples/μm as well.

In all above described experiments, the methane flow was kept constant; therefore, the variation of the hydrogen flow is the factor that apparently affects the density of the formed ripples. **Figure 16** shows the Raman measurements of the different hydrogen flows. In **Figure 16a**, we see the full width half maximum (FWHM) of the 2D peak versus the hydrogen flow. Despite some extreme values, mainly in the samples grown at 25 and 30 sccm of hydrogen flow, the majority of the spectra presents an FWHM close to the value of 28 cm^{-1}, which is the reference value of the 2D peak FWHM for single-layer graphene [62]. Narrower 2D bands are correlated with flat and undoped regions on the Cu (100) and (110) surfaces. The generally compressed (0.3% of strain) and n-doped (Fermi level shift of 250 meV) graphene on Cu (111) shows the 2D band FWHM minimum of 20 cm^{-1}. In contrast, graphene grown on Cu foil under the same conditions reflects the heterogeneity of the polycrystalline surface and its 2D band is accordingly broader with FWHM > 24 cm^{-1} [63]. The authors have previously studied the polycrystalline nature of the copper surface by electron backscatter diffraction (EBSD) mapping. After the heating up to 1040°C and posterior cooling, the domains increase significantly in size. According to these observations, graphene domains grow forming different morphologies which depend on the copper lattice orientation, in accordance to previous

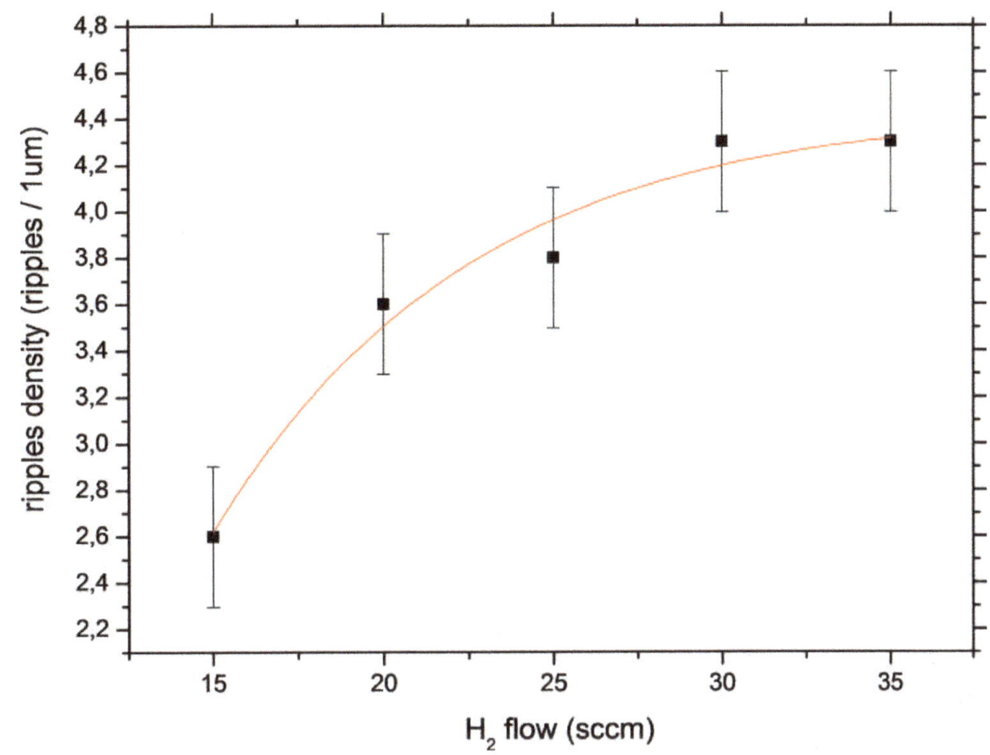

Figure 15. Ripples density of graphene for various hydrogen flows.

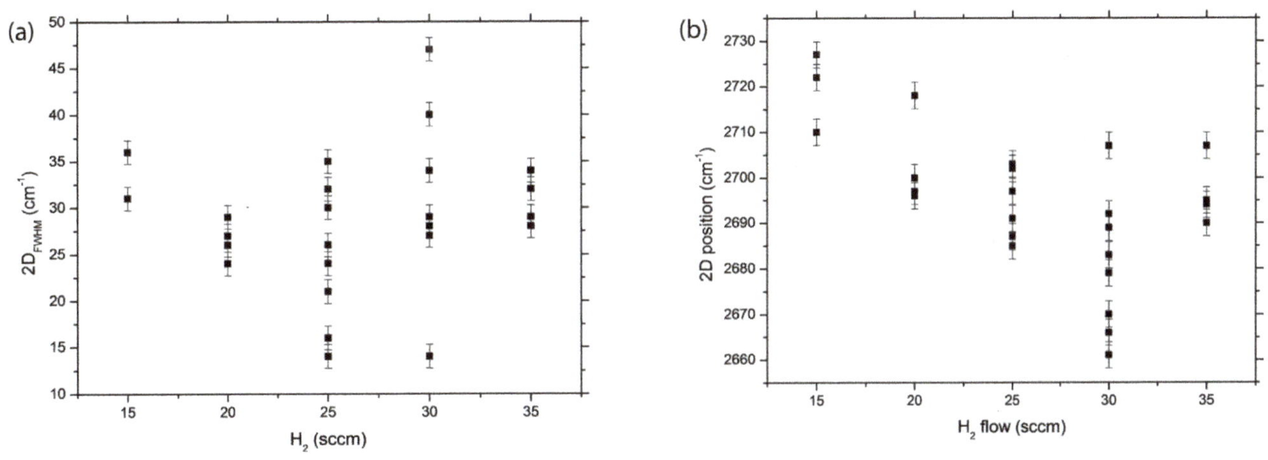

Figure 16. (a) Plot of the 2D FWHM with respect to the different hydrogen flows. The blue line corresponds to the reference value of the 2D peak FWHM for single-layer graphene, (b) plot of the 2D position with respect to the different hydrogen flows. The blue line at 2680 cm^{-1} corresponds to the position of relaxed graphene.

theoretic and experimental results [45]. In concrete, we have observed how graphene domains with a four-lobe morphology grow over the Cu (100) plane, while graphene domains with a butterfly-like morphology grow over the Cu (221) plane.

By the shift in the 2D peak, we calculate the graphene strain. **Figure 16b** shows the position of the 2D peak versus the hydrogen flow. The position of the 2D peak shifts from 2719 to 2680 cm^{-1} with the increase of the hydrogen flow from 15 to 30 sccm. The values presented in **Table 1** regarding the 2D FWHM and the 2D position for the different hydrogen flows are the average values resulting from those presented in the plots of **Figure 16a** and **b**.

The results reveal an average red-shift of the 2D peak following the hydrogen flow increase. We have associated this shift with the strain introduced to the graphene as a result of the ripple's formation and the partial ablation of graphene. Such shifting of the 2D peak between suspended and supported graphene has been observed before [64]. Rao et al. have observed a shifting of the order of 10 cm^{-1}. In our case, we propose to explain the decrease of the compressive strain effect as the origin of this shift. As mentioned earlier, the graphene replicates the morphology of the copper substrate. In order to understand this phenomenon, we propose a mechanism based on thermal mobility of Cu adatoms. Surface adatoms can take energy from hydrogen bombardment, which facilitates their migration and ripple's formation.

Theory of thermal grooving was first developed by Mullins et al. [65]. In this work, surface migration is proposed as the most probable mechanism of groove formation. The origin of the grooves can be due to evapouration or due to surface diffusion of the copper adatoms. As temperature falls, surface diffusion is favoured thanks to its small activation energy, in compare to the enthalpy of evapouration. Although, in their model, Mullins et al. consider that the grooves are formed during the heating of the surface. As the development of the groove is proportional to $t^{1/4}$, where t is the annealing time, they come in accordance with experimental results which indicate that the grooves did not develop more after some hours of annealing.

Today we know thank to *in situ* observations that the grooves, or ripples as we refer to in our text, develop during the cooling step. As observed by *in situ* scanning electron microscopy [66], ripples are formed in the copper surface during the cooling step, in concrete in the range between 600 and 700°C. Surface re-construction of the pre-melted copper occurs in this range of temperatures.

The model proposed by Mullins allows us to calculate the depth of the groove when we know the separation of two consecutive maxima.

$$\frac{s}{d} = \frac{4.73}{m} \tag{6}$$

where s is the separation between the two maxima, d is the groove depth and $m = \tan b = 0.10$ (b is the equilibrium angle).

As we know the separation of the two maxima from the SEM images (~0.5μm), we calculate the depth d = ~10 nm. Values of this order are very similar to the ones observed by other authors [67].

The increase in the hydrogen bombardment increases the adatoms density and, therefore, the ripple's density. From molecular dynamic simulations that have been performed by Rosen et al. [68], it has been calculated that the positive energy transfer that can occur during the bombardment of metallic clusters by a gas. This energy can promote the diffusion of the metallic adatoms, promoting the formations of ripples. Graphene grows overlapping the copper ripples. The migration of the copper adatoms results in the formation of convex regions where copper is disordered. Ordered copper adatoms remain in the concave regions (**Figure 17a**) [69]. The ordered placement of the copper adatoms in these regions favours the matching with the graphene.

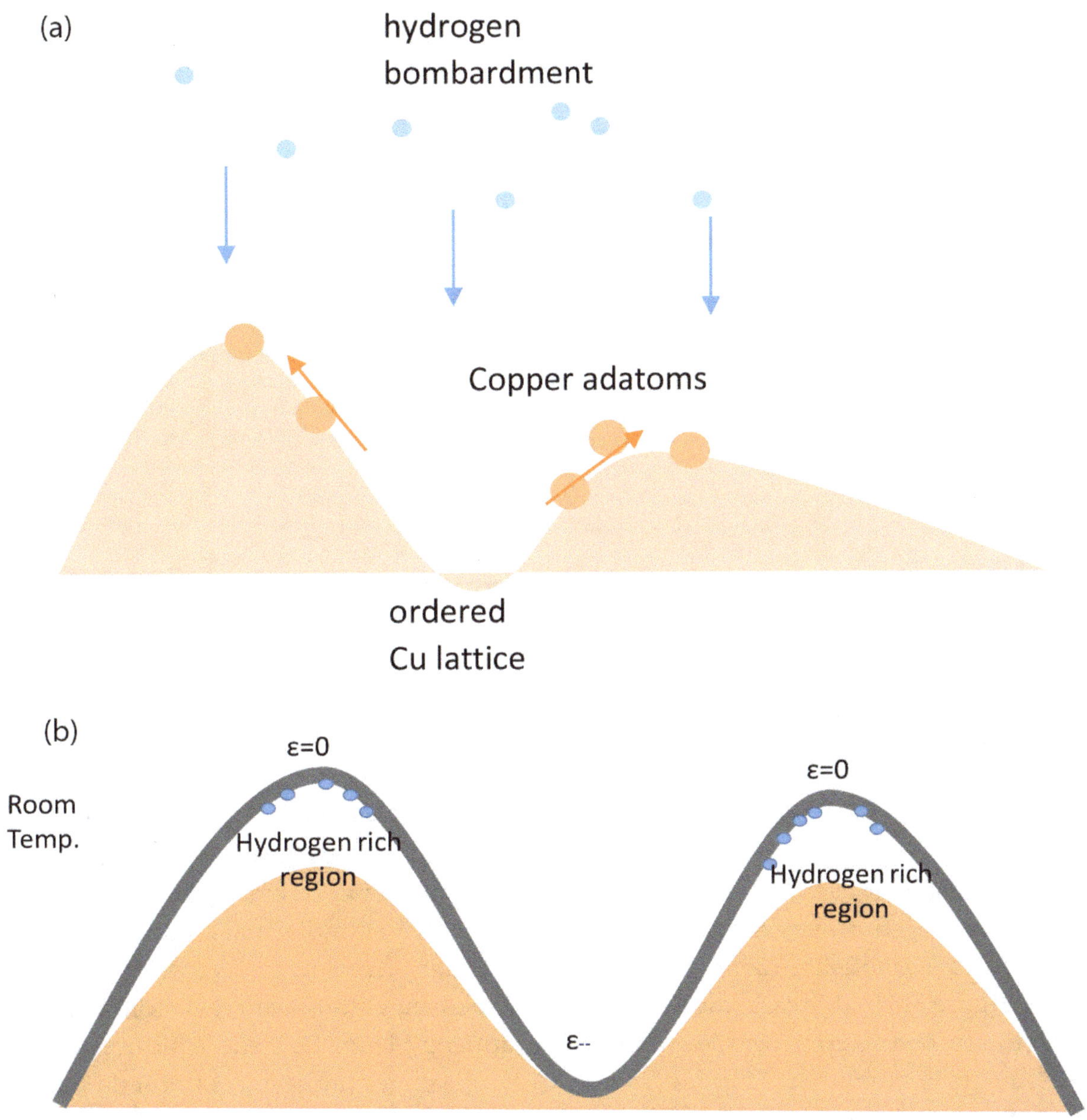

Figure 17. (a) Illustration of the ripple's formation as a result of the hydrogen bombardment. (b) Aspect of the graphene film after its ablation from the concave regions.

As explained above, while cooling, the copper foil contracts. This induces an increase in the compressive stress of the graphene in the concave region. In the convex region, the graphene detaches from the copper and remains suspended and relaxed. This detachment probably is the reason that graphene in the convex regions favours the hydrogen storage. As explained in previous works by experimental and density functional theory calculations [69], graphene chemisorption of atomic hydrogen is energetically favourable in the convex regions (**Figure 18b**). In that work, scanning tunnelling microscopy (STM) images have shown a large increase in corrugation due to the C–H bonds on the convex areas of the graphene surface after the exposure of pristine graphene to atomic hydrogen. DFT simulations have observed and resolved hydrogen dimers and tetramers on top of the carbon atoms, on distances that do not extend beyond 4 Å between them.

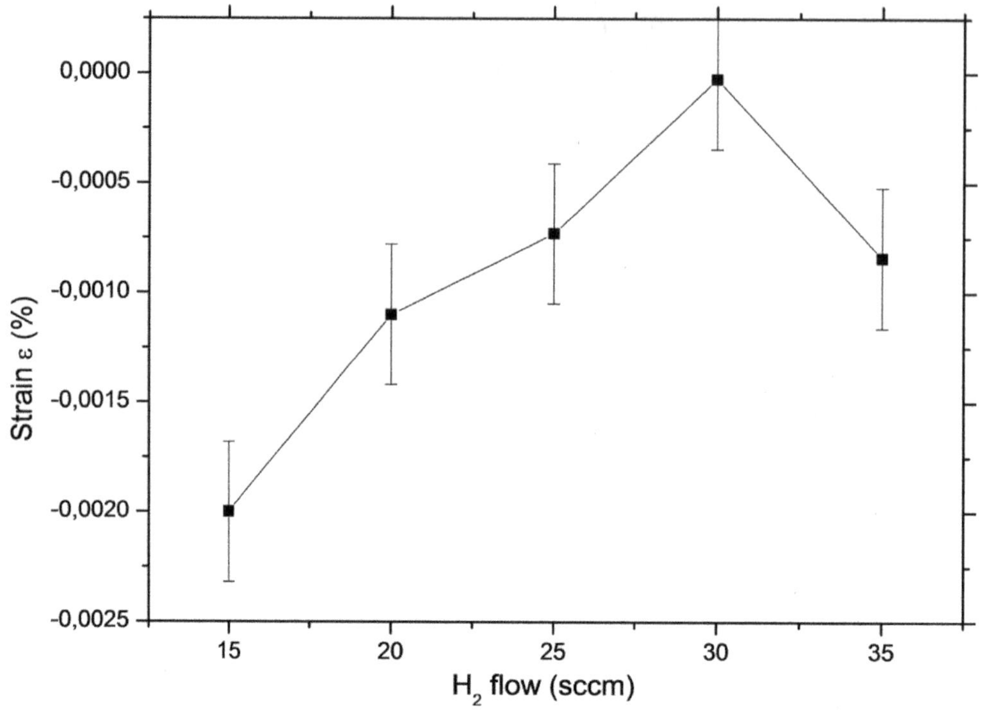

Figure 18. Calculated compressive strain (Eq. 7) of graphene layer for the different hydrogen flow rates from the 2D Raman peak shift.

According to Goler et al. [69] calculation and using 650°C (~930 K) as the approximate temperature of the hydrogen desorption from the puckered graphene, the only positions where hydrogen is stable at elevated temperatures, a desorption energy barrier of 1.4 eV is calculated. The hydrogen dimers are calculated to be as stable as molecular hydrogen. This is a result of the favourable hydrogen chemisorption on the convex areas, like we propose in the model of **Figure 5b**. In this case, the local curvature increases after the first H atom is adsorbed because the carbon atom protrudes out of the graphene plane. This effect induces adhesion of subsequent H atoms. In accordance with Goler et al., the adhesion of atomic hydrogen becomes thus barrierless.

In our work, we consider that the red-shift of the 2D peak is a result of the reduction/absence of the compressive strain with the increase in the hydrogen flow. With the increase in the ripples density, more regions where graphene is relaxed are contained in each spectrum, resulting in reinforcing of the red-shift of the 2D peak.

From the shift in the 2D peak, we can calculate the strain of the graphene. This is calculated by equation

$$\varepsilon = \frac{-\Delta u_{2D}}{2\, u_{2D}^{0}\, \gamma_D} \tag{7}$$

where ε is the value of the strain, $\gamma_D = 3.55$ is the Gruneisen parameter and Δu_{2D} denotes the shift of the frequency of the 2D phonon mode with respect to $u_{2D}^{0} = 2680\ \mathrm{cm^{-1}}$ (which is the position of the 2D peak for unstrained pristine graphene) [70]. The compressive strain is reduced with the increase of the hydrogen flow from $\varepsilon = -0.2\%$ for 15 sccm of hydrogen to $\varepsilon = -0.026\%$ for 30 sccm of hydrogen flow rate (**Figure 18**).

4. Conclusions

In the present chapter, we study the CVD growth of graphene over copper foils. We are interested in the effect of the growth parameters, gases mixtures, pressure, temperature in the graphene synthesis. The characterization of SEM images gives information about the effect of hydrogen in the process. The lower nucleation density results in formation of larger crystals but lower total coverage. Considering the coverage, the etching effect of hydrogen in the graphene film plays an important role. In addition, we underline the linear dependence between the C/L^2 and the nucleation density. We highlight the increase of hydrogen partial pressure under stable flow and the fact that the relative pressure $\langle P_{H2} \rangle / \langle P_{CH4} \rangle$ ratio diagram follows the same variation as the percent coverage diagram.

We study the kinetics, from a perspective of temperature dependence, of CVD-grown graphene in low pressure and over polycrystalline copper substrate. Growth temperature ranges between 970 and 1070°C. The growth takes place in a methane/hydrogen atmosphere. In all samples, high-quality, defect-free single-layer graphene is grown, as revealed by Raman spectroscopy. SEM images analysis provide information considering the nucleation density, in the range between 0.5 and 8 nuclei/1000 μm^2, and the growth rate, in the range between 150 and 1400 μm^2/20 min, of the graphene domains. Activation energy of 3.01 eV has been derived from the Arrhenius equation. Finally, we study the profile of the ripples formed in the graphene surface. Ripples of 0.2 μm and 4 nm height are formed, introducing an intrinsic strain in the graphene. The geometry of the ripples is not affected by the growth temperature.

Through the control of the flow, the ripple density is regulated. An increase in the flow of hydrogen can reduce the compressive strain, as a result of the levitation of graphene from the copper substrate in the concave regions. However, the introduced hydrogen flow should be selected with care, as high flow rates can lead to anisotropic etching of the graphene film.

Acknowledgements

The first author was founded by the Greek State Scholarships Foundation (IKY). The authors would like to thank the CCiT-UB for help with the structural and morphological characterization. This work was developed in the frame of the project 2014SGR984 of AGAUR from the Generalitat de Catalunya and the projects MAT2010-20468 and ENE2014-56109-C3 1-R of MICINN from the Spanish Government.

Author details

Stefanos Chaitoglou*, Enric Bertran and Jose Luis Andujar

*Address all correspondence to: schaitog@gmail.com

FEMAN Group, IN2UB, Department of Applied Physics, University of Barcelona, Barcelona, Catalonia, Spain

References

[1] Li X et al. Large-area synthesis of high-quality and uniform graphene films on copper foil. Science 2009;**324**:1312–4.

[2] Bae S et al. Roll-to-roll production of 30-inch graphene films for transparent electrodes. Nat. Nanotechnol. 2010;**5**:574.

[3] Han G H, Shin H-J, Kim E S, Chae S J, Choi J-Y and Lee Y H. Poly(ethylene co-vinyl acetate)-assisted one-step transfer of ultra large graphene. Nano 2011;**6**:59.

[4] Miseikis V et al. Rapid CVD growth of millimetre-sized single crystal graphene using a cold-wall reactor. 2D Mater. 2010;**2**:014006.

[5] Yazyev O V and Louie S G. Topological defects in graphene: dislocations and grain boundaries. Phys. Rev. B 2010;**81:**195420.

[6] Li X F, Wang L L, Chen K Q and Luo Y. Effects of interface roughness on electronic transport properties of nanotube–molecule–nanotube junctions. J. Phys. Chem. C 2010;**115**:12616–24.

[7] Kumar S B and Guo J. Strain-induced conductance modulation in graphene grain boundary. Nano Lett. 2012;**12**:1362.

[8] Yazyev O V and Louie S G. Electronic transport in polycrystalline graphene. Nat. Mater. 2010;**9**:806–9.

[9] Gran Table R, Shenoy V B and Ruoff R S. Anomalous strength characteristics of tilt grain boundaries in graphene. Science 2010;**330:**946–8.

[10] Kim K, Artyukhov V I, Regan W, Liu Y, Crommie M F, Yakobson B I and Zettl A. Ripping graphene: preferred directions. Nano Lett. 2012;**12**:12293–7.

[11] Vlassiouk I et al. Graphene nucleation density on copper: fundamental role of background pressure. J. Phys. Chem.C 2013;**117:**18919–26.

[12] Li X, Magnuson C W, Venugopal A et al. Large-area graphene single crystals grown by low-pressure chemical vapour deposition of methane on copper. J. Am. Chem. Soc. 2011;**133**:2816–9 .

[13] Bhaviripudi S, Jia X, Dresselhaus M S and Kong J. Role of kinetic factors in chemical vapour deposition synthesis of uniform large area graphene using copper catalyst. Nano Lett. 2010;**10**:4128–33.

[14] Li X, Magnuson C W, Venugopal A. et al. Graphene films with large domain size by a two-step chemical vapour deposition process. Nano Lett. 2010;**10**:4328–34.

[15] Wu B, Geng D, Xu Z, Guo Y, Huang L, Xue Y, Chen J, Yu G and Liu Y. Self-organized graphene crystal patterns. NPG Asia Mater. 2013;**5:**36.

[16] Zhang Y, Li Z, Kim P, Zhang L and Zhou C. Anisotropic hydrogen etching of chemical vapour deposited graphene. ACS Nano 2011;**6:**126–32.

[17] Geng D,Wu B, Guo Y, Luo B, Xue Y, Chen J, Yu G and Liu Y. Controllable synthesis of submillimeter single-crystal monolayer graphene domains on copper foils by suppressing nucleation. J. Am. Chem. Soc. 2013;**135**:6431–4.

[18] Vlassiouk I, Regmi M, Fulvio P, Dai S, Datskos P, Eres G and Smirnov S. Role of hydrogen in chemical vapour deposition growth of large single-crystal graphene. ACS Nano 2011;**5**:6069–76.

[19] Zhang X, Ning J, Li X, Wang B, Hao L, Liang M, Jin M and Zhi L. Hydrogeninduced effects on the CVD growth of high-quality graphene structures. Nanoscale 2013;**5**:8363–6.

[20] Ma T, Ren W, Zhang X, Liu Z, Gao Y, Yin LC, Ma XL, Ding F and Cheng H M. Edge-controlled growth and kinetics of single crystal graphene domains by chemical vapour deposition. Proc. Natl Acad. Sci. USA 2013;**110**:20386–91.

[21] Mohsin A et al. Synthesis of millimeter-size hexagon-shaped graphene single crystals on resolidified copper. ACS Nano 2013;**7**:8924.

[22] Yan Z, Lin J, Peng Z, Sun Z, Zhu Y, Li L, Xiang C, Samuel E L, Kittrell C and Tour L M. Toward the synthesis of wafer-scale single crystal graphene on copper foils. ACS Nano 2012;**6**:9110–7.

[23] Wu Y A, Fan Y, Speller S, Creeth G L, Sadowski J T, He K, Robertson A W, Allen C S and Warner J H. Large single crystals of graphene on melted copper using chemical vapour deposition. ACS Nano 2012;**6**:5010–7.

[24] Geng D,Wu B, Guo Y, Huang L, Xue Y, Chen J, Yu G, Jiang L, Hu W and Liu Y. Uniform hexagonal graphene flakes and films grown on liquid copper surface. Proc. Natl Acad. Sci. USA 2012;**109**:7992–6.

[25] Wang H, Wang G, Bao P, Yang S, Zhu W, Xie X and Zhang W J. Controllable synthesis of submillimeter single-crystal monolayer graphene domains on copper foils by suppressing nucleation. J. Am. Chem. Soc. 2012;**134**:3627–30.

[26] Wu T, Ding G, Shen H, Wang H, Sun L, Jiang D, Xie X and Jiang M. Triggering the continuous growth of graphene toward millimeter-sized grains. Adv. Funct. Mater. 2013;**23**:198–203.

[27] Brown L et al. Polycrystalline graphene with single crystalline electronic structure. Nano Lett. 2014;**14**:5706.

[28] Faggio G, Capasso A, Messina G, Santangelo S, Dikonimos T, Gagliardi S, Giorgi R, Morandi V, Ortolani L and Lisi N. High temperature growth of graphene films on copper foils by ethanol chemical vapour deposition. J. Phys. Chem. C 2013;**117**:21569–76.

[29] Hao Y et al. The role of surface oxygen in the growth of large single-crystal graphene on copper. Science 2013;**342**:720.

[30] Gan L and Luo Z. Turning off hydrogen to realize seeded growth of subcentimeter single-crystal graphene grains on copper. ACS Nano 2013;**7**:9480.

[31] Li J, Wang X-Y, Liu X-R, Jin Z, Wang D and Wan L-J. Facile growth of centimeter-sized single-crystal graphene on copper foil at atmospheric pressure. J. Mater. Chem. C 2015;**3**:3530.

[32] Wang Y, Zheng Y, Xu X, Dubuisson E, Bao Q, Lu J and Loh K P. Electrochemical delamination of cvd grown graphene film: toward the recyclable use of copper catalyst. ACS Nano 2011;**5**:9927–93.

[33] Deng S, Berry V. Wrinkled, rippled and crumpled graphene: an overview of formation mechanism, electronic properties, and applications. Mater. Today 2016;**19**:197–212.

[34] Zhu W, Low T, Perebeinos V, Bol A, Zhu Y, Yan H, Tersoff J and Avouris P. Structure and electronic transport in graphene wrinkles. Nano Lett. 2012;**12:**3431–6.

[35] Duan W, Gong K and Wang Q. Controlling the formation of wrinkles in a single layer graphene sheet subjected to in-plane shear. Carbon 2011;**49**:3107–12.

[36] Park W. H. Electrical performance of chemical vapour deposition graphene on PET substrate tailored by Cu foil surface morphology. Eur. Phys. J. Appl. Phys. 2014;**67**:30701.

[37] Park W H, Jo I, Hong B H and Cheong H. Controlling the ripple density and heights: a new way to improve the electrical performance of CVD-grown graphene. Nanoscale 2016;**8**:9822.

[38] Kato R, Tsugawa K, Yamada T, Ishihara M and Hasegawa M. Improvement of multilayer graphene synthesis on copper substrate by microwave plasma process using helium at low temperatures. Japanese J. Appl. Phys. 2014;**53**:015505.

[39] Hussain S, Amade R, Moreno H and Bertran E. RF PECVD growth and nitrogen plasma functionalization of CNT son copper foil for electrochemical applications. Diam. Relat. Mater. 2014;**49**:55–61.

[40] Chaitoglou S, Pascual E, Bertran E and Andujar J L. Effect of a balanced concentration of hydrogen on graphene CVD growth. J. Nanomater. 2016;**2016**:9640935.

[41] Chaitoglou S and Bertran E. Effect of pressure and hydrogen flow in nucleation density and morphology of graphene bidimensional crystals. Mater. Res. Express 2016;**3**:075603.

[42] Raskova Z, Krcma F, Klima M and Kousal J. Diagnostic of plasma chemical treatment of archaeological artefacts. Czech. J. Phys. 2011;**52:**D927–32.

[43] Chang Y M, Leu J, Lin B H, Wang Y L and Cheng Y L. Comparison of H2 and NH3 treatments for copper interconnects. Adv. Mater. Sci. Eng. 2013;**7**:825195.

[44] Chen S, Brown L, Levendorf M et al. Oxidation resistance of graphene-coated Cu and Cu/Ni alloy. ACS Nano 2011;**5**:1321–7.

[45] Meca E, Lowengrub J, Kim H, Mattevi C and Shenoy V B. Epitaxial graphene growth and shape dynamics on copper: phase-field modeling and experiments. Nano Lett. 2013;**13**:5692–7.

[46] Hong J, Park M K, Lee E J, Lee D E, Hwang D S and Ryu S. Origin of new broad Raman D and G peaks in annealed graphene. Sci. Rep. 2013;**3**:2700.

[47] Wu P, Zhai X, Li Z and Yang J. Bilayer graphene growth via a penetration mechanism. J. Phys. Chem. C 2014;**118**:6201–6.

[48] Wood J D, Schmucker S W, Lyons A S, Pop E and Lyding J W. Effects of polycrystalline Cu substrate on graphene growth by chemical vapour deposition. Nano Lett. 2011;**11**:4547–54.

[49] Costa S D, Righi A, Fantini C, Hao Y, Magnuson C, Colombo L, Ruoff R S and Pimenta A. Resonant Raman spectroscopy of graphene grown on copper substrates. Solid State Commun. 2012;**152**:1317–20.

[50] Das A, Chakraborty B and Sood A. Raman spectroscopy of graphene on different substrates and influence of defects. Bull. Mater. Sci. 2008;**31**:579–84.

[51] Kim H, Mattevi C, Reyes Calvo M, Oberg J, Artiglia L, Agnoli S, Hirjibehedin C, Chhowalla M and Saiz E. Activation energy paths for graphene nucleation and growth on Cu. ACS Nano 2012;**6**:3614–23.

[52] Loginova E, Norman C B, Peter J F and Kevin F M. Evidence for graphene growth by C cluster attachment. New J. Phys. 2008;**10**:093026.

[53] Robinson V N E and Robins J L. Nucleation kinetics of gold deposited onto UHV cleaved surfaces of NaCl and KBr. Thin Solid Films 1974;**20**:155–75.

[54] Xing S, Wu W, Wang Y, Bao J and Pei S. Kinetic study of graphene growth: temperature perspective on growth rate and film thickness by chemical vapour deposition. Chem. Phys. Lett. 2013;**580**:62–6.

[55] Hayashi K, Sato S, Ikeda M, Kaneta C and Yokoyama N. Selective graphene formation on copper twin crystal. J. Am. Chem. Soc. 2012;**134**:12492.

[56] Lucchese M M, Stavale F, Martins Ferreira E H, Vilani C, Moutinho M V O, Capaz R, Achete C and Jorio A. Quantifying ion-induced defects and Raman relaxation length in graphene. Carbon 2010;**48**:1592–7.

[57] Eckmann A, Felten A, Verzhbitskiy I, Davey R and Casiraghi C. Raman study on defective graphene: effect of the excitation energy, type, and amount of defects. Phys. Rev. B 2013;**88**:035426.

[58] Cancado L G, Jorio A, Ferreira E H M, Stavale F, Achete C A, Capaz R B, Moutinho M V O, Lombardo A, Kulmala T S and Ferrari A C. Quantifying defects in graphene via Raman spectroscopy at different excitation energies. Nano Lett. 2011;**8**:3190.

[59] Paronyan T M, Pigos E M, Chen G and Harutyunyan A R. Formation of ripples in graphene as a result of interfacial instabilities. ACS Nano 2011;**5**:9619–27.

[60] Yoon D, Son Y and Cheong H. Negative thermal expansion coefficient of graphene measured by Raman spectroscopy. Nano Lett. 2011;**11**:3227–31.

[61] Tapasztó L, Dumitrica T, Kim S, Nemes-Incze P, Hwang C and Biró L P. Breakdown of continuum mechanics for nanometre-wavelength rippling of graphene. Nat. Phys. 2012;**8**:739–42.

[62] Lin Z, Ye X, Han J, Che Q, Fan P, Zhang H, Xie D, Zhu H and Zhong M. Precise control of the number of layers of graphene by picosecond laser thinning. Sci. Rep. 2015;**5**:11662.

[63] Frank O, Veravova J, Hol V, Kavan L and Kalbac M. Interaction between graphene and copper substrate: the role of lattice orientation. Carbon 2014;**68**:440–51.

[64] Rao R, Chen G, Arava L M, Kalaga K, Ishigami M, Heinz T F, Ajayan P M and Harutyunyan A R. Graphene as an atomically thin interface for growth of vertically aligned carbon nanotubes. Sci. Rep. 2013;**3**:1891.

[65] Mullins W W. Theory of thermal grooving. J. Appl. Phys. 1957;**28**:333.

[66] Wang Z, Weinberg G, Zhang Q, Lunkenbein T, Klein-Hoffmann A, Kurnatowska M, Plodinec M, Li Q, Chi L, Willinger M et al. Direct observation of graphene growth and associated copper substrate dynamics by in situ scanning electron microscopy. ACS Nano 2015;**9**:1506–19.

[67] Meng L, Su Y, Geng D, Yu G, Liu Y, Dou R, Nie J and He L. Hierarchy of graphene wrinkles induced by thermal strain engineering. Appl. Phys. Lett. 2013;**103**:251610.

[68] Westergren J, GrSnbeck H, Rosh A and Nordbolm S. Molecular dynamic simulations of metal cluster cooling and heating in noble gas atmosphere. Nanostruct. Mater. 1999;**12**:281–6.

[69] Goler S, Coletti C, Tozzini V, Piazza V, Mashoff T, Beltram F, Pellegrini V and Heun S. Influence of graphene curvature on hydrogen adsorption: Toward hydrogen storage devices. J. Phys. Chem. C 2013;**11**:11506–13.

[70] Troppenz G V, Gluba M A, Kraft M, Rappich J and Nickel N H. Strain relaxation in graphene grown by chemical vapour deposition. J. Appl. Phys. 2013;**114**:214312.

Permissions

All chapters in this book were first published in GMSPM, by InTech Open; hereby published with permission under the Creative Commons Attribution License or equivalent. Every chapter published in this book has been scrutinized by our experts. Their significance has been extensively debated. The topics covered herein carry significant findings which will fuel the growth of the discipline. They may even be implemented as practical applications or may be referred to as a beginning point for another development.

The contributors of this book come from diverse backgrounds, making this book a truly international effort. This book will bring forth new frontiers with its revolutionizing research information and detailed analysis of the nascent developments around the world.

We would like to thank all the contributing authors for lending their expertise to make the book truly unique. They have played a crucial role in the development of this book. Without their invaluable contributions this book wouldn't have been possible. They have made vital efforts to compile up to date information on the varied aspects of this subject to make this book a valuable addition to the collection of many professionals and students.

This book was conceptualized with the vision of imparting up-to-date information and advanced data in this field. To ensure the same, a matchless editorial board was set up. Every individual on the board went through rigorous rounds of assessment to prove their worth. After which they invested a large part of their time researching and compiling the most relevant data for our readers.

The editorial board has been involved in producing this book since its inception. They have spent rigorous hours researching and exploring the diverse topics which have resulted in the successful publishing of this book. They have passed on their knowledge of decades through this book. To expedite this challenging task, the publisher supported the team at every step. A small team of assistant editors was also appointed to further simplify the editing procedure and attain best results for the readers.

Apart from the editorial board, the designing team has also invested a significant amount of their time in understanding the subject and creating the most relevant covers. They scrutinized every image to scout for the most suitable representation of the subject and create an appropriate cover for the book.

The publishing team has been an ardent support to the editorial, designing and production team. Their endless efforts to recruit the best for this project, has resulted in the accomplishment of this book. They are a veteran in the field of academics and their pool of knowledge is as vast as their experience in printing. Their expertise and guidance has proved useful at every step. Their uncompromising quality standards have made this book an exceptional effort. Their encouragement from time to time has been an inspiration for everyone.

The publisher and the editorial board hope that this book will prove to be a valuable piece of knowledge for researchers, students, practitioners and scholars across the globe.

List of Contributors

Rebeca Ortega-Amaya
SEES, Electrical Engineering Department, Center for Research and Advanced Studies of the National Polytechnic Institute, Mexico City, Mexico

Yasuhiro Matsumoto and Mauricio Ortega-López
SEES, Electrical Engineering Department, Center for Research and Advanced Studies of the National Polytechnic Institute, Mexico City, Mexico
Nanoscience and Nanotechnology Program, Center for Research and Advanced Studies of the National Polytechnic Institute, Mexico City, Mexico

Esteban Díaz-Torres and Manuel Alejandro Pérez-Guzmán
Nanoscience and Nanotechnology Program, Center for Research and Advanced Studies of the National Polytechnic Institute, Mexico City, Mexico

Claudio Davet Gutierrez-Lazos
Center for Research in Physical Mathematical Sciences, Faculty of Mathematical Physical Sciences, The Autonomous University of Nuevo León, San Nicolás de los Garza, Nuevo León, Mexico

Ulises Antonio Méndez Romero and Miguel Ángel Velasco Soto
Center for Advanced Materials Research S.C. (CIMAV), Unidad Monterrey, Apodaca, NL, Mexico

Liliana Licea Jiménez, Jaime Álvarez Quintana and Sergio Alfonso Pérez García
Center for Advanced Materials Research S.C. (CIMAV), Unidad Monterrey, Apodaca, NL, Mexico
GENES—Group of Embedded Nanomaterials for Energy Scavenging, Apodaca, NL, Mexico

Alexander López
Escuela Superior Politécnica del Litoral, ESPOL, Facultad de Ingeniería Mecánica y Ciencias de la Producción, Campus Gustavo Galindo Km 30.5 Vía Perimetral, Guayaquil, Ecuador

Benjamin Santos
INRS-EMT, Université du Québec, Varennes, Québec, Canada

Janusz Edward Jacak
Department of Quantum Technologies, Faculty of Fundamental Problems of Technology, Wroclaw University of Science and Technology, Wroclaw, Poland

Maurizio Galimberti, Vincenzina Barbera and Annalisa Sironi
Department of Chemistry, Materials and Chemical Engineering "G. Natta", Polytechnic University of Milan, Milan, Italy

Prashantha Kumar H.G. and Anthony Xavior M.
School of Mechanical Engineering, VIT University, Vellore, Tamil Nadu, India

Petr Machac
Department of Solid State Engineering, University of Chemistry and Technology, Prague, Czech Republic

Norio Tsubokawa, Takeshi Yamauchi and Shingo Tamesue
Graduate School of Science and Technology, Niigata University, Niigata, Japan

Kazuhiro Fujiki
Niigata Institute of Technology, Kashiwazaki, Niigata, Japan

Hui Li, Yifan Li and Wei Chen
Key Laboratory for Liquid-Solid Structural Evolution and Processing of Materials, Ministry of Education, Shandong University, Jinan, People's Republic of China

Stefanos Chaitoglou, Enric Bertran and Jose Luis Andujar
FEMAN Group, IN2UB, Department of Applied Physics, University of Barcelona, Barcelona, Catalonia, Spain

Index

R

S

T

V

W